# Business Math For Dummies

**Cheat Sheet**

## Fraction and Decimal Equivalents

In the following table, you find some of the more commonly used fractions and their decimal equivalents. An ellipsis (three dots) after a number means that the digit continues on forever.

| Fraction | Decimal | Fraction | Decimal |
|---|---|---|---|
| $\frac{1}{16}$ | 0.0625 | $\frac{5}{9}$ | 0.555 . . . |
| $\frac{1}{9}$ | 0.111 . . . | $\frac{9}{16}$ | 0.5625 |
| $\frac{1}{8}$ | 0.125 | $\frac{3}{5}$ | 0.6 |
| $\frac{3}{16}$ | 0.1875 | $\frac{5}{8}$ | 0.625 |
| $\frac{1}{5}$ | 0.2 | $\frac{2}{3}$ | 0.666 . . . |
| $\frac{2}{9}$ | 0.222 . . . | $\frac{11}{16}$ | 0.6875 |
| $\frac{1}{4}$ | 0.25 | $\frac{3}{4}$ | 0.75 |
| $\frac{5}{16}$ | 0.3125 | $\frac{7}{9}$ | 0.777 . . . |
| $\frac{1}{3}$ | 0.333 . . . | $\frac{4}{5}$ | 0.8 |
| $\frac{3}{8}$ | 0.375 | $\frac{13}{16}$ | 0.8125 |
| $\frac{2}{5}$ | 0.4 | $\frac{7}{8}$ | 0.875 |
| $\frac{7}{16}$ | 0.4375 | $\frac{8}{9}$ | 0.888 . . . |
| $\frac{4}{9}$ | 0.444 . . . | $\frac{15}{16}$ | 0.9375 |
| $\frac{1}{2}$ | 0.5 | | |

## Doubling Your Money

If you deposit money at a particular interest rate, you see here how long it takes for your investment to double in size. For interest rates that fall between the rates in the chart, remember that the time will fall in between also. (See Chapter 22 for more on how to interpolate with interest rates.)

| Interest Rate Compounded Quarterly | Amount of Time Needed to Double |
|---|---|
| 2% | ≈ 34.7 years |
| 3% | ≈ 23.2 years |
| 4% | ≈ 17.4 years |
| 5% | ≈ 13.9 years |
| 6% | ≈ 11.6 years |
| 7% | ≈ 10.0 years |
| 8% | ≈ 8.8 years |
| 9% | ≈ 7.8 years |
| 10% | ≈ 7.0 years |
| 11% | ≈ 6.4 years |
| 12% | ≈ 5.9 years |
| 13% | ≈ 5.4 years |
| 14% | ≈ 5.0 years |
| 15% | ≈ 4.7 years |

## Equivalent Distances and Acreage

1 mile = 5,280 feet = 1,760 yards

1 rod = 5½ yards = 16½ feet

1 yard = 3 feet = 36 inches

1 foot = 12 inches

1 meter ≈ 39.36 inches

1 inch ≈ 2.54 centimeters

1 acre = 43,560 square feet = 4,840 square yards

## The Most Common Area and Perimeter Formulas

The formulas shown here are some of the more frequently used ones when determining the areas and perimeters of rooms and tracts of land.

| Figure | Area Formula | Perimeter Formula | Variables |
|---|---|---|---|
| Square | $A = s^2$ | $P = 4s$ | $s$ = length of side |
| Rectangle | $A = lw$ | $P = 2(l + w)$ | $l$ = length $w$ = width |
| Triangle | $A = \frac{1}{2}bh$ | $P = a + b + c$ | $a, c$ = sides $b$ = base $h$ = height |
| Circle | $A = \pi r^2$ | $P = 2\pi r$ | $r$ = radius |

*For Dummies: Bestselling Book Series for Beginners*

# Business Math For Dummies®

## Financial Formulas

For your quick reference, here are some of the most popular financial formulas that you may need.

| Name | Formula | Variables |
|------|---------|-----------|
| Simple interest | $I = Prt$ | $P$ = principal <br> $r$ = interest rate <br> $t$ = time in years |
| Compound interest | $A = P\left(1 + \dfrac{r}{n}\right)^{nt}$ | $P$ = principal <br> $r$ = rate <br> $n$ = compoundings <br> $t$ = years |
| Effective rate | $\left(1 + \dfrac{r}{n}\right)^{n} - 1$ | $r$ = interest rate <br> $n$ = compoundings |
| Amortized loan payment | $R = \dfrac{Pi}{1 - (1 + i)^{-n}}$ | $P$ = amount borrowed <br> $i$ = interest rate per period <br> $n$ = number of payments |
| Remaining balance | $B = R\left[\dfrac{1 - (1 + i)^{-(n-x)}}{i}\right]$ | $R$ = regular payment <br> $i$ = interest rate per period <br> $n$ = number of payments <br> $x$ = number of payments already made |

## Important Angle Measures in Degrees

These sketches of the more commonly used angle measures can help you estimate angles when doing all sorts of things, such as creating lot lines or figuring fencing:

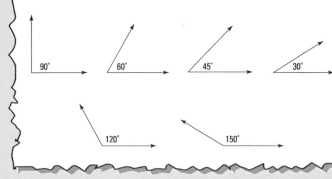

90°   60°   45°   30°

120°   150°

## Common Financial Acronyms

The financial world is chock-full of terms and acronyms. You may not be familiar with all of them, so take a glance through the following list. In no time, you'll be speaking like a financial pro.

**ABC**  activity based costing

**AP**  accounts payable

**APR**  annual percentage rate

**AR**  accounts receivable

**ARM**  adjustable rate mortgage

**BFY**  budget fiscal year

**CD**  certificate of deposit

**CDSC**  contingent deferred sales charge

**CFO**  chief financial officer

**CFP**  certified financial planner

**CIO**  chief information officer

**DB**  defined benefit (retirement plan)

**EFT**  electronic funds transfer

**EPS**  earnings per share

**FASB**  Financial Accounting Standards Board

**FBMS**  Financial and Business Management System

**FDIC**  Federal Deposit Insurance Corporation

**FTE**  full-time employee

**FIFO**  first in, first out

**FY**  fiscal year

**GDP**  gross domestic product

**GNP**  gross national product

**IPO**  initial public offering

**IR**  invoice receipt

**IRA**  Individual Retirement Account

**IRS**  Internal Revenue Service

**JTWROS**  joint tenants with rights of survivorship

**LIFO**  last in, first out

**LOI**  letter of intent

**LWOP**  leave without pay

**NASDAQ**  National Association of Securities Dealers Automated Quotation

**NAV**  net asset value

**NYSE**  New York Stock Exchange

**OWCP**  Office of Workers Compensation Programs

**P-card**  purchase card

**PE**  price to earnings (ratio)

**PO**  purchase order

**PR**  purchase requisition

**PSP**  profit sharing plan

**REIT**  real estate investment trust

**ROE**  return on equity

**ROI**  return on investment

**S&P**  Standard & Poor's

**SEC**  Securities Exchange Commission

**SEP**  simplified employee pension

**TSA**  tax-sheltered annuity

**YTM**  yield to maturity

### For Dummies: Bestselling Book Series for Beginners

*Business*
*Math*
FOR
DUMMIES®

# Business Math

## FOR

# DUMMIES®

## by Mary Jane Sterling

**WILEY**

Wiley Publishing, Inc.

**Business Math For Dummies**®

Published by
**Wiley Publishing, Inc.**
111 River St.
Hoboken, NJ 07030-5774
www.wiley.com

Copyright © 2008 by Wiley Publishing, Inc., Indianapolis, Indiana

Published simultaneously in Canada

For general information on our other products and services, please contact our Customer Care Department within the U.S. at 800-762-2974, outside the U.S. at 317-572-3993, or fax 317-572-4002.

For technical support, please visit www.wiley.com/techsupport.

Wiley also publishes its books in a variety of electronic formats. Some content that appears in print may not be available in electronic books.

Library of Congress Control Number: 2008927911

ISBN: 978-0-470-23331-3

Manufactured in the United States of America

10  9  8  7  6  5  4  3  2  1

WILEY

# About the Author

**Mary Jane Sterling** is the author of four other *For Dummies* titles: *Algebra For Dummies, Algebra II For Dummies, Trigonometry For Dummies,* and *Math Word Problems For Dummies.* She has honed her math-explaining skills during her years of teaching mathematics at all levels: junior high school, high school, and college. She has been teaching at Bradley University, in Peoria, Illinois, for almost 30 of those years.

When not teaching or writing, Mary Jane keeps busy by working with her Kiwanis Club, advising Bradley University's Circle K Club, and working with members of the Heart of Illinois Aktion Club (for adults with disabilities). All the volunteer projects taken on for these clubs help keep her busy and involved in the community.

# Dedication

I dedicate this book to my two business math inspirations. First, to my son, Jon, who came up with the idea of writing a book for business professionals who needed a refresher — specifically for his real estate workforce. And second, to my husband, Ted, who is the financial guru of the family and has had years of experience in the business and financial arena.

# Author's Acknowledgments

I give a big thank you to Stephen Clark, who took on this project and never ceased to be encouraging and upbeat. Also to Jessica Smith, a wonderful wordsmith and cheerful copy editor; her wonderful perspective and insights will pay off big time for the readers. Thank you to Technical Editor Benjamin Schultz who helped out so much by keeping the examples in this book realistic and the mathematics correct. Thank you to Lindsay Lefevere who took the initial suggestion and molded it into a workable project. And thank you to Composition Services, who took some pretty complicated equations and laid them out in the most readable forms.

## Publisher's Acknowledgments

We're proud of this book; please send us your comments through our Dummies online registration form located at www.dummies.com/register/.

Some of the people who helped bring this book to market include the following:

*Acquisitions, Editorial, and Media Development*

**Project Editor:** Stephen R. Clark

**Acquisitions Editor:** Lindsay Sandman Lefevere

**Copy Editor:** Jessica Smith

**Editorial Program Coordinator:** Erin Calligan Mooney

**Technical Editor:** Benjamin Schultz, MA, Lecturer, Department of Business Communication, Kelley School of Business

**Editorial Manager:** Christine Meloy Beck

**Editorial Assistants:** Joe Niesen, David Lutton

**Cartoons:** Rich Tennant (www.the5thwave.com)

*Composition Services*

**Project Coordinator:** Katie Key

**Layout and Graphics:** Carrie A. Cesavice, Stephanie D. Jumper, Julia Trippetti

**Proofreaders:** Melissa D. Buddendeck, Jessica Kramer

**Indexer:** Christine Spina Karpeles

*Special Help*
David Nacin, PhD

---

**Publishing and Editorial for Consumer Dummies**

    **Diane Graves Steele,** Vice President and Publisher, Consumer Dummies

    **Joyce Pepple,** Acquisitions Director, Consumer Dummies

    **Kristin A. Cocks,** Product Development Director, Consumer Dummies

    **Michael Spring,** Vice President and Publisher, Travel

    **Kelly Regan,** Editorial Director, Travel

**Publishing for Technology Dummies**

    **Andy Cummings,** Vice President and Publisher, Dummies Technology/General User

**Composition Services**

    **Gerry Fahey,** Vice President of Production Services

    **Debbie Stailey,** Director of Composition Services

# Contents at a Glance

# Table of Contents

# Introduction

*B*usiness and mathematics — they just seem to go together. Of course, I could make a case for *any* topic to successfully mesh with math. Then again, I'm a bit prejudiced. But you have to agree that you can't do much in the home or in the business or real estate worlds without a good, solid mathematical background.

I became interested in writing *Business Math For Dummies* because of Jon, one of my children. He works with real estate agents and managers, and over time he heard a recurring complaint: This or that agent hadn't done math in a long time and couldn't remember how to do the proper computations. Sure, computers and calculators do the actual arithmetic, but you have to know what to enter into the computer or calculator. Finding the area of a circular garden isn't terribly difficult if you know how to use the formula for the area of a circle, but it can be tricky if you haven't used that formula in a long time.

For instance, you may not remember what *pi* is (and I don't mean the apple or cherry variety). And who would have thought that you'd ever need to solve a proportion after you grew up and got a job? The math in the business world isn't difficult, but it does take a refresher or a bit of relearning. And here you find it — all in one spot.

In this book, you find tons of mathematical explanations coupled with some sample business or financial problems. I show you the steps used in the math and explain how the answers help you complete a transaction or continue on with a project.

## About This Book

In your business, you probably have all sorts of specialized procedures and formulas. And you're probably pretty good at what's particular to your situation. So what you find in this book is all the background info for more specialized material. I show you how to do financial computations that use simple formulas along with other more-involved formulas. You'll find geometric structures and patterns and uses of fractions, decimals, and percents.

Having said all that, I doubt you want to read this book from cover to cover. After all, no snooze alarm is loud enough to rouse you if you attempt a front-to-back read. But if you love math that much, please feel free to make this a pleasure read. However, most folks simply will go to those topics that interest or concern them.

This modularity is what's so great about this book. You can jump backward or forward to your heart's delight. In fact, I refer to earlier material in later chapters (if I think it would help explain the topic at hand), and I titillate you early on with promises of exciting applications further on as you read. In other words, you'll find the organization to be such that you can quickly turn to the pages you need as you need them.

Please use this book as a reference or a study guide. You don't want to memorize the formulas and procedures shown here. If you did, your head would surely burst.

# Conventions Used in This Book

This book is designed to be user friendly. I'm guessing that you're too busy to have to hunt up a dictionary if a word isn't familiar or to poke around for a formula (and how to use it) if you need to determine an item's markup or a loan's payment amount.

So as you read this book, you'll see important terms highlighted with *italics,* and I always include a definition with these terms. **Boldfaced** text highlights keywords of bulleted lists and the actions you must take in numbered steps. Also, when a process requires a formula, I state the formula and identify all the letters and symbols.

# What You're Not to Read

Occasionally in this book you'll see some material accompanied by a Technical Stuff icon. And you'll come across some sidebars, which are those gray-shaded boxes with interesting but nonessential information. You'll find these items when I just couldn't help myself — I had to tell you about some mathematical property, geometric eccentricity, or financial feature because I found it to be so fascinating. Do you need the information in order to understand the process? No. Will it enlighten you and make your world a better place? Well, of course. You can take them or leave them; pass them by today and come back to them in the future — or not.

# Foolish Assumptions

I find that separating my foolish assumptions from my realistic assumptions is really difficult. After all, what you may consider to be foolish, I may consider to be realistic (and vice versa). But let me tell you what I think, and you can do the sorting.

I'm assuming that you have access to at least a simple scientific calculator. You really can't do some of the problems involving powers, radicals, and huge numbers without the help of a calculating device. So, you're really handy with an abacus? All the more power to you — but I think you still want to invest in a calculator. Similarly, computers and spreadsheet software programs aren't a necessity, but I assume that you can use a computer, even if you don't have your own. After all, today's financial world is handled pretty much using spreadsheets with formulas embedded into them.

Most of the math in this book is arithmetic, but you also find some basic algebra. Here's the deal: You don't need to solve tricky quadratic equations or problems involving calculus, but I do assume that you understand how to solve an equation such as $1,000 = 3x - 8$ (by finding the value of $x$). Don't panic! I show you how to do these problems, but I assume that you've at least seen the algebraic process before and just need a reason to use it.

# How This Book Is Organized

Mathematics and business just seem to go together naturally, but I do try to organize the different topics and techniques in this book, and I give you some practical perspective of how they relate. Even when I start with the mathematics in the beginning, you see how those basic math operations and techniques relate to business topics; otherwise you wouldn't be inspired to continue reading the material! In all the chapters, I reference other topics that are closely related to what you're reading to save you time in your quest for the information you need.

## Part I: Reviewing Basic Math for Business and Real Estate Transactions

The basic math of business is really the basic math of life. So, the chapters in this part take you through the fundamentals of dealing with fractions, decimals, percents, and proportions. You see how the different forms that a percent can take are used to do computations necessary to be successful in business. A little algebra is even thrown in to make the topics interesting. Algebra is a rather cryptic language, but a simple translation is all you need to make your way.

# Part II: Taking Intriguing Math to Work

Formulas make the world go 'round — literally. But the formulas in this section mainly deal with money and financial situations, measurements, and statistics. In this part, you see how to use formulas in spreadsheets, and you find your way around statistical information.

# Part III: Discovering the Math of Finance and Investments

The chapters in this part are, well, *interesting*. In other words, simple interest and compound interest are used to a large extent in this part. You use formulas involving interest to determine how much an investment grows over time. And you use interest to determine the extra expense involved when borrowing money over different periods of time. Plus, no discussion of finance would be complete without talking about stocks and bonds. You find the ins and outs of their mathematics in this part. And, as promised, you find information on promissory notes as an option for borrowing.

# Part IV: Putting Math to Use in Banking and Payroll

Bank accounts are places to accumulate money and grow interest. You also use them to make payments to suppliers or employees. In the chapters in this part, you see variations on budgets and payroll management. Insurance and valuation of assets both play a big part in the overall financial picture of a business, which is why I include them in the general picture of managing the income and outgo involved in running a business successfully.

# Part V: Successfully Handling the Math Used in the World of Goods and Services

Whether you're on the buying end or the selling end of a business, you need to understand that discounts, markups, overhead, and depreciation all play a part in your business's profits, revenue, and costs. So in this part, I show you the mathematics involved and give you some examples. And because managing inventory is critical to the health of a company and its revenue figures, I explain methods for turnover and valuation.

## Part VI: Surviving the Math for Business Facilities and Operations

When purchasing property, you need accurate measurements, a fair assessment of the value, and a reasonable mortgage payment. Also, you need to secure a profitable rental amount if you're leasing the property to others. You find the math necessary for all the aspects of measuring, borrowing money for, purchasing, and renting property in this part's chapters.

## Part VII: The Part of Tens

A *For Dummies* book isn't complete without the Part of Tens. In this part, I include two top ten lists that give you quick, fun, and relevant information. For instance, I provide ten tips for managing leases and rental property, as well as a list of the ten main things to look for when reading a financial form.

# Icons Used in This Book

In the margins of this book, you see little pictures, which are called icons. These icons highlight specific types of information. They aren't meant to distract you. No, instead they're meant to attract your attention to items that warrant notice. Here are the icons I include and what they mean:

This icon calls your attention to situations and examples where the business math that's discussed is used in a sample situation.

This icon flags important information that you should keep in mind as you read or solve a problem. I often use this icon to highlight a rule or formula that you use in the current discussion. These rules or formulas are necessary for the mathematical computation and are applicable to other problems related to the topic.

When you see this icon, you'll know that you're coming upon a helpful or time-saving tip.

You'll find this icon attached to interesting but nonessential information. The information next to this icon is loosely related to the discussion at hand but isn't necessary to your understanding of the topic.

This icon points out a particularly sticky problem or common misunderstanding. I use this icon sparingly so that, when you do see it, you know I'm serious about the information.

## Where to Go from Here

The range of topics in business mathematics (and therefore in this book) is pretty broad. However, because of the large scope of material shown here, you'll be able to find what you need in one place or another. If you need some help with the basics, feel free to start with the first part. Then jump to the chapters that relate to your specific business situation. You just need to pick and choose what you need from the different parts of this book (and then use them or abuse them to your heart's content).

As you're reading through the book, remember that if you come across a particular mathematical process that you can't quite remember, you can be sure I refer you to the chapter that covers the process or something very much like it. You can then apply the concepts or processes to your particular situation. Feel free to use this book as a reference more than a guide. It's here to clear up the puzzling points and get you set in the right direction.

# Part I

# Reviewing Basic Math for Business and Real Estate Transactions

The 5th Wave          By Rich Tennant

"David's using algebra to calculate the tip. Barbara— would you mind being a fractional exponent?"

# In this part . . .

The mathematics of fractions, decimals, and percents is fully dissected and then put back together in these chapters. Why? Well, the world of business is tied to these topics. So, in this part, you investigate percent increases and percent decreases, and you see how the math goes right along with fractions and proportions. You also discover how to create new ways of using percents and proportions that you haven't even considered.

# Chapter 1

# Starting from the Beginning

* * *

## In This Chapter

▶ Aligning fractions, decimals, and percents to their place in business

▶ Understanding how formulas can help you solve business math problems

▶ Using exponents in everything financial

▶ Dealing with the math for inventory, overhead, and depreciation

▶ Picturing business scenarios using tables, charts, and graphs

* * *

*B*usiness mathematics involves a lot of arithmetic, some algebra, a touch of geometry, and dibs and dabs of other mathematical topics. But the major portion of mathematics that's found in business is arithmetic.

Getting you off to a good start is the goal of this chapter. You may be looking for answers to some deep, dark mathematical secrets; this chapter helps you light the way toward realizing that the basic math involved in business was never meant to be kept a secret. You may not see the relevance in some mathematical processes. But this chapter makes the necessity for mathematics abundantly clear.

On the other hand, you may have a firm grasp on the business math basics; you're looking for more — for some explanations as to *why*, not just *how*. So, in this chapter, I show you many of these *whys*, and I direct you to even more complex mathematics when the occasion allows.

 Most of the math in business isn't compartmentalized into one section or another. Fractions and decimals are found in all applications. Proportions and percents are rampant. And measurements are necessary for many different business processes. In other words, the math of business involves computations shared by all the different aspects. The main trick to doing the math is to know when to apply what. Use the material in this book to help yourself become comfortable with when to use what and how to use it successfully.

# Fracturing the Myths about Fractions, Decimals, and Percents

It could be that adding fractions is something that you do every day; in that case, fractions are fresh in your mind, and you find them easy to deal with. On the other hand, you may not have found a common denominator in years (and hope never to have to again). You also may have made it a point not to deal with decimals. But fractions and decimals can't be ignored; they need to be embraced — or, at least, tolerated.

It's an undeniable fact that fractions, decimals, and percents form the basis of much of the math in business. Before you run away, screaming and sobbing, let me tell you that in this book, I ease you into some of the less popular mathematical subjects when they arise. I show you the way to deal with the math in the most quick and efficient way possible.

In Part II of this book, you find fractions cropping up in formulas and with various measurement situations. You probably never complained much when told that you had to take half of an amount to complete a computation. After all, a half is the simplest fraction. But, if you can manage one fraction, you can manage them all. In Chapter 2, I go over some of the operations needed with fractions, and I show you how to change them to percents and back again.

Decimals are the middle ground between fractions and percents. You really can't get around them — nor do you really want to avoid them. When figuring percent increases or decreases in Chapter 3, you see how the change from percents to decimals is necessary. Discounts and markups are often confusing to understand and compute, but Chapter 17 shows you how to handle the ups and downs correctly using our good friends, the percent and decimal.

And how in the world can you deal with the interest earned on your account unless you haul out those delightful decimals? When you compute simple interest, you use the formula $I = Prt$. Nestled between the money amount ($P$ stands for principal) and the number of years involved ($t$ stands for time) is the rate of interest, $r$, which is given as a percent and changed to a decimal in order to do the computation.

You find percents and decimals cropping up throughout this book. If you come across a conundrum (okay, even just a little challenge), you can always refer to the chapters in Part I, which deal with fractions, decimals, percents, and their basic applications and computations.

# Capitalizing on Patterns in Formulas

A *formula* is nothing more than a relationship between values that always works and is always true. Wouldn't it be nice if formulas worked for people, too? But that's the big difference between people and numbers. Numbers are known to behave much better and more predictably than people, which is why you can embrace formulas with such confidence.

Formulas have been around since the beginning of recorded history, but in the Middle Ages, formulas weren't in their current neat-and-compact forms. It was only a couple hundred years ago that algebraic notation became popular and formulas such as $P = 2(l + w)$ and $a^2 + b^2 = c^2$ became a part of mathematical history.

The trick to working with any formula is knowing what the different variables stand for and how to perform the mathematical operations involved. For example, you should know that $A = \frac{1}{2}bh$ gives you the area of a triangle. You simply multiply one side of the triangle (the base, *b*) by the height of the triangle (*h*, which is measured from the base) and then find $\frac{1}{2}$ of the product. (Check out Chapter 21 for more on the area formulas.)

In Chapter 5, I go over the different rules for performing more than one operation in an expression. In all of Part II, you see how to use formulas in different settings. It's a splendid situation when you can put a formula into a computer spreadsheet and let the technology do repeated computations for you. So in Chapter 5, I give you some spreadsheet guidance.

Get ready, because you can find some pretty impressive formulas in Parts III and VI, where loans, mortgages, and other financial manipulations play a big part in the discussion. You don't want to memorize the formulas for annuities and sinking funds, however. You just need to become comfortable using the formulas and have confidence in your answers. You can always flip to the actual formula when it comes time to use it.

You probably remember how to do a *mean average,* but do you know whether the *median* or *mode* would be a better measure of the middle in your particular situation? Just use the formulas or rules in this book to find out. For instance, when you're managing rental properties or another business, you need to do some comparisons from month to month and year to year. Your statistics skills, which you can gain in Chapter 8, will put you in good stead for the computations needed. You don't find any heavy-duty statistics or statistical formulas in this book. For further investigations, check out *Statistics For Dummies* (Wiley).

# Finding the Power in Exponents

An *exponent* is a power. In other words, it's shorthand algebra that tells you to multiply something by itself over and over again. Exponents are fundamental in compound interest formulas and amortized loan formulas. In fact, they're great for everything financial.

For example, do you want to know what an investment of $10,000 will be worth in 5 years? Part III presents lots of options for dealing with your money and for doing the computations necessary. Mortgage rates change constantly, and so you often need to make decisions on your mortgages and loans. Those decisions rely heavily on computations using exponents. You find the lowdown in Part III and, again, in Part VI.

Are you a little shaky on the use of exponents? Check out Table 1-1, which gives you a quick reminder.

| Table 1-1 | The Rules for Operations Involving Exponents | |
|---|---|---|
| *Operation* | *How to Handle* | *Example* |
| Multiplying the same bases | Add the exponents. | $a^x \cdot a^y = a^{x+y}$ |
| Dividing the same bases | Subtract the exponents. | $\dfrac{a^x}{a^y} = a^{x-y}$ |
| Raising a power to a power | Multiply the exponents. | $(a^x)^y = a^{xy}$ |
| Handling negative exponents | Move the power to the denominator (the bottom part of the fraction) and change the sign. | $a^{-x} = \dfrac{1}{a^x}$ |
| Dealing with roots | Roots change to fractional exponents. | $\sqrt[x]{a} = a^{1/x}$ |

Use the rules involving multiplication and roots to simplify $9^{3/2}\sqrt{9}$.

First off, change the radical to a fractional exponent. Then multiply the two numbers together by adding the exponents like this:

$$9^{3/2} \cdot 9^{1/2} = 9^{3/2\,+1/2} = 9^2 = 81$$

Okay, that was fun. But you may not be convinced that exponents are all that important. What if I asked you to do a quick comparison of how much more money you accumulate if you invest a certain lump sum for 10 years instead of just 5 years? I show you how the exponents work in the following example. (You can go to Chapter 9 to get more of the details.)

Compare the amount of money accumulated if you deposit an insurance settlement for 10 years in an account earning 8% interest compounded quarterly versus just leaving that money in the account for 5 years (at the same interest rate).

The formula for compound interest is:

$$A = P\left(1 + \frac{r}{n}\right)^{nt}$$

To do the comparison of how much money accumulates, write a fraction with the 10 years compounding in the *numerator* (the top part of the fraction) and the 5 years compounding in the *denominator* (the bottom). Here's what your math should look like:

$$\frac{\text{Compounded 10 years}}{\text{Compounded 5 years}} = \frac{P\left(1 + \frac{0.08}{4}\right)^{4\,(10)}}{P\left(1 + \frac{0.08}{4}\right)^{4\,(5)}}$$

$$= \frac{P(1.02)^{40}}{P(1.02)^{20}}$$

$$= \frac{\cancel{P}(1.02)^{\cancel{40}\,20}}{\cancel{P}(1.02)^{\cancel{20}}} = (1.02)^{20} \approx 1.49$$

Performing some reducing of fractions and operations on exponents, you see that the amount of money accumulated is about 1½ times as much as if it's left for 10 years (rather than for 5). You can also use the formula to get the respective amounts of money for the two different investment times, but this equation shows you the power of a power for any amount of money invested at that rate and time.

# Doing Some Serious Counting

Keeping track of your inventory means more than taking a clipboard to the storeroom and counting the number of boxes that are there. Even a smaller business needs a systematic way of keeping track of how many items it has, how much the different items cost, and how quickly the items are used and need to be replaced.

So, in Part V, you find information on inventory, overhead, and depreciation. The amount of inventory on hand affects the cost of insurance, and, in turn, the cost of insurance affects the overall profit. *Profit* is determined by finding the difference between revenue and cost. And each part of the profit equation involves its own set of computations.

One of the best ways of keeping track of inventory and the related costs and revenue is to use a computer spreadsheet. You get some formal explanations on spreadsheets in Chapter 5. I also tell you throughout the book when the use of spreadsheets is possible.

Other types of counting or tallying come in the form of measurements — linear measurements, area, volume, and angles. You find uses for measuring lengths and widths and areas when you're building something new or renovating something old. You need accurate area computations when you're planning for the space needed for production or a particular volume. That way you can store what you've produced.

In these cases and more, you need to be able to switch from one measurement unit to another — from feet to yards, for example — and apply the measures to the correct situations. In Chapter 7, you find the basics of measuring, and you apply these measures in the chapters dealing with insuring spaces, renting properties, figuring acreage, and computing depreciation, just to name a few.

## Painting a Pretty Picture

Many people are visual — they learn, understand, and remember better if they have a picture of the situation. I'm one of those people — and proud of it. Present a problem to me, and I'll try to draw a picture of the situation, even if it means drawing stick figures of Ted, Fred, and Ned to do a comparison of their salaries.

Pictures, charts, graphs, and tables are extremely helpful when trying to explain a situation, organize information, or make quick decisions. You find tables of values throughout this book. A *table* is a rectangular arrangement of information with columns of items sharing some quality and rows of items moving sequentially downward. Tables of information can be transformed into spreadsheets or matrices so you can do further computations. (Matrices aren't covered in this book, but if you'd like, you can find information on them in *Algebra II For Dummies*, published by Wiley.)

Charts and graphs are quick, pictorial representations of a bunch of numbers or other numerical information. You don't get exact values from charts or graphs, but you get a quick, overall picture of what's going on in the business. You can find pie charts and line graphs and more in Chapter 6.

# Chapter 2

# Fractions, Decimals, and Percents

● ● ● ● ● ● ● ● ● ● ● ● ● ● ● ● ● ● ● ● ● ● ● ● ● ● ● ● ● ● ● ● ● ● ● ● ● ● ● ● ● ● ● ● ●

## In This Chapter

▶ Converting fractions to decimals and vice versa

▶ Relating percents to their decimal equivalents

▶ Performing basic operations on fractions

● ● ● ● ● ● ● ● ● ● ● ● ● ● ● ● ● ● ● ● ● ● ● ● ● ● ● ● ● ● ● ● ● ● ● ● ● ● ● ● ● ● ● ● ●

*F*ractions, decimals, and percents are closely related to one another. Every fraction has a percent equivalent, but you have to go through decimals to get from fractions to percents or from percents to fractions. And, although percents may be the most easy to understand of the three different forms, you can't really perform many operations involving percents without changing them to one of the other forms. You may want to just do away with all but one of these formats to avoid having to go through the hassle of doing all the conversions. However, you really wouldn't want to do that because each format has a place in everyday life and business.

In this chapter, I show you the ins and outs of changing from one form to another — fractions to decimals to percents, and back again. You'll see why the different formats are necessary and useful (and not evil).

## Changing from Fractions to Decimals

Everyone loves fractions. In fact, you probably find it hard to believe that anyone would ever want fractions to change! Okay. I jest. The practical reason for changing from fractions to decimals is to aid in computations and get a sense of the value. It's easy to trip up when entering a fraction into an arithmetic operation. But decimals work nicely in calculators. Also, the decimal equivalents of many fractions are easier to compare. For example, which is larger, $\frac{23}{57}$ or $\frac{26}{67}$? The decimal equivalent of $\frac{23}{57}$ is about 0.404, and the equivalent of $\frac{26}{67}$ is about 0.388. Much easier to compare now, aren't they?

Changing a fraction to a decimal is pretty straightforward — you just divide until the division comes to a screeching halt or until you see a pattern repeating over and over. One or the other (ending or repeating) will always occur when you start with a fraction.

## Considering the two types of decimals

The two types of decimal numbers that arise from dividing a fraction's numerator by its denominator are terminating and repeating decimals. A *terminating decimal* is just what it sounds like: it's a decimal that comes to an end. You may need to do a bit of dividing to come to that end, but you will find the end. A *repeating decimal* is one that never ends. Instead, it repeats itself over and over in a distinctive pattern. The pattern may contain one digit, two digits, or a huge number of digits.

### Terminating decimals

Terminating decimals are created from fractions whose denominators have the following:

- ✓ Factors (divisors) of only 2 and 5
- ✓ Powers of 2 and 5
- ✓ Multiples of powers of 2 and 5

For example, the fraction ⅝ has a terminating decimal, because the number 8 is $2^3$, a power of 2. And the fraction $\frac{137}{250}$ has a terminating decimal, because the number 250 is equal to $2 \cdot 5^3$, the product of 2 and a power of 5.

In general, you divide as many times as 2 or 5 are factors of the denominator. For example, if the denominator is 16, which is the 4th power of 2, you divide 4 times. If the denominator is 3,125, the 5th power of 5, you divide 5 times. If the denominator contains both powers of 2 and powers of 5, then the number that has the higher power (the most factors) determines how many times you divide.

### Repeating decimals

When the denominator of a fraction has factors other than 2 or 5, the decimal equivalent of the fraction always repeats. Always. For example, the fraction ⅓ has a decimal equivalent of 0.3333 . . . . The 3s go on forever. You indicate a repeating decimal by showing enough of the numbers that it's apparent what's repeating and then writing an *ellipsis* ( . . . ). Or, if you prefer, you can draw a line across the top of the repeating digits.

Here's what I mean: The fraction ⅖ has a repeating decimal, $.\overline{285714}$. All six digits repeat; if you didn't use the bar across the top, you'd write the decimal equivalent as 0.285714285714 . . . . Doing so shows that the six digits repeat over and over.

Find the decimal equivalent of the fraction $\dfrac{7}{375}$.

Divide 7 by 375, and continue until you find the repeating pattern. How do I know it repeats? Because the denominator, 375, is divisible by 3 and 5 — the 3 factor causes the repeating part. When dividing the problem, your math will look like this:

$$
\begin{array}{r}
.01866 \\
375\,\overline{)7.00000} \\
375 \\ \hline
3250 \\
3000 \\ \hline
2500 \\
2250 \\ \hline
2500 \\
2250 \\ \hline
250
\end{array}
$$

You see that, after the first few divisions, the remainder 250 keeps appearing every time. The first three digits of the decimal don't repeat, just the 6s. You'd continue to get only 6s, so the decimal equivalent is written 0.018666 . . . or $0.018\overline{6}$.

## Rounding decimals up or down

When you change the fraction ¼ to its decimal equivalent, you end up with a nice, tame decimal of 0.25. But the fraction ⅙ has a repeating decimal equivalent that never ends — 0.1666 . . . . You can even have terminating decimals that get a little long-winded. For example, the fraction $\dfrac{79}{3125}$ has a decimal value of 0.02528, which isn't all that long, but you may have more digits than you think is necessary for the situation at hand.

*Rounding off* the extra decimal digits takes care of overly long numerical values. Of course, rounding a decimal changes the number slightly. You're eliminating a more exact value when you drop pieces from the number. But rounding decimals under the right circumstances is expedient and useful.

Suppose, for example, that you're planning to cut a piece of wood into four equal pieces. So you measure the wood, divide by four, and cut the pieces to size. Consider a piece of wood measuring 6 feet, 3⅜ inches. One-fourth of the length of the board is 1 foot, 6.84375 inches. (Chapter 7 shows you how to deal with feet and inches, and later in this chapter I show how to divide fractions.) With that measurement in hand, you're all set to create four pieces of wood that measure 1 foot, 6.84375 inches each, right? Of course not. You know that the sawdust accounts for at least the last three decimal digits of the number. So, to make things a bit easier, you round off the measure to about 1 foot, 6.8 inches, which is equal to about 1 foot, 6⅞ inches.

When rounding decimals, you decide how many decimal places you want in your number, and then you round to the nearer of that place and lop off the rest of the digits. You can also use the Rule of 5 when rounding. I explain the different ways to round in the following sections.

### Rounding to the nearer digit

The rule for rounding is that you choose how many decimal digits you want in your answer, and then you get rid of the excess digits after rounding up or down. Before discarding the excess digits, use the following steps:

1. **Count the number of digits to be discarded, and think of the power of 10 that has as many zeros as digits to discard.**

2. **Use, as a comparison, half of that power of 10.**

   For example, half of 10 is 5, half of 100 is 50, and so on.

3. **Now consider the discarded digits.**

   If the amount being discarded is smaller than 5, 50, or 500 (and so on), you just drop the extra digits. If the amount being discarded is larger than 5, 50, or 500 (and so on), you increase the last digit of the number you're keeping by 1 and discard the rest. If you have exactly 5, 50, or 500 (and so on), you use a special rule, the *Rule of 5,* which I cover later in this chapter.

In the following examples, I talk about rounding a number to the "nearer tenth" or "nearer hundredth." This naming has to do with the number of placeholders present. To read more about this naming convention, check out the later section, "Ending up with terminating decimals."

Round the number 45.63125 to the nearer hundredth (leave two decimal places).

You want to keep the whole number, 45, and the first three digits to the right of the decimal point — the 6 and, possibly, the 3. The digits left over form the number 125. Because 125 is less than 500, just drop off those three digits. So your answer is that 45.63125 rounds to 45.63 when it's rounded to the nearer hundredth.

Round the number 645.645645 . . . to the nearer thousandth (leave three decimal places).

You want to keep the whole number, 645, to the left of the decimal point and at least the first two digits to the right of the decimal point. You keep the 6 and 4 and make a decision about the 5. Because the numbers to be dropped off represent 645 (or 6,456, 64,564, and so on), you see that the numbers to be dropped off represent a number bigger than 500 (or 5,000, 50,000, and so on). So you add 1 to the 5, the last digit, forming the rounded number 645.646 to replace the repeating decimal part of the original number.

Round the number 1.97342 to the nearer tenth (leave one decimal place).

When you round, you'll be dropping off the digits 7342, which represent a number bigger than 5,000. So you add 1 to the 9. But 1 + 9 = 10, so you have to carry the 1 from the 10 to the digit to the left of the decimal point. Your resulting number is 2.0 after dropping the four digits off. You leave the 0 after the decimal point to show that the number is correct (has been rounded) to the nearer tenth.

### Rule of 5

When you're rounding numbers and the amount to drop off is exactly 5, 50, or 500 (and so on), you round either up or down, depending on which choice creates an even number (a number ending in 0, 2, 4, 6, or 8). The reason for this rule is that half the time you round up and half the time you round down — making the adjustments due to rounding more fair and creating less of an error if you have to repeat the process many times.

Round the number 655.555 to the nearer hundredth (leave two decimal places).

Because there are only three digits, only the last digit needs to be dropped off. And because you have exactly 5, you round the hundredth place to a 6, giving you 655.56. Why? Rounding up gives you an ending digit that's an even number.

# Converting Decimals to Fractions

Decimals have a much better reputation than fractions. Why, for the life of me, I can't understand. But many people do prefer the decimal point followed by a neat row of digits. For most applications, the decimal form of a number is just fine. For many precision computations, though, the *exact form,* a number's fractional value, is necessary. Those working with tools know that fractions describe the different sizes (unless you're working with the metric system).

Changing a decimal into a fraction takes a decision on your part. First, you decide whether the decimal is terminating or repeating. Then, in the case of the repeating decimal, you decide whether all the digits repeat or just some of them do. What if the decimal doesn't terminate or repeat and just goes on forever and ever? Then you're done. Only decimals that terminate or repeat have fractional equivalents. The decimals that neither terminate nor repeat are called *irrational* (how appropriate).

In the following sections, I explain how to change terminating and repeating decimals into fractions. (For more information on irrational numbers, check out the nearby sidebar "Discovering irrational numbers.")

## Ending up with terminating decimals

*Terminating decimals* occur when a fraction has the following in its denominator:

- ✔ Only factors of 2 or 5
- ✔ Powers of 2 or 5
- ✔ Products of powers of 2 or 5

A terminating decimal has a countable number of decimal places, and the number formed has a name based on the number of decimal places being used. The names of the first eight decimal places with a 1 place holder are:

| | |
|---|---|
| 0.1 | one tenth |
| 0.01 | one hundredth |
| 0.001 | one thousandth |
| 0.0001 | one ten-thousandth |
| 0.00001 | one hundred-thousandth |
| 0.000001 | one millionth |
| 0.0000001 | one ten-millionth |
| 0.00000001 | one hundred-millionth |

## Discovering irrational numbers

It is our famous friend, Pythagoras, who's often given credit for discovering irrational numbers. Even though the ancient Babylonians have surviving documents that suggest that they were also onto the idea, it was the Pythagoreans who formalized the discovery with the irrationality of the square root of 2. The proof involves taking a square, drawing its diagonal, and setting up a proportion between the sides of the square and the diagonal. From the proportion, the Pythagoreans produced the equation $2a^2 = b^2$ and were able to argue that the equation has no solution if $a$ and $b$ are integers. (Integers, by the way, are positive and negative whole numbers and 0.) Other references to the irrational number proof show up in Aristotle's works and Euclid's *Elements*. The irrational square root of 2 has a decimal that neither terminates nor repeats; here are the first 60 or so decimal places of the square root of 2: 1.41421356237309504880168872420969807856967 18753769480731766797 3799 . . . . If you want more decimal places than appear here, go to www.rossi.com\sqr2.htm.

So, for example, you read the number 0.003456 as *three thousand four hundred fifty-six millionths.* The *millionths* comes from the position of the last digit in the number. And, as a bonus, the names of the decimal places actually tell you what to put in the denominator when you change a terminating decimal to a fraction. How?

When changing a terminating decimal to its fractional equivalent, you place the decimal digits over a power of 10 that has the same number of 0s as the number of decimal digits. Then you reduce the fraction to lowest terms. Check out the following example to see what I mean.

Rewrite the decimal 0.0875 as a fraction in its lowest terms.

The number 0.0875 is eight hundred seventy-five ten-thousandths. To rewrite this number as a fraction, you put the digits 875 (you don't need the lead zero because 875 and 0875 are the same number) in the numerator of a fraction and 10,000 in the denominator (you use 10,000 because there are 4 digits to the right of the decimal point and 4 zeros in 10,000 ). Now reduce the fraction. To do so, first divide the numerator and denominator by 25, and then reduce the numerator and denominator by 5. The math should look like this:

$$\frac{875}{10,000} = \frac{\overset{35}{\cancel{875}}}{\underset{400}{\cancel{10,000}}} = \frac{\overset{7}{\cancel{35}}}{\underset{80}{\cancel{400}}} = \frac{7}{80}$$

Yes, I know that I could have reduced the fraction in one step by dividing both the numerator and denominator by 125, but I don't really know my multiples of 125 that well, so I chose to reduce the fractions in stages.

# Dealing with repeating decimals

Repeating decimals come from fractions. But unlike terminating decimals, you can't put the decimal part over a power of 10, because the decimal never ends. In other words, it doesn't settle down to a particular power of 10, so you'd never stop putting zeros in the fraction.

*Repeating decimals* separate into two different groups or types. The first type is the one in which all the digits repeat. You don't have any start-off digits that don't fit in the repeating pattern. The second type of repeating decimal consists of those that have one or more digits that appear just once and don't fit into the repeating pattern. Each group has its own way of converting from decimals to fractions. I explain in the following sections.

## Catching all the digits repeating

A decimal number in which all the digits repeat looks something like these:

> 0.88 . . .
>
> 0.123123 . . .
>
> 0.14851485 . . .

It doesn't matter how many digits are in the repeating pattern. As long as all the digits are a part of the pattern, the number falls into this first group of repeating decimals.

To find the fractional equivalent of a repeating decimal in which all the digits repeat, put the repeating digits into the numerator of a fraction, and put as many 9s as there are digits in the repeating pattern in the denominator of the fraction. Then reduce the fraction to lowest terms. Sounds pretty simple, doesn't it? Try your hand at the following example.

Find the fractional equivalents of 0.88 . . . and 0.123123 . . .

For the decimal 0.88 . . ., put an 8 in the numerator and a 9 in the denominator. You get the fraction ⁸⁄₉, which is already in reduced form.

For the decimal 0.123123 . . ., put 123 in the numerator and 999 in the denominator. You reduce the fraction by dividing the numerator and denominator by 3. Here's what your simplifying should look like:

$$\frac{123}{999} = \frac{\cancel{123}^{41}}{\cancel{999}_{333}} = \frac{41}{333}$$

### *Working with digits that don't repeat in the pattern*

Repeating decimals with nonrepeating parts look like the following:

0.00345345 . . .

0.12699999 . . .

0.32517842842842 . . .

The nonrepeating parts never show up again after the decimal starts repeating, but that part does affect how you determine the corresponding fraction. You can't put the nonrepeating part over 9s and you can't put it over 0s. The actual technique involves a bit of a compromise.

To find the fractional equivalent of a repeating decimal with one or more nonrepeating digits, follow these steps:

1. **Set the decimal equal to $N$.**

2. **Multiply both sides of the equation by the power of 10 that has as many zeros as there are nonrepeating digits.**

3. **Multiply both sides of the new equation by the power of 10 that has as many zeros as there are repeating digits.**

4. **Subtract the equation in Step 2 from the equation in Step 3, and solve for $N$.**

Put these steps to use with the following example problem.

Find the fractional equivalent of the repeating decimal 0.123555555555 . . . .

You can solve this problem using the steps as previously given:

1. **Set the decimal equal to $N$ to get $N = 0.123555555555$ . . . .**

2. **Multiply both sides by 1,000 (because the 123 doesn't repeat), which gives you $1,000N = 123.555555555$ . . . .**

3. **Multiply each side of the equation in Step 2 by 10 (because only one digit, the 5, keeps repeating), and you get $10,000N = 1235.555555555$ . . . .**

4. **Subtract the equation in Step 2 from the equation in Step 3, and then solve for N:**

$$10{,}000N = 1235.555555555\ldots$$
$$-\ 1000N = \phantom{1}123.555555555\ldots$$
$$9000N = 1112$$

$$N = \frac{1112}{9000}$$

$$= \frac{\overset{139}{\cancel{1112}}}{\underset{1125}{\cancel{9000}}} = \frac{139}{1125}$$

I divided the numerator and denominator by 8 to reduce the fraction to lowest terms.

# Understanding the Relationship between Percents and Decimals

Percents are found daily in newspaper ads, financial statements, medical reports, and so on. However, percentages aren't really the same as the rest of our numbers. You need to change from a percent to a decimal before doing any computations involving percents of other quantities.

A *percent* is a number that represents *how many compared to 100.* Percents are usually much quicker and easier to understand or know the worth of than their fractional counterparts. Sure, you have some easier equivalents, such as ½ and 50% or ¾ and 75%. But a fraction's worth may not jump out at you. For instance, you may not immediately realize that $\frac{7}{20}$ is 35%.

In the earlier section, "Changing from Fractions to Decimals," I show you how to get from a fraction to its decimal equivalent. In this section, you see how to change from decimals to percents and from percents to decimals.

Percents are easy to understand, because you can quickly compare the percent amount to 100. If you're told that 89% of the people have arrived at an event, you know that most of the people have come. After all, 89 out of 100 is a lot of those expected. Similarly, if you're told that the chance of rain is 1%, you know that the chances are slim-to-none that you'll have precipitation, because 1 out of 100 isn't a large amount.

Decimals are fairly easy to deal with, but you may have more trouble wrapping your head around how much 0.125 of an estate is. Did you realize that 0.125 is the same as 12.5%? If so, you're probably in the minority. But good for you for knowing!

## Transforming from percents to decimals

A *percentage* is recognizable by the % symbol, or just by the word, *percent*. Percents are sort of pictures of amounts, and so they need to be changed to a decimal or fractional equivalent before combining them with other numbers. If I told you that 89% of the people had arrived at a meeting, and you're expecting 500 people, you can't just multiply 89 by 500 to get the number of people who have arrived. Why? Well, $89 \times 500 = 44{,}500$. So you can see that the answer doesn't make any sense. Instead, you need to change percents to decimals before trying to do computations.

To change a percent to its decimal equivalent, move the decimal point in the percent two places to the left. Moving the decimal point two places to the left gives you the same result as multiplying the percent amount by 0.01 (one hundredth). Why are you multiplying by 0.01? Because percents are comparisons to 100; you're changing the percent to how many out of 100.

The decimal equivalent of 12.5% is 0.125, and the decimal equivalent of 6% is 0.06. Notice that with 6%, I had to add a 0 in front of the 6 to be able to move the decimal point the two decimal places. You assume that there's a decimal point to the right of the 6 in 6%.

If you're told that 89% of the people have arrived at the company-wide meeting, and you're expecting 500 people, how many people have arrived?

To find out, simply change 89% to its decimal equivalent, 0.89. Then multiply $0.89 \times 500 = 445$ people.

When checking on the latest shipment of eggs, your dairy manager tells you that ½% of them are cracked. The shipment contained 3,000 dozen eggs. How many were cracked?

First, change 3,000 dozen to an actual number of eggs by multiplying $3{,}000 \times 12 = 36{,}000$ eggs. Next, change ½% to a decimal. The fraction ½ is 0.5 as a decimal, so ½% is 0.5%. Now, to change 0.5% to a decimal, move the decimal point two places to the left to get 0.005. Multiply 36,000 by 0.005, and you discover that 180 eggs are cracked. That's 15 dozen eggs in all.

## Moving from decimals to percents

When you change a number from a decimal to a percent, it's probably because you have a task in mind. Most likely, you've started with a fraction and are using the decimal as the *transition number* — the number between fractions and percents. If the decimal terminates, you can just carry all the digits along in the percent, or you can round off to a predetermined number of digits. If the decimal repeats, you want to decide how many decimal points you want in the percent value.

To change a decimal to its percent equivalent, move the decimal point in the decimal two places to the right.

If 13 out of 20 people received a flu shot, what percentage of people have received the shot?

To find the percentage, first write 13 out of 20 as a fraction. Then you can divide 13 by 20 to get the decimal equivalent: 0.65. (You can see how to change fractions to decimals in the earlier section "Changing from Fractions to Decimals.") Now change 0.65 to a percent by moving the decimal point two places to the right, which gives you 65%. (I don't show the decimal point to the right of the 5 in 65%, because it's understood to be there when it's not otherwise shown.)

The evening crew has completed 67 out of the 333 packets that need to be filled for the big order. What percentage of the packets has been completed?

The decimal equivalent of the fraction $\frac{67}{333}$ is 0.201201 . . . . As you can see, this is a repeating decimal. Moving the decimal point two places to the right gives you 20.1201201 . . . . Now you need to decide how many decimal places to keep.

Let's say that you want just the nearer percent. In that case, you round off all the digits to the right of the decimal point. The number 1,201,201 is less than half of 5,000,000, so you can just drop all the digits to the right of the decimal point and call the percentage 20%. (See "Rounding decimals up or down," earlier in this chapter, for more on how to round numbers and drop digits.)

## Dealing with more than 100%

Sometimes you may find it difficult to get your head around the report that there's a 300% increase in the number of travelers in a company or that you're getting a 104% increase in your salary. Just what do the percents really mean, and how do you work with them in computations?

First, think about 100% of something. For instance, if you're told that 100% of a job is complete, you know that the whole thing is wrapped up. But it doesn't make sense that 200% of the job is complete — you can't do more than all of the job. So, instead, think of 100% in terms of money. If I pay you 100% of what I owe you, say $800, I pay you all the money — $800. But if I pay you 200% of what I owe you, I pay you all of it twice — or $1,600. (Now, why would I do that? Because I really like you!)

For percents that exceed 100%, apply the same rule for changing percents to decimals. The equivalent of 200% is 2.00, and the equivalent of 104% is 1.04.

Try this example to better understand how to work with percentages that are equal to more than 100. Suppose you're promised a 104% increase in salary if you agree to stay in your position for another 5 years. If you're currently earning $87,000, what will your salary be for the next 5 years?

To start, multiply 1.04 by $87,000 to get $90,480. Now the question is: Does the $90,480 actually represent an *increase* — an amount to be added to your current salary? If so, you'll be earning $87,000 + $90,480 = $177,480. Or did the person offering the increase actually mean to just increase your salary by 4%, in which case you'd be earning the $90,480? You need to be sure you understand the terms exactly. In Chapter 3, you can see how to be sure of your figures when talking about percentages and their ups and downs.

# Coming to Grips with Fractions

I'll bet that fractions are your favorite things. Everyone I know just loves fractions. Oh, who am I kidding? I didn't have you fooled for a minute. So fractions are way down on your list of things to talk about at a party, but they can't be ignored. Fractions are a way of life. You cut a pie into eighths, sixths, or fourths (or, heaven forbid, sevenths). You whip out your ⅞-inch wrench. You measure a board to be 5 feet and ⁹⁄₁₆ inches long.

See what happened when the United States decided not to go metric? We got stuck using all these fractions when computing. So, until that happy day when the country changes its mind regarding the metric system, you *can* deal with fractions and the operations that go with them.

Fractions are still useful for describing amounts. They're exact numbers, so a sale proclaiming ⅓ off the original price tells you to divide by 3 and take one of them away from the original price. Do you get the same thing changing the fraction to a decimal? No, not really. The decimal for ⅓ is 0.3333 . . ., which approximates ⅓ but isn't exactly the same. You don't really notice the difference unless you're dealing with large amounts of money or assets.

## Adding and subtracting fractions

Fractions can be added and subtracted only if they have the same denominator. So before adding or subtracting them, you have to find equivalent values for the fractions involved so that they have the same denominator. One fraction is equivalent to another if the numerator and denominator of the one fraction are the same multiple as the numerator and denominator of the other fraction. For example, the following three fractions are equivalent to the fraction $\frac{6}{18}$:

$$\frac{6}{18} \cdot \frac{5}{5} = \frac{30}{90}, \ \frac{6}{18} \cdot \frac{8}{8} = \frac{48}{144}, \ \frac{6}{18} \cdot \frac{\frac{1}{6}}{\frac{1}{6}} = \frac{1}{3}$$

You could find infinite numbers of fractions that are equivalent to the given fraction. The three I chose are found by multiplying the numerator and denominator by 5, 8, and ⅙.

As I note earlier, you can only add or subtract fractions if they have the same denominator. So if two fractions that need to be added together don't have the same denominator, you have to change the fractions until they do. You change either one or both of the fractions until they have the same (*common*) denominator. For example, if you want to add ⅗ and ⁷⁄₁₀, you change the ⅗ to an equivalent fractions with 10 in the denominator. The fraction equivalent to ⅗ is ⁶⁄₁₀, which has the same (common) denominator as the fraction ⁷⁄₁₀.

Find the sum of ¾ and ⅜.

Change ¾ to a fraction with 8 in the denominator. Then add the two numerators together. Your math should look like this:

$$\frac{3}{4} \cdot \frac{2}{2} = \frac{6}{8}$$
$$\frac{6}{8} + \frac{3}{8} = \frac{9}{8}$$

The sum of the two fractions, ⁹⁄₈, is an *improper fraction* — it's bigger than 1. So, you need to rewrite it as 1⅛.

To write an improper fraction as a mixed number, you divide the denominator into the numerator. The number of times that the denominator divides is the whole number, in front. The remainder goes in the numerator of the fraction that's left. If you need to borrow from the whole number in order to subtract, you add the equivalent of 1 to the fraction.

Find the difference between 9⅓ and 4⅞.

The common denominator of the two fractions is 24. So write the two as equivalent fractions and then subtract. Here's what the math looks like:

$$9\frac{1}{3} \cdot \frac{8}{8} = \ 9\frac{8}{24}$$
$$-4\frac{7}{8} \cdot \frac{3}{3} = -4\frac{21}{24}$$

You borrow 1 from the 9, which you write as $\frac{24}{24}$ before adding it to the top fraction, like so:

$$\begin{aligned} {}^{8}\!\!\not{9}\,\frac{8}{24} + \frac{24}{24} &= \quad 8\,\frac{32}{24} \\ -4\,\frac{21}{24} \quad\quad &= -4\,\frac{21}{24} \\ &\quad\quad\; 4\,\frac{11}{24} \end{aligned}$$

Whenever you have to borrow from a whole number in order to subtract fractions, you change the 1 that's borrowed from the whole number to a fraction with the common denominator in both the top and bottom of the fraction that's added.

## Multiplying and dividing fractions

Adding and subtracting fractions is generally more complicated than multiplying and dividing them. The nice thing about multiplication and division of fractions is that you don't need a common denominator. And you can reduce the fractions before ever multiplying to make the numbers smaller and more manageable. The challenge, however, is that you have to change all mixed numbers to improper fractions and change whole numbers to fractions by putting them in the numerator with a 1 in the denominator.

Multiply 5⅓ by 4⅕.

To find the product of this problem, first change the mixed numbers to improper fractions. To do so, multiply the whole number by the denominator, add the numerator, and then write the sum over the denominator, like this:

$$5\,\frac{1}{3} = \frac{5 \cdot 3 + 1}{3} = \frac{16}{3}$$
$$4\,\frac{1}{5} = \frac{4 \cdot 5 + 1}{5} = \frac{21}{5}$$

You're almost ready to multiply the two improper fractions together. But first, remember that it helps to reduce the multiplication problem. Do so by dividing the 3 in the denominator of the first fraction and the 21 in the numerator of the second fraction by 3:

$$\frac{16}{3} \cdot \frac{21}{5} = \frac{16}{\underset{1}{\not{3}}} \cdot \frac{\not{21}^{\,7}}{5}$$

After you do that, you're ready to multiply the two numerators and two denominators together. Your math will look like this:

$$\frac{16}{1} \cdot \frac{7}{5} = \frac{112}{5} = 22\frac{2}{5}$$

The final answer, 22⅖, is obtained by dividing the numerator by the denominator and writing the quotient and remainder as a mixed fraction.

To divide one fraction by another, you change the problem to multiplication by flipping the second fraction. In other words, division is actually multiplication by the *reciprocal* of a number. The reciprocal of 2, for example, is ½, and the reciprocal of ⅞ is 2/7.

Divide 8¾ by 3⅛.

To begin, change the mixed numbers to improper fractions. Then change the division to multiplication by multiplying the first fraction by the reciprocal of the second fraction. Your math will look like this:

$$8\frac{3}{4} \div 3\frac{1}{8} = \frac{35}{4} \div \frac{25}{8} = \frac{35}{4} \cdot \frac{8}{25}$$

Now reduce the multiplication problem by dividing by 5 in the upper left and lower right and dividing by 4 in the other two numbers. Then multiply across the top and bottom and write the final answer as a mixed number. Here's what the math will look like:

$$\frac{35}{4} \cdot \frac{8}{25} = \frac{\overset{7}{\cancel{35}}}{\underset{1}{\cancel{4}}} \cdot \frac{\overset{2}{\cancel{8}}}{\underset{5}{\cancel{25}}} = \frac{14}{5} = 2\frac{4}{5}$$

# Chapter 3

# Determining Percent Increase and Decrease

· · · · · · · · · · · · · · · · · · · · · · · · · · · · · · · · · · · · · · · · · · · · · · · ·

## In This Chapter

▶ Applying percentages to increases and decreases

▶ Deciphering the percent change language

▶ Working backward from the percent changes to the original prices

· · · · · · · · · · · · · · · · · · · · · · · · · · · · · · · · · · · · · · · · · · · · · · · ·

*I*nformation involving percentages is fairly easy to decipher. If you're 45 percent finished with a project, for instance, you know that you're almost half done. Percents are based on the idea *how many out of 100,* so the comparison of the percentage to the total amount is quick and universally understood.

You're probably confronted with percent increases and decreases on a daily basis. For example, you may deal with salary increases or decreases. Citing these changes as percentages is easier when it comes to comparing the changes. One person's increase of $2,000 per year, for instance, is much different from another person's increase of the same amount — especially when one person currently earns $10,000 and the other earns $100,000.

In this chapter, you find methods for determining the percent increase or decrease so that you can do comparisons. You also find out how to work backward from an increased amount to see what the original figure was. If you're a little shaky on percents and doing computations with them, flip to Chapter 2. There you can get reacquainted with the ins and outs of percentages.

## Working with Percent Increase

A *percent increase* isn't just an increase in the original amount; it's a comparison to the original amount. When you get a 5% increase in salary, you get what you originally earned plus an additional 5% of that original amount. If you're told the dollar amount of your increase and want to determine what percent that increase is, you compare the amount of the increase by the original amount. The original amount, the new amount, the increase, and the percent increase are all entwined with one another and allow you to solve for any one of the values when given the others. I explain how in the following sections.

## Computing new totals with increases

You can add or subtract percents to get their sums or differences; the answers in these cases are in percents. But when mixing percents with numerical amounts, such as money or time, you have to change the percent to a decimal before multiplying. (I discuss the percent-to-decimal procedure in detail in Chapter 2.)

When figuring the amount resulting from a particular percent increase, you change the percent to a decimal and then multiply by the original amount. After you have the amount of the increase, you add that amount to the original number to get the new total. Or, if you want one-stop computing, you can add the percentage value of the increase to 100%. Then you change the total percentage to a decimal, multiply by the original amount, and have the new total amount as a final result.

When you have amount $A$ and want to determine a percent increase of $p$%, you find the

- Percent increase by multiplying $A \times 0.01p$.
- New total by adding 100% + $p$% and then multiplying $A \times 0.01(100 + p)$.

**Tip:** Multiplying the percent by 0.01 is the same as moving the decimal point two places to the left.

Find the percent increase and new salary if your employer is going to give you a 5% increase on your current salary of $48,200.

To find the percent increase, multiply 0.05 by $48,200 to get $2,410. The new salary is the sum of the increase and the original salary: $2,410 + $48,200 = $50,610. You also can get the new total by multiplying the original salary by 105% (add the 5% increase to 100%): $48,200 × 1.05 = $50,610.

## Determining the percent increase

When given the percentage that some number is to be increased, you change the percentage to a decimal and multiply by the original amount. So, for example, a 3% increase on 200 items is $0.03 \times 200 = 6$. As you can see, it takes multiplication to find the percent increase, so it takes division to find out the percent if you only know the increase.

To find the percentage amount that something has increased, you divide the amount of the increase by the original value. The formula you use looks like this:

$$\text{percent increase} = \frac{\text{amount of increase}}{\text{original amount}}$$

$$= \frac{\text{increased amount} - \text{original amount}}{\text{original amount}}$$

Try out the following example to see how this formula works.

Suppose that your best salesman sold 400 cars last year and 437 cars this year. What was the percent increase of his sales?

To find the percent increase, first subtract 400 from 437 to get an increase of 37 cars this year. Now divide 37 by the original amount (last year's number) to get the decimal value that you can then change to a percent. Here's what the math looks like:

$$\text{percent increase} = \frac{437 - 400}{400} = \frac{37}{400} = 0.0925$$

As you can see, the decimal answer is 0.0925, which is equal to 9.25%. (Refer to Chapter 2 if you need a reminder on how to change from a decimal to a percent.)

## Solving for the original amount

With the basic equation used to solve for the percent increase (refer to the previous section), you can solve for either the percent increase, the increased amount, or the original amount — as long as you're given the other two values. I could give you three different formulas or equations, one to solve for each of the values, but it's easier to just remember one formula and perform some simple algebra to solve for what's missing.

Imagine you're told that all the prices in a store increased by 4½% last year. You have a list of all the current prices and want to determine the original price of an item. To do so, you just have to use the formula for percent increase (provided in the previous section). You fill in the percentage and the new price, let the original price be represented by $x$, and then solve for $x$. Check out the following example for practice.

Say that a hardware store is selling a power washer for $208.95. This current price is 4½% higher than the price for the same power washer last year. What was the cost of the washer last year?

To find out, plug all the numbers into the equation for finding the percent increase:

$$\text{percent increase} = \frac{\text{increased amount} - \text{original amount}}{\text{original amount}}$$

$$4\frac{1}{2}\% = \frac{208.95 - x}{x}$$

$$0.045 = \frac{208.95 - x}{x}$$

Now multiply each side of the equation by the denominator on the right, which is $x$. Then add $x$ to each side and divide by the coefficient of $x$. Your math should look like this:

$$x \cdot 0.045 = \frac{208.95 - x}{\cancel{x}} \cdot \cancel{x}$$

$$0.045x = 208.95 - x$$

$$+\, 1x = \qquad\quad +\, x$$

$$1.045x = 208.95$$

$$\frac{\cancel{1.045}x}{\cancel{1.045}} = \frac{208.95}{1.045}$$

$$x \approx 199.95$$

Last year's price for the power washer was $199.95.

# Looking into Percent Decrease

Computing *percent decrease* has a lot of similarities to computing percent increase — you change percents to decimals, you multiply by the original amount, and then you refer to the original amount. However, sometimes the math is just a bit trickier, because you subtract rather than add (subtraction is a more difficult operation for most folks). And the amount of the decrease often gets obscured (you end up computing with a number that wasn't in the original problem) when you find the reduced amount directly, without subtracting. I explain everything you need to know about computing percent decreases in the following sections.

For many of you, the best method for determining whether you've made an error in computation is just common sense. You don't have to be a mathematician to have a feel for the answer. If you're figuring the percent increase in your salary and you get a 400% increase, you'll probably realize that this isn't realistic. (If it *is* realistic, then we need to talk.) Basically, you see if the answer resembles what you've expected; most of the errors come from using the wrong operation, not in the actual arithmetic (because most people use a calculator, anyway).

## Finding new totals with decreases

Percent decreases play a huge role in the world of discount pricing. You see advertisements by merchants announcing that certain items, or even all items, are marked down by a particular percent of the original value of the item. The percent decrease is determined by multiplying the decimal equivalent of the percent decrease by the original amount.

When you have amount $A$ and want to determine a percent decrease of $p\%$, you find the

- ✔ Percent decrease by multiplying $A \times 0.01p$.
- ✔ New total by subtracting $100\% - p\%$ and multiplying $A \times 0.01(100 - p)$.

Multiplying by 0.01 has the same effect as moving the decimal point two places to the left. (You can see more on changing percents to decimals in Chapter 2.)

***Tip:*** For the store that has to determine all the new prices of all the red-tagged items, a nice spreadsheet does the job quickly and accurately. In Chapter 5, you see some suggestions for using a computer spreadsheet.

This example should get you going in the right direction with percent decreases: Say that the after-the-holidays sale at a furniture store features 40% off all red-tagged items. What's the new price of a $3,515 sofa after the 40% decrease in price?

The amount of the decrease is 40%, which is 0.40 as a decimal. First multiply the original amount by the percentage in decimal form: $\$3,515 \times 0.40 = \$1,406$. Then subtract that product from the original amount to get the new price: $\$3,515 - \$1,406 = \$2,109$.

## Figuring out the percent decrease

Say you purchase several thousand dollars' worth of office supplies and are given a volume discount on the entire order. Your bill just shows the total before the discount and the net cost after the discount. You can determine the percentage of the discount by using the formula for percent decrease.

To find the percent decrease, you divide the amount of the decrease by the original value. Here's the formula:

$$\text{percent decrease} = \frac{\text{amount of decrease}}{\text{original amount}}$$

$$= \frac{\text{original amount} - \text{reduced amount}}{\text{original amount}}$$

Using this formula, try out the following example problem.

What's the percent discount if you paid $1,418.55 for an order whose original cost (before the discount) was $1,470?

To solve, simply plug all the numbers into the formula for percent decrease, like so:

$$\text{percent decrease} = \frac{1470.00 - 1418.55}{1470.00} = 0.035$$

The decimal 0.035 is equivalent to 3.5%.

## Restoring the original price from a decreased price

If you purchase a piece of machinery after getting the frequent customer discount of 10%, you may want to determine the original cost for insurance purposes. To find out that original cost, you use the formula for determining the percent decrease. You let the original amount be represented by $x$, and then you solve for $x$ in the equation after replacing the other quantities with their respective values.

For example, suppose you're such a good customer at Tracy's Tractors that you're given a 10% discount on your purchase of a new tractor. You have to pay only $7,688.70. You need to insure the tractor at its replacement cost. What did the tractor cost before the discount?

Using the percent discount formula, replace the percent with 10% and the discounted price with $7,688.70. Let the original price be represented by $x$ and solve for $x$. Here's what the new equation will look like:

$$\text{percent decrease} = \frac{\text{original amount} - \text{reduced amount}}{\text{original amount}}$$

$$10\% = \frac{x - 7688.70}{x}$$

$$0.10 = \frac{x - 7688.70}{x}$$

Now multiply each side of the equation by $x$. Then subtract $x$ from each side and divide by the coefficient of $x$. Your math will look like this:

$$x \cdot 0.10 = \frac{x - 7688.70}{\cancel{x}} \cdot \cancel{x}$$

$$0.10x = x - 7688.70$$

$$-1x = -x$$

$$-0.90x = -7688.70$$

$$\frac{\cancel{-0.90}x}{\cancel{-0.90}} = \frac{-7688.70}{-0.90}$$

$$x = 8,543$$

As you can see, the tractor originally cost $8,543.

# Chapter 4

# Dealing with Proportions and Basic Algebra

· · · · · · · · · · · · · · · · · · · · · · · · · · · · · · · · · · · · · · · · · · · · · · · · · ·

## In This Chapter

▶ Working with proportions

▶ Writing and solving linear equations from algebra

▶ Considering direct and indirect variation

· · · · · · · · · · · · · · · · · · · · · · · · · · · · · · · · · · · · · · · · · · · · · · · · · ·

*O*ne of the most useful structures in mathematics is the proportion. You use it to parcel out amounts in fair and equal portions and to determine equivalent lengths, volumes, or amounts of money. A proportion is nothing more than two fractions set equal to one another, but it's beautifully symmetric and functionally versatile.

Working with proportions helps you solve simple algebraic equations. An *algebraic equation* has at least one unknown variable lurking in its depths. You've solved an algebraic equation when you figure out what a missing variable is worth. The rules for solving simple, linear algebraic equations make a lot of sense and are easy to remember. The key is to keep the equation balanced.

I also cover direct and indirect variation in this chapter. Both of these are actually comparisons of things. Two items vary directly with one another if one of the items is tied to the other by some multiple. For example, if Henry always runs twice as fast as Henrietta, then when Henrietta runs at 3 meters per second, you know that Henry runs at 6 meters per second. Indirect variation takes a bit more explanation, but it's basically an opposite type of relationship — as one thing gets bigger, the other gets smaller.

You won't get as much exposure to solving equations in this chapter as you would if you were reading *Algebra For Dummies* or *Algebra II For Dummies* (both from Wiley), but do know that I show you the basics and all you need to handle equations special to business math.

# Setting Up Proportions

A *proportion* is an equation in which one fraction is set equal to another fraction. The following are examples of proportions:

$$\frac{3}{6} = \frac{25}{50}$$
    Two fractions equaling ½

$$\frac{1 \text{ mile}}{5,280 \text{ feet}} = \frac{\frac{1}{4} \text{ mile}}{1,320 \text{ feet}}$$
    Equating miles and feet

$$\frac{1 \text{ hour of work}}{10 \text{ hours of work}} = \frac{\$15.80}{\$158.00}$$
    Equating hourly rates with hours

When setting up a proportion, you keep the same units either across from one another or in the same fraction — above and below one another. Try your hand at proportions with the following example.

Set up a proportion involving the fact that 1 pound equals 16 ounces and 80 ounces equals 5 pounds.

To set up this proportion, you either have to write the pounds across from one another or above and below one another. Also, you have to keep the equivalences across from one another or above and below one another. The following are just four of the different ways that the proportion involving pounds and ounces can be written correctly:

$$\frac{1 \text{ pound}}{16 \text{ ounces}} = \frac{5 \text{ pounds}}{80 \text{ ounces}} \qquad \frac{1 \text{ pound}}{5 \text{ pounds}} = \frac{16 \text{ ounces}}{80 \text{ ounces}}$$

$$\frac{80 \text{ ounces}}{16 \text{ ounces}} = \frac{5 \text{ pounds}}{1 \text{ pound}} \qquad \frac{5 \text{ pounds}}{1 \text{ pound}} = \frac{80 \text{ ounces}}{16 \text{ ounces}}$$

A property common to all proportions is that the *cross products* (the diagonal products) are always equal. This commonality makes checking and solving proportions even easier. When you multiply the top left number by the bottom right number, you get the same result as multiplying the bottom left number by the top right number.

For example, in the equation involving fractions equaling ½ (shown at the beginning of this section), you see that the cross products are both 150. The cross products in the proportion involving miles and feet have cross products of 1,320. You can check the other proportions so far and see that the cross products always come out the same.

Proportions have several properties that make them easy to deal with and solve. Given the proportion $\frac{a}{b} = \frac{c}{d}$, the following are always true:

✔ The cross products are equal: $ad = bc$.

✔ The *reciprocals* (the upside-down versions of fractions) also form a proportion: $\frac{b}{a} = \frac{d}{c}$.

✔ You can reduce either fraction vertically, dividing by a common factor:

$$\frac{a}{b} = \frac{e \cdot f}{e \cdot g}, \frac{a}{b} = \frac{\cancel{e} \cdot f}{\cancel{e} \cdot g}, \frac{a}{b} = \frac{f}{g}.$$

✔ You can reduce horizontally, across the top or bottom, dividing by a common factor: $\frac{a}{e \cdot f} = \frac{c}{e \cdot g}, \frac{a}{\cancel{e} \cdot f} = \frac{c}{\cancel{e} \cdot g}, \frac{a}{f} = \frac{c}{g}.$

Ready for some practice? Simplify the following proportion by reducing the fractions: $\dfrac{6 \text{ days}}{20 \text{ days}} = \dfrac{144 \text{ hours}}{480 \text{ hours}}$

You have many choices for reducing the fractions. Take the most obvious first, and divide the two bottom numbers by 10. Then divide the numbers in the left fraction by 2. Here's what your reductions should look like so far:

$$\frac{6}{\underset{2}{\cancel{20}}} = \frac{144}{\underset{48}{\cancel{480}}}$$

$$\frac{\overset{3}{\cancel{6}}}{\underset{1}{\cancel{2}}} = \frac{144}{48}$$

$$\frac{3}{1} = \frac{144}{48}$$

Next, divide the two top numbers by 3. Then divide the two numbers in the right fraction by 48:

$$\frac{\overset{1}{\cancel{3}}}{1} = \frac{\overset{48}{\cancel{144}}}{48}$$

$$\frac{1}{1} = \frac{\overset{1}{\cancel{48}}}{\underset{1}{\cancel{48}}}$$

$$\frac{1}{1} = \frac{1}{1}$$

As you can see, you end up with 1 = 1. So what does that tell you? Nothing, really, except that proportions always act that way and give you a true statement. The real power and versatility of proportions come into play when you solve for some unknown value within the proportion.

# Solving Proportions for Missing Values

Proportions and their properties find their way into many financial and scientific applications. For instance, determining doses of medicine incorporates proportions involving a person's weight. Vegetation and animal habitats are in a balance expressed with proportions as well. And figuring a seller's share of the real estate taxes involves ratios and proportions.

# Setting up and solving

The biggest challenge when solving a proportion for an unknown value is in setting the proportion up correctly. (I detail the correct way to set up a proportion in the earlier section, "Setting Up Proportions.") When you have an unknown value, you let that value be represented by a variable, usually $x$, you cross-multiply, and then you solve the resulting equation for that variable. The following example gives you a chance to solve a proportion for a missing value.

A distributor of garbage disposals has been getting complaints that pieces are missing from the boxes of new disposals. In the past month, the distributor has sold 143 disposals and has received 15 complaints. He has 200 more of the disposals in his warehouse. If the complaint pattern holds, how many of the remaining disposals can he expect to have missing parts?

To find out, set up one fraction with the 143 disposals in the *numerator* (the top part of the fraction) and the 15 complaints in the *denominator* (the bottom part). The other fraction in the proportion needs to have the 200 disposals in the numerator — opposite the 143 disposals — and an $x$ in the denominator, representing the unknown number of complaints. Here's what the proportion looks like:

$$\frac{143 \text{ disposals}}{15 \text{ complaints}} = \frac{200 \text{ disposals}}{x \text{ complaints}}$$

Solve for $x$ by cross-multiplying and then dividing each side of the equation by the coefficient of $x$. Take a look at the math:

$$\frac{143}{15} = \frac{200}{x}$$
$$143x = 3{,}000$$
$$\frac{143x}{143} = \frac{3{,}000}{143} \approx 20.979$$

So, the distributor can expect to have about 21 more complaints — unless he decides to open the boxes beforehand to check whether the parts are all there.

# Interpolating when necessary

In Chapter 15, you find many charts and tables with rows and columns full of numbers. When reading tables and using the numbers in computations or reports, it's sometimes necessary to read between the lines, or, mathematically speaking, to interpolate.

*Interpolation* is basically inserting numbers between entries in a table. You assume that the change from one entry to another is pretty uniform and that you can find a number halfway between two numbers on the table by halving the difference. Proportions allow you to find numbers between the entries that are more or less than halfway between.

Table 4-1 shows the cumulative sales of a realty company for the first few months of the year. (The word *cumulative* means that it's the sum of everything up to that point.) On March 1, the total number of homes sold by the realty company was 50. On April 1, the total number of homes sold was 100. If you assume that the sales were relatively steady throughout the month of March, how many homes had been sold by March 20?

| Table 4-1 | Cumulative Number of Sales |
| --- | --- |
| *Month* | *Number of Sales* |
| January | 10 |
| February | 20 |
| March | 50 |
| April | 100 |
| May | 180 |

To interpolate, set up a proportion in which one fraction has the number of days involved and the other fraction has the number of homes. Now you need to do the following three things to solve the entire problem:

1. **Determine how many homes were sold from March 1 through April 1.**

2. **Solve for the number of homes sold by March 20.**

3. **Add the result from Step 2 to the 50 homes that were sold by March 1.**

First, 50 – 20 = 30 tells you that 30 homes were sold in March. Let the fraction involving the days have 20 in the numerator and 31 in the denominator. ("Thirty days hath September . . .") Put the *x* opposite the 20 and the 30 opposite the 31. So your proportion looks like this:

$$\frac{20 \text{ days}}{31 \text{ days}} = \frac{x \text{ homes}}{30 \text{ homes}}$$

Now cross-multiply and divide each side of the equation by the coefficient of *x*:

$$\frac{20}{31} = \frac{x}{30}$$
$$600 = 31x$$
$$\frac{600}{31} = \frac{\cancel{31}x}{\cancel{31}}$$
$$x \approx 19.354$$

It looks like about 19 homes were sold between March 1 and March 20, so the year-to-date number on March 20 is 20 + 19 = 39 homes. Great work!

When using proportions to interpolate, be careful not to assume a relatively equal rate between the consecutive entries in the table. In the practice problem, if I would have tried to interpolate starting from the beginning of the year, I would have received a completely different (and incorrect) number. Only 20 homes were sold during the first two months and then 30 more during the third month. The average change is different for the different months.

# Handling Basic Linear Equations

When solving proportions for an unknown value, and when performing many mathematical computations involving measures, money, or time, you'll likely find yourself working with an equation that has a variable in it. You solve for the value of the unknown by applying basic algebraic processes. These processes keep the integrity of the equation while they change its format so you can find the value of the unknown.

When solving a basic linear equation of the form $ax + b = c$, where $x$ is an unknown variable and $a$, $b$, and $c$ are constants, you apply the following rules (not all will be used in every equation — you pick and choose per situation):

- ✔ Add the same number to each side of the equation.
- ✔ Subtract the same number from each side of the equation.
- ✔ Multiply both sides of the equation by the same number (but don't multiply by 0).
- ✔ Divide each side of the equation by the same number (but don't divide by 0).
- ✔ Distribute (multiply) each term in parentheses by the number outside the parentheses so that you can drop the parentheses.
- ✔ Check your answer in the original equation to be sure it makes sense.

Here's an example you can try: Solve this equation for the value of $x$: $3(2x - 7) = 4x - 13$.

First distribute the 3 over the terms in the parentheses by multiplying each term by 3. You get $6x - 21 = 4x - 13$.

Now subtract $4x$ from each side and add 21 to each side. Doing so allows you to get the $x$'s on one side of the equation and the numbers without $x$'s on the other side. Here's how your math should look:

$$6x - 21 = 4x - 13$$

$$\begin{array}{rcr} -4x & & -4x \\ \hline 2x - 21 = & & -13 \\ +21 & & +21 \\ \hline 2x & = & +8 \end{array}$$

Now divide each side of the equation by 2, and you get that $x = 4$. Always check your answer. If you replace the $x$ with 4, you get $3(2\times4 - 7) = 4\times4 - 13$. In the parentheses, you get $8 - 7 = 1$. So the equation now reads $3(1) = 16 - 13$, or $3 = 3$. When you see a true statement — one in which a number equals itself — you know that you've done the process correctly.

**GO FIGURE**

Just for kicks, try out another practice problem. Solve the following proportion for the value of $x$: $\dfrac{200}{400} = \dfrac{2x}{x + 30}$

First reduce the fraction on the left by dividing both the numerator and denominator by 200. Then cross-multiply. Your math looks like this so far:

$$\frac{\cancel{200}^{1}}{\cancel{400}^{2}} = \frac{2x}{x + 30}$$

$$\frac{1}{2} = \frac{2x}{x + 30}$$

$$1(x + 30) = 2(2x)$$

$$x + 30 = 4x$$

Now solve the equation for the value of $x$.

$$x + 30 = 4x$$

$$\begin{array}{rcr} -x & & -x \\ \hline 30 = & & 3x \end{array}$$

$$\frac{30}{3} = \frac{\cancel{3}x}{\cancel{3}}$$

$$10 = x$$

Check to be sure that $x$ does equal 10. By checking your answer to see if it makes sense, you're more apt to catch silly mistakes. Go back to the original proportion and replace the $x$'s with 10s, like so:

$$\frac{200}{400} = \frac{2(10)}{10 + 30}$$

$$\frac{200}{400} = \frac{20}{40}$$

You can stop with the check at this stage, noting that both fractions are equal to ½. Or you can cross-multiply to get 8,000 on each side of the equation. You also can reduce both fractions to get the same numbers. In any case, you've shown that the 10 does represent the *x* in the proportion.

# Comparing Values with Variation

What is *variation?* It's how one entity varies with relation to another. Variation comes in two varieties: direct and indirect. Consider the following statements:

- ✔ The pressure of a gas is indirectly proportional to its volume.
- ✔ The productivity of the team is indirectly proportional to the number of absences.
- ✔ The revenue earned is directly proportional to the hours worked.
- ✔ The speed of the truck is directly proportional to the acceleration.

I explain the differences between the two types of variation in the following sections.

## Getting right to it with direct variation

When one quantity varies directly with another, one quantity is directly proportional to the other. An algebraic expression of how the two quantities *a* and *b* vary directly is: $a = kb$, where *k* is the *constant of proportionality,* which is the multiplier that connects the two quantities. If you divide each side of the equation by *b*, you get the following proportion:

$$\frac{a}{b} = \frac{k}{1}$$

And this shows you how the proportion and variation are tied together. When quantities vary directly, one quantity is a multiple of the other. If you know at least one of the quantities in the relationship, you can solve for the other (if it's unknown). Check out the following example problem to see exactly what I mean.

The revenue earned by the sales office varies directly with the number of hours spent out in the field. Last month, the sales team earned $450,000 and spent a total of 1,200 hours meeting with customers. How much will the team earn if it spends 2,000 hours out in the field?

Using direct variation, you need to find the value of $k$, the constant of proportionality. Write the equation $450,000 = k \times 1,200$ for the relationship between the quantities. Divide each side of the equation by 1,200, and you get that $k = 375$. Now, to solve for the amount earned if the team is in the field for 2,000 hours, plug the numbers into the formula and solve: $375 \times 2,000 = \$750,000$.

What does the $k$, the constant of proportionality, really represent in this problem? You can interpret $k$ to be the number of dollars earned per hour spent with a customer — a sort of average for the business.

## Going the indirect route with indirect variation

Two quantities vary *inversely,* or indirectly, if, as one quantity increases, the other decreases proportionately. Instead of a constant multiplier of $k$, like you find with direct variation, the equation tying the two quantities together is:

$$a = \frac{k}{b}$$

Try your hand at the following indirect variation example.

When a 2 x 4 pine board is suspended between two buildings (with just the ends of the board at the building edges), the maximum weight that the board can support varies indirectly with the distance between the buildings. If the distance between the buildings is 10 feet, the board can support 480 pounds. What's the maximum amount of weight that such a board can support if the buildings are 20 feet apart? How about 30 feet apart?

First write the relationship between the 10-foot distance and the 480 pounds so you can solve for $k$ by multiplying each side of the equation by 10:

$$480 = \frac{k}{10}$$
$$k = 4,800$$

Next, solve for the amount of weight that can be supported, putting 4,800 for the value of $k$ and the distance of 20 feet in the denominator:

$$\frac{4,800}{20} = 240$$

The board can support half as much at that distance — or 120 fewer pounds when the distance is increased by 10 feet.

Now, using the equation and replacing the denominator with 30, you get:

$$\frac{4,800}{30} = 160$$

Increase the distance by another 10 feet, and the amount of weight decreases by another 80 pounds.

# Part II
# Taking Intriguing Math to Work

The 5th Wave                                          By Rich Tennant

Well, obviously one of the cells in the navigational spreadsheet is corrupt!

# In this part . . .

Whether found in a graph, a table, or a computer spreadsheet, a formula is a powerful and dependable mathematical structure. You can find formulas for measurements of physical structures and formulas for determining averages and other statistical values. This part helps you master all of these intriguing mathematical formulas.

# Chapter 5

# Working with Formulas

## In This Chapter

▶ Making sense of all the letters and symbols in a formula

▶ Solving formulas using the correct order of operations

▶ Using calculators and spreadsheets to simplify your math

**W**hether you want them to be or not, formulas are a part of your life. For instance, when you pay your income tax, the amount you owe is determined with a formula. Yes, I know that you may use a table of values, but where do you think those values came from? Why, a formula, of course! When the meteorologist on television talks about the heat index or wind chill factor, the numbers come from a formula. And when you order new carpeting for your office, you take measurements and determine the amount needed using a formula.

You use lots and lots of formulas at work and at home. Like many folks, you may not have realized how much formulas are a part of your life. So, as you can see, formulas are our friends. They offer organization and structure in a chaotic world. A formula qualifies as such when it consistently gives you correct results and answers to questions.

In this chapter, you see how to use formulas properly when you have more than one choice of operation involved. You also see how to simplify the complex parts of a formula. I even show you some of the ins and outs of using calculators to help you solve mathematical expressions. Finally, I show you the beauty of entering a formula into a computer spreadsheet, such as Microsoft Excel, and seeing hundreds of computations magically appear before your very eyes.

# Familiarizing Yourself with a Formula

Formulas are usually long-established relationships that you use in your calculator or computer spreadsheet (or, heaven forbid, with a pencil and some paper). You find formulas in textbooks, almanacs, mathematical tables, and, best of all, on the Internet.

Having a formula such as $I = Prt$ or $S = \frac{1}{2}n(n+1)$ is fine and dandy as long as you know what to do with it. Otherwise, it may as well be Greek. So the first course of action to use when confronted with a formula is to identify all the players. What do the $I$, $P$, $r$, $t$, $S$, and $n$ stand for in the formulas? Lots of times the letters used in a formula have the same beginning letter as what the letters represent — but not always. After you've conquered the identification hurdle, you need to find numbers to replace the letters within the formula. Next, you check to see that the numbers you're using are in the same units — you can't mix feet and inches, pounds and tons, or apples and oranges. Finally, you get to the fun part: performing the mathematical operations. I explain each of these steps in the following sections.

## Identifying variables and replacing them correctly

A formula is essentially an equation in which some relationship between the values in the equation is always true. For example, the area of a rectangle is always equal to the product of the rectangle's length multiplied by the rectangle's width. So the formula for the area of a rectangle is written $A = lw$. (I introduce this formula in Chapter 21.)

When using a formula, you first need to know what each letter represents, and then you need to replace the corresponding letters with the correct values from the problem. Consider, for example, the formula $I = Prt$. This is the simple interest formula. (You find more info on interest in Chapter 9.) The letter $I$ represents the amount of interest earned when the principal, $P$, is deposited at the rate of interest, $r$, for a certain number of years, $t$. Here's another way to view this formula:

Interest = Principal $\times$ Rate $\times$ Time

In the simple interest formula, you have three values being multiplied by one another on the right side of the equation. So if you're told that $5,000 is invested for 6 years at 4% interest, you have to know that the $5,000 replaces the $P$; the decimal equivalent of 4%, 0.04, replaces the $r$; and 6 years replaces the $t$.

After the letters are replaced by numbers, you have to resort to another notation to show the original multiplication. If you just fill in the numbers with no grouping symbols, you get $I = 5,0000.046$, which hardly makes sense. Instead, you use parentheses to separate the values: $I = 5,000(0.04)(6)$.

Now, as another example, consider the formula for computing compound interest:

$$A = P\left(1 + \frac{r}{n}\right)^{nt}$$

The letters in this formula are

   *A* for the total amount of money accumulated

   *P* for the principal or initial money deposited

   *r* for the interest rate as a decimal

   *n* for the number of times each year the interest is compounded

   *t* for the number of years involved

Note that the letters in the formula pretty much represent the first letter of each word involved. It isn't always possible to match each value with its corresponding letter, but, when possible, it's most helpful to have the coordination.

In the compound interest formula, if you're told to find the interest rate that results in an accumulation of $150,000 when $100,000 is deposited for 5 years and the interest is compounded quarterly, you need to know how to replace the variables with the appropriate numbers. In this example, the $150,000 replaces the *A*, the $100,000 replaces the *P*, the letter *n* is replaced with a 4, and *t* is replaced with 5. When you replace the letters in the compound interest formula, you get the following equation:

$$150,000 = 100,000\left(1 + \frac{r}{4}\right)^{4\,(5)}$$

In real-life situations, the substitutions aren't going to be given to you as $A = 150,000$, $r = 4.5\%$, and so on. Instead, you have to read, interpret, substitute, simplify, and then solve.

## *Adjusting for differing units*

Formulas dealing with perimeter or area need input values in inches, feet, yards, or some other linear measure. Formulas dealing with interest need input values involving money, percentages, and time. But you can't just enter values willy-nilly without considering the units involved. Just because the *t* stands for time doesn't mean that you get to put in days or months. You only get to replace with those units if the interest is coordinated with that time measure.

The same goes for area and perimeter formulas. For example, say that you're asked to find the area of a rectangle. If you're given the length in feet and the width in inches, you need to change one or the other so that the measures use the same units.

In general, to change units, you start with the known equivalence between two different units, and then perform multiplication or division to create the numbers that you need. The following examples provide you some practice with changing from one unit to another.

Find the area of a factory floor space that's 40 feet, 8 inches wide by 26 yards, 2 feet, and 9 inches long.

The measurements given here are a conglomeration of units, so you need to change all of them to the same unit. However, if you change everything to inches, the numbers will get awfully big. Changing to yards, on the other hand, involves fractions of yards as well as fractions of feet (which I guarantee won't be fun). So a good compromise is to change everything to feet before computing the area.

Here are some tips to help with this problem: The area of a rectangle is found by multiplying its length by its width. (Refer to Chapter 21 for lots more information on area, perimeter, and measures.) Also, keep in mind these equivalents:

   1 yard = 3 feet

   1 foot = 12 inches

So to change the width of 40 feet, 8 inches to a measure in feet only, you have to divide the 8 inches by 12. This calculation gives you ⅔ foot. So the width is 40⅔ feet.

To change 26 yards, 2 feet, 9 inches to feet, first multiply 26 × 3 = 78 feet. Then divide the 9 inches by 12 to get ¾ foot. The length is 78 feet + 2 feet + ¾ foot = 80¾ feet.

Now, to find the area of this rectangle, you just have to multiply the length by the width, like this:

$$80\frac{3}{4} \times 40\frac{2}{3} = \frac{323}{4} \times \frac{122}{3} = \frac{39,406}{12} = 3,283\frac{10}{12} = 3,283\frac{5}{6}$$

So the area is 3,283⅚ feet, or 3,283 feet, 10 inches. You change the ⅚ feet to inches by multiplying ⅚ by 12.

Here's another example to help you practice unit conversion: What's the interest on a loan of $21,000 at 9% annual interest if you're borrowing it for 90 days?

Interest is found with the formula $I = Prt$, where $P$ is the principal or amount borrowed, $r$ is the annual rate of interest (written as a decimal), and $t$ is the number of years you're borrowing the money.

To use the interest formula, you need to change the 90 days into a part of a year. The simplest method is to use *ordinary* interest (where you use 360 for the number of days in a year) so that when you divide 90 by 360 you get ¼ year. (In Chapter 9, you can find an explanation of the difference between *ordinary* and *exact* interest.) The annual rate of interest percentage, written as a decimal, is 0.09. And the fraction that will fill in for time, ¼, is equal to 0.25 in decimal form. So, in this problem, $I = \$21,000(0.09)(0.25) = \$472.50$. You pay almost $500 for the privilege of borrowing the money for 90 days.

## *Recognizing operations*

A mathematical operation tells you to do something to a number or a set of numbers. The four basic operations are: +, −, ×, and ÷. These four basic operations are called *binary* operations, because they're performed using two numbers. For example, you add 3 + 5. You can't add 4 + . A second number is needed to complete the task. "Aren't all operations binary?" you ask. No. And thank goodness they aren't. Too many other tasks need to be performed — and many of them on one number at a time. Some of the nonbinary operations are $\sqrt{\ }$ for finding a square root, ! for finding a factorial, and $\left[\kern-0.15em\left[\ \ \right]\kern-0.15em\right]$ indicating to round back to the greatest integer.

While adding and subtracting are indicated only by the + and − signs, this isn't the case for multiplying and dividing. You can denote multiplication and division several different ways. Read on for details.

### *Multiple ways of showing multiplication*

Formulas containing the multiplication operation seldom have the multiplication sign, ×, written in them. Because formulas have letters representing numerical values, a multiplication sign often gets confused with the letter $x$. The letter $x$ is the most frequently used letter in algebra equations, too, so a multiplication sign isn't used there, either. Instead of the traditional multiplication sign, multiplication in these cases is indicated with a dot, a grouping symbol, or with nothing at all.

For example, four ways of writing *a* times *b* are: *ab*, *a·b*, *a*(*b*), and *a*[*b*]. The simple interest formula *I* = *Prt* shows three values multiplied together on the right. And the formula for the sum of integers, *S* = ½*n*(*n* + 1), has ½ multiplying *n*, which then multiplies the results inside the parentheses.

### Dividing and conquering

Division is another operation that's seldom shown with its traditional division sign, ÷, in a formula. The two most commonly used methods of showing division in a formula are to use a slash (/) or to write the numbers involved in the division as a fraction. So ½ means to divide 1 by 2. And in the formula for finding the sum of squares, $S = \dfrac{n(n+1)(2n+1)}{6}$, you divide the product (result of multiplying the numbers in the numerator) by 6.

# Simplifying and Solving a Formula

After you've determined which formula to use, what each letter in the formula represents, and where to substitute the various numbers (see the previous section for details), you get to simplify the formula and compute for the unknown value. In other words, you get to solve the problem.

Most times, you'll be solving for the lone letter on one side of the equation. But occasionally (when you're so lucky!), you get to solve for a letter or quantity that's embedded in the operations of the formula. Now you can't just go in and start ripping a formula apart. The rules governing solving equations ensure that you get the intended answer. Everyone uses the same rules, so everyone gets the same answer (if, of course, they do their arithmetic correctly). I explain these rules in the following sections.

## Operating according to the order of operations

Centuries ago, before algebra was done symbolically with letters, exponents, and operations, the processes in math were written out in words and explanations. While this method of using words may have made the directions pretty clear, the method was long and cumbersome, and it wasted space.

Along came the algebraic symbols, and the mathematics became neat, tidy, and concise. In addition to these shorthand symbols came some rules that were established so that everyone knew what the symbols meant. This way folks had a road map for following the processes indicated by the symbols.

One important part of this list of rules is the *order of operations.* The order of operations kicks in when you have more than one operation to perform in an expression. Just as the name implies, the operations of $+, -, \times, \div, \sqrt{\phantom{x}}$, and the exponents (powers) have to line up in order to be performed.

When grouping symbols appear in an equation, perform the operations within the grouping symbols first. Grouping symbols include ( ), [ ], { }, | |, and fraction lines. The grouping symbols supersede any operations, because their contents need to be simplified and changed into a single value. After the groupings are dealt with, you proceed with the rest of the process.

The order of operations dictates that, if no grouping symbols interfere, you perform operations in the following order, moving from left to right:

1. **Take roots and raise to powers.**

2. **Multiply and divide.**

3. **Add and subtract.**

Try your hand at the order of operations with the example problems in the following sections.

## *Example 1*

Simplify the following expression using the order of operations:

$$6 + 2^3 - 4 \cdot 3 + \frac{18}{2}\sqrt{121}$$

Because the expression has no grouping symbols, you know that you first need to deal with the power and root. So raise 2 to the third power and find the square root of 121. Be sure to put a dot or parenthesis to indicate multiplication between the fraction and the answer to the root of 121. After the radical is dropped, you lose the grouping symbol that indicates multiplication. Here's what your new expression looks like:

$$6 + 8 - 4 \cdot 3 + \frac{18}{2} \cdot 11$$

Now, moving left to right, multiply the 4 and 3, divide the 18 by 2, and then take the result of the division and multiply it by 11. After all these calculations, you get the following:

$$6 + 8 - 12 + 9 \cdot 11 = 6 + 8 - 12 + 99$$

Now, add the 6 and 8 to get 14. Subtract the 12 to get 2, and add the 2 to 99. The final answer is 101.

### Example 2

Evaluate numbers in the following formula for compound interest (see Chapter 9):

$$A = P\left(1 + \frac{r}{n}\right)^{nt}$$

$$= 10,000\left(1 + \frac{0.06}{4}\right)^{4\,(25)}$$

***Note:*** In this version of the formula, the rate, $r$, is replaced with 6% (or 0.06), the $n$ is replaced with 4 to represent quarterly compounding, and the $t$ is replaced with 25 for that many years.

Start by simplifying the fraction in the parentheses and adding it to 1. Then raise the result to the appropriate power and multiply the result by 10,000. Here's what your math should look like:

$$A = 10,000\,(1 + 0.015)^{4\,(25)}$$

$$= 10,000\,(1.015)^{100}$$

$$= 10,000\,(4.43204565)$$

$$= 44,320.4565 \approx 44,320.46$$

### Example 3

Simplify the numbers in the following formula for the payment amount of an amortized loan (see Chapter 12 for more on amortized loans) where $120,000 is borrowed at 0.08 interest for 60 months:

$$R = \frac{120,000\,(0.008)}{1 - (1 + 0.008)^{-60}}$$

You start by first working on the denominator. Add the numbers in the parentheses and then raise the result to the appropriate power. Subtract that result from 1. Multiply the two numbers in the numerator together, and then divide the sum by what's in the denominator. Your math should look like this:

$$R = \frac{120,000\,(0.008)}{1 - (1.008)^{-60}}$$

$$= \frac{120,000\,(0.008)}{1 - 0.61996629}$$

$$= \frac{120,000\,(0.008)}{0.38003371}$$

$$= \frac{960}{0.38003371} \approx 2,526.09$$

## Making sure what you have makes sense

When simplifying expressions involving lots of numbers and operations, it's easy to make a simple arithmetic error. You may have a mental meltdown, or you may enter the numbers into the calculator incorrectly. For these and several other reasons, it's always a good idea to have an approximate answer in mind — or, at least, a general range of possible answers.

For example, say you want to find the interest earned on a deposit of $10,000 for 10 years. If you come up with an answer of $100,000 in interest, hopefully you'll realize that the answer isn't reasonable. After all, that's more interest than the initial deposit! Unheard of! (Or, if it's correct, tell me what bank you're using because I'm going to sign up!)

Here's another example: If you're determining the monthly payments on a $4 million house and come up with $500 per month, you should be a bit suspicious of your answer (and ever hopeful). Why? Well, if you think about it, $500 per month is awfully low for such a large home — and it would take almost 700 years to pay back just the principal at this rate.

True, you'll come across some situations where you have no clue what the answer should be. I'm faced with them all the time. You just work as carefully as you can, and perhaps check with someone else, too. In the big picture of mathematics, though, using common sense goes a long way toward accuracy.

# Computing with Technology

Gone are the days of the abacus, the rolled-up sleeves, the green visor, and the quill pen. Okay, I've really mixed up some computing centuries, but you probably get my point. Technology is here to stay. And that technology can make your life much easier. However, you also can get into trouble much more quickly when using technology; in a complex spreadsheet, a simple error in only one cell can create calculating havoc.

In this section, I don't try to sell you any particular brand of calculator or any particular spreadsheet product. I'm just going to whet your appetite — tease you with some neat features that technology brings to your business math table. The directions and suggestions are as general as possible. So it's up to you to check out your own personal calculator or computer spreadsheet to find the specific directions and processes necessary.

# Calculators: Holding the answer in the palm of your hand

No matter how much you love to crunch numbers in your head, you eventually come across calculations that are too big or too difficult. In that case, you look to a calculator for help. You can work most of these computations with a simple scientific calculator. A scientific calculator raises numbers to powers and finds roots and the values of logarithms.

*TIP*

You can go fancy-schmantzy and invest in a graphing calculator (or even in one that does calculus), but be aware that the more features you have in your calculator, the more opportunities you have of going astray. Besides, if you don't need to do fancy calculations, why spend more money on a fancy calculator?

*REMEMBER*

The main challenge of using a calculator is to direct the calculator to do what you *mean.* The calculator computes with a certain set of rules — mostly based on the order of operations (see the earlier section, "Operating according to the order of operations," for more). So you need to push the right button, use enough parentheses, and interpret the resulting answer. That way your calculator computes the right information and you answer your question correctly.

The four basic operations each have their own button on a calculator. When you hit the ⊞ or ⊟ button, you see the + or – right on the screen. When you hit ×, most calculators show the multiplication as an asterisk (*). The ÷ button usually conjures a slash (/).

## Wielding the power of exponents

Many financial and geometrical formulas involve exponents (powers) of the values in the expression. Some examples of these types of formulas include the following:

**Area of a circle:** $A = \pi r^2$

**Accumulated money from compound interest:** $A = P\left(1 + \dfrac{r}{n}\right)^{nt}$

**Regular payment amount in a sinking fund:** $R = \dfrac{Ai}{(1+i)^n - 1}$

You access the power (exponent) button on a scientific calculator in one of several ways. If you want to square a value (raise it to the second power), you hit the button that looks like this:

## Taking advantage of hand-held and online calculators

Hand-held calculators have been around since the 1960s. Some of the first calculators did little more than add, subtract, multiply, and divide. You had the choice between answers correct to two decimal places or the expanded five decimal places. Wow! We've come a long way, baby. Nowadays, you have more choices than you can ever fully investigate. For under $10, you can do exponential, logarithmic, and trigonometric computations to an impressive degree of accuracy.

When deciding on a calculator, your main considerations are the following:

✔ What do you need the calculator for?

✔ What type of power source do you prefer (solar or battery, for example)?

✔ What type and how large a display do you want?

✔ How big do you want the keys on the keypad (some of us have big fingers)?

The big names in calculators are Texas Instruments, Hewlett-Packard, and Casio, but you'll find more brands if you hunt around enough. The prices vary, and you can get spectacular deals on eBay.

If you don't want to invest in a hand-held calculator, you can find many online calculators that are quite good. These online calculators are usually geared to a specific chore: computing a mortgage payment, figuring percentages, determining a rental amount, converting temperatures or times, computing areas, and so on. In some respects, online calculators make a lot of sense, because technology is changing so rapidly. After all, the calculator or computer you bought yesterday is already obsolete.

The square, or second power, is usually the only power to get its own button on a calculator. Too many other powers are used in computing, so the rest are taken care of with a general power button, which looks like one of the following:

The two buttons with $x$ raised to the $y$ or $y$ raised to the $x$ are sort of scripted. You have to enter the numbers exactly in the correct order for the calculator to compute what you mean. To use this button, you put in the $x$ value first, you hit the button, and then finally you put in the $y$ value.

I'm sure you're dying to try out these new tricks on your calculator. As practice, determine how to enter the following expressions in a calculator:

$7^2 + 1$

$3^5 - 4^{1/3}$

$2(5^8)$

Here's how you input each of the exponents into your calculator:

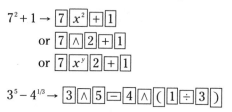

$$7^2 + 1 \rightarrow \boxed{7}\boxed{x^2}\boxed{+}\boxed{1}$$
$$\text{or } \boxed{7}\boxed{\wedge}\boxed{2}\boxed{+}\boxed{1}$$
$$\text{or } \boxed{7}\boxed{x^y}\boxed{2}\boxed{+}\boxed{1}$$

$$3^5 - 4^{1/3} \rightarrow \boxed{3}\boxed{\wedge}\boxed{5}\boxed{-}\boxed{4}\boxed{\wedge}\boxed{(}\boxed{1}\boxed{\div}\boxed{3}\boxed{)}$$

Notice here that the fractional exponent has to have parentheses around the numbers. Negative exponents also require parentheses around them.

$$2\left(5^8\right) \rightarrow \boxed{2}\boxed{(}\boxed{5}\boxed{\wedge}\boxed{8}\boxed{)}$$
$$\text{or } \boxed{2}\boxed{\times}\boxed{5}\boxed{\wedge}\boxed{8}$$

The order of operations dictates that the 5 in this expression is raised to the power first, and then the multiplication takes place. You don't need to input into your calculator the parentheses to be assured of having the operations done in the correct order. (But they don't hurt either.)

### Distinguishing between subtraction and negativity

The subtraction symbol (–) is understood as being an operation. Subtraction is one of the four basic binary operations. And, in algebra, students are told that *subtract, minus, negative, opposite,* and *less* are all indicated with the same symbol: the subtraction sign. The algebra ruling works fine when dealing with algebraic expressions. But calculators are a bit fussy and make a distinction between the operation of subtraction (*minus* and *less*) and the condition of being negative (*opposite*). Most scientific calculators have a subtraction button. You also find a separate negative button, which is distinguished from the subtraction button by parentheses: (–).

To get familiar with subtraction and negativity, determine how to enter the following expressions in a calculator:

$$16 - 18$$
$$-16 - (-18)$$
$$-3^4$$

Here's how to enter the previous expressions in your calculator:

$$16 - 18 \rightarrow \boxed{16}\boxed{-}\boxed{18}$$
$$-16 - (-18) \rightarrow \boxed{(-)}\boxed{16}\boxed{-}\boxed{(-)}\boxed{18}$$

You can type in the parentheses, but they aren't really necessary in this case. But when in doubt, always use the parentheses to avoid errors. Look at the difference between not using parentheses around a negative number being raised to a power and then using the parentheses. You get two different answers:

$$-3^4 \rightarrow \boxed{(-)}\boxed{3}\boxed{\wedge}\boxed{4}$$

vs.

$$(-3)^4 \rightarrow \boxed{(}\boxed{(-)}\boxed{3}\boxed{)}\boxed{\wedge}\boxed{4}$$

The answer to this first expression is –81, because the calculator raises the 3 to the fourth power and then changes the number to the opposite. If you want the number –3 raised to the fourth power, you have to put parentheses around both the negative and the 3. Notice that the answer is positive. Why? Raising a negative number to the fourth power gives you a positive result. It's all tied to the order of operations.

### *Grouping operations successfully*

Usually you won't run into any difficulties if you use more parentheses than necessary. I'm usually pretty heavy-handed with parentheses in math and commas in writing (but my editor takes care of that). The parentheses help you say what you mean — mathematically. Parentheses are needed if you have more than one term in the denominator (or bottom) of a fraction or more than one term in a radical. They're supposed to make your intent clearer.

Using the grouping info I provide, determine how to enter the following expressions in a calculator:

$$\frac{28}{3^2+5} \text{ and } \sqrt{18-2^2}$$

You input the first expression like this:

$$\frac{28}{3^2+5} \rightarrow \boxed{28}\boxed{\div}\boxed{(}\boxed{3}\boxed{\wedge}\boxed{2}\boxed{+}\boxed{5}\boxed{)}$$

The parentheses ensure that the power and sum are performed in the denominator, and then the result divides the 28.

And here's how to enter the second expression:

$$\sqrt{18-2^2} \rightarrow \boxed{\sqrt{\phantom{x}}}\boxed{(}\boxed{18}\boxed{-}\boxed{2}\boxed{\wedge}\boxed{2}\boxed{)}$$

You want the root of the result under the radical.

### Going scientific with scientific notation

*Scientific notation* is used to write very large or very small numbers in a useful, readable, compact form.

A number written in scientific notation consists of

[*a number between 1 and 10*] × [*a power of 10*]

For example, the number 234,000,000,000,000,000,000,000,000 written in scientific notation looks like this: $2.34 \times 10^{26}$

The power of 26 on the 10 tells you that the decimal point was moved 26 places, from the end of the last zero to directly behind the 2. The number 2.34 is between 1 and 10.

Now here's an example at the opposite end of the spectrum — it includes a number that's quite small. Writing the number 0.000000000000000000000000000000000234567 in scientific notation gives you: $2.34567 \times 10^{-37}$.

A negative power is used in this instance, because very small numbers require that the decimal point moves to the right to get the nonzero digit part (the first number) to form a number between 1 and 10.

Calculators go into scientific notation mode when the result of a computation is too large for the screen or has more digits than the calculator can handle. Instead of displaying scientific notation like I do here, calculators show an *E* and then, usually, the exponent. So, if you multiply numbers together and get the result 4.3E16, the calculator is reporting that the answer is written in scientific notation and is $4.3 \times 10^{16} = 43,000,000,000,000,000$.

## Repeating operations: Simplifying your work with a computer spreadsheet

Computer spreadsheets are truly wonderful tools. You not only get orderly reports of numbers all typed out neatly and in regular rows and columns, but you also have computing capability that's equal to a calculator's capability. For instance, you can direct the computer spreadsheet to do all sorts of computations, such as adding all the numbers in a column and performing multiplications and additions on selected numbers. And here's the best thing: You can then tell the spreadsheet to repeat these same operations over and over on lots and lots of numbers. Now you can spend all that extra time lying on a beach somewhere . . . .

In this section, I don't go into too much detail on how to enter formulas or values, because different computer packages have different rules. But after you see how tables of data are produced, hopefully you'll be inspired to check into the particular spreadsheet program that you have and find the correct commands to type in.

In the following sections, I show you two examples of spreadsheets that you can create. One shows you how to sum up the revenue from different sources, and another determines the payment of an amortized loan.

### Summing across rows or down columns

A useful type of table is one that shows the revenue from sales, where some of the money from the month's sales come in that month, another percentage of those sales comes in the second month, and the rest (that actually does come in) is received in the third month. Every month, the total amount of revenue has to be tallied from the different sources — from previous months and from the current billing.

Guess what? You can use a spreadsheet to determine the amounts of money coming in each month as a percentage of particular sales. You also can use the spreadsheet to find the sum of all the revenue sources for the month.

Show the entries for the columns of a spreadsheet where you expect to collect 50% of the revenue from sales during the first month, 45% of the revenue during the second month, 4% during the third month, and write off the last 1% as a bad debt. Table 5-1 shows the setup. You copy the three percentage entries into the corresponding cells for each month. If you type in formulas and refer to sales amounts in the first column, the spreadsheet does the computations for you.

| **Table 5-1** | | **Creating a Spreadsheet** | | |
|---|---|---|---|---|
| *Projected Sales* | *January* | *February* | *March* | *April* |
| January: $400,000 | 0.5($400,000) = $200,000 | 0.45($400,000) = $180,000 | 0.04($400,000) = $16,000 | |
| February: $500,000 | | = 0.5($500,000) = $250,000 | = 0.45($500,000) = $225,000 | = .04($500,000) = $20,000 |
| March: $700,000 | | | = 0.5($700,000) = $350,000 | = .45($700,000) = $315,000 |
| April: $900,000 | | | | = 0.5($900,000) = $450,000 |

The table, which extends down for each month's revenue, could be extended across for a full year or for several years. The spreadsheet does the computation for you, rounded to the number of decimal places you indicate in your setup.

With one of the table commands available, you can find the sum of the numbers going across the rows of the table or those going down the columns. Just highlighting and dragging across the values you want usually does the trick.

So the sum of the entries in the April column will come out to be 4% of February's revenue, 45% of March's revenue, and 50% of April's revenue. If you change the sales amount (you find an extra $1,000 in April perhaps), the formula structure of the table will not only adjust the percentage amount across the row, but it also takes care of all the columns that are affected, too.

### Creating an amortized loan schedule

You can determine the monthly payment of a particular loan using one of the appropriate formulas (see Chapter 12 for more on loan formulas). But it gets a bit tedious if you want to find the monthly payments involved in more than one or two scenarios — where you change the interest rate a bit or the length of the loan a bit. You don't want to have to type the numbers into the calculator over and over again.

Thankfully, a computer spreadsheet makes a table of loan payments quickly and accurately. You do have to type in the equation and set up the input values, but after doing the preliminaries, you can immediately copy, drag, and observe all the possibilities. The following example will walk you through the process.

Create a chart of the monthly loan payments on a $100,000 loan where you compare interest rates of 8%, 8.25%, 8.5% and 8.75%. Compare these rates at 15 years, 20 years, 25 years, and 30 years.

You start with a spreadsheet that has the interest rates, as decimals, at the top of consecutive columns and the years at the beginning of rows. You can see what I mean in Figure 5-1.

| | A | B | C | D | E | F |
|---|---|---|---|---|---|---|
| 1 | | 0.08 | 0.0825 | 0.085 | 0.0875 | |
| 2 | 15 | | | | | |
| 3 | 20 | | | | | |
| 4 | 25 | | | | | |
| 5 | 30 | | | | | |
| 6 | | | | | | |
| 7 | | | | | | |
| 8 | | | | | | |

**Figure 5-1:** Setting up a spreadsheet for loan payments.

TIP

I typed in the interest rates and years, but you can have the computer spreadsheet fill in a bunch of consecutive amounts for you. By simply entering in a command to add a certain amount to each successive value, you save yourself a lot of typing.

For instance, in Figure 5-1 you enter the 15 for 15 years. Then in the cell directly below that one, you enter a command such as "= A2+5." The number 20 should appear in that cell. Now you can copy the command and drag down through as many cells as you want, and each will show a number that's 5 more than the previous cell. The same thing works when moving across the interest rates. You can increase by 0.25%, as shown in the figure, or you can pick some other increment. Also, you can format the cells to produce just about any number of decimal points. For example, you can set the format to show you four decimal places; the numbers are then rounded to that many places.

Now you're ready for the fun part: entering the formula referencing the cell positions. The variables in your formula will be entries such as A2 or B3, telling the formula to look at that particular row and column. To begin, remember that the formula for the monthly payment amount of an amortized loan is

$$R = \frac{P\left(\frac{r}{12}\right)}{1 - \left(1 + \frac{r}{12}\right)^{-12t}}$$

where $R$ is the amount of the regular payment, $P$ is the amount borrowed, $\frac{r}{12}$ is the interest rate each month, and $t$ is the number of years. (You can find all the details on this formula in Chapter 12.)

Now you type the formula into cell B2 of the table. The following is one possibility for the format of the formula (which is the one I use in my spreadsheet). However, you need to check the instructions and help menu with your particular computer spreadsheet. Here's the formula I used:

$$= \left(100{,}000 *\ (B1/12)\right) / \left(1 - (1 + B1/12) \wedge (-12*15\ )\right)$$

Yes, this all fits into the one tiny cell. Actually, what you'll see is just the numerical answer in the cell. The formula should be available in the editing box. You can then copy that cell and hold and drag to the right until the whole row is highlighted. The spreadsheet picks up the interest values from the respective column heads.

Figure 5-2 shows what your row in the spreadsheet should look like.

|  | A | B | C | D | E | F |
|---|---|---|---|---|---|---|
| 1 |  | 0.08 | 0.0825 | 0.085 | 0.0875 |  |
| 2 | 15 | 955.6521 | 970.1404 | 984.7396 | 999.4487 |  |
| 3 | 20 |  |  |  |  |  |
| 4 | 25 |  |  |  |  |  |
| 5 | 30 |  |  |  |  |  |
| 6 |  |  |  |  |  |  |
| 7 |  |  |  |  |  |  |
| 8 |  |  |  |  |  |  |

**Figure 5-2:**
Part of the
loan
payment
schedule.

When it comes to spreadsheets, you really need to just play around with them. Say to yourself, "I wonder if I can get the spreadsheet to . . ." And, amazingly, you usually *can* get the spreadsheet to do what you want. When in doubt, consult the program manual for help.

# Chapter 6

# Reading Graphs and Charts

. . . . . . . . . . . . . . . . . . . . . . . . . . . . . . . . . . . . . . . . . . . . . . . .

## In This Chapter

▶ Making sense of a scatter plot

▶ Discovering line graphs

▶ Reading and creating histograms

▶ Charting portions of a whole with a pie chart

. . . . . . . . . . . . . . . . . . . . . . . . . . . . . . . . . . . . . . . . . . . . . . . .

Some people love numbers and find it easy to interpret pages full of them. But others are intimidated by them. How can numbers be presented so that they make sense to those who aren't so great with numbers? Through visuals such as graphs and charts.

Graphs and charts provide lots of information quickly. It's true that a picture is worth a thousand words. After all, you can quickly grasp the financial situation of a business when you see a line on a graph that represents earnings sloping down.

These visuals don't give you detailed information, but they do set the scene as a whole. After you have the situation pictured in your mind, you can then decide how far you need to drill down into the material in order to make an informed decision.

The more common types of graphs are scatter plots, line graphs, and bar graphs (or histograms). *Scatter plots* are numbers represented by dots distributed about a rectangular area. A *line graph* connects points to one another. Trends and items that are connected — that affect one another — are best shown with line graphs. In the section on line graphs, you see how to get around the need for a uniform labeling. You also see how graphs can be abused.

A *bar graph* is good for showing volumes or frequencies of items that may or may not have a connection to one another. Other graphs you may use include *pie charts,* where the different-sized pieces represent different portions of the whole scene, and *pictorial graphs,* which catch your eye and make a statement at the same time.

The purpose of graphs and charts is to show as much information as quickly, efficiently, and accurately as possible. The labels on the axes are an equal distance apart, and they have numbers that are the same distance apart. With the information in this chapter, you can become better at choosing the best type of graph or chart for your situation. This way you can easily get your information to those folks who need it.

# Organizing Scattered Information with a Scatter Plot

A *scatter plot* may look just like what its name implies: a bunch of dots scattered all over the place. A scatter plot is used to see whether the data that it represents has any visible trend or pattern. If the dots are scattered all over the chart with no rhyme or reason, you can safely conclude that there's no correlation between the values that are being plotted. However, if you see the plotted dots lying more in a clump or in an upward-leaning gathering, you may determine that some connection or trend is taking place between the values.

Scatter plots, line graphs, and histograms are drawn with respect to two *axes*. A horizontal axis and a vertical axis cross one another, and the intersection acts as a starting point for each set of information.

For example, you may keep records of the days' temperatures and make a chart of how many customers you have in your ice cream store on a day with that certain temperature. Or you may be keeping track of the number of hours a machine is run each day and how much it costs per hour to run the machine (averaging in all the labor, materials, utilities, and so on).

In Figure 6-1, you see two scatter plots that are based on the previously mentioned examples. The scatter plot on the left-hand side represents customers and temperatures; on the right-hand side, hours and average cost are plotted. Keep in mind that a scatter plot is designed to help you determine whether you see some cause and effect — some pattern or trend.

**Figure 6-1:**
Scatter
plots
showing
trends.

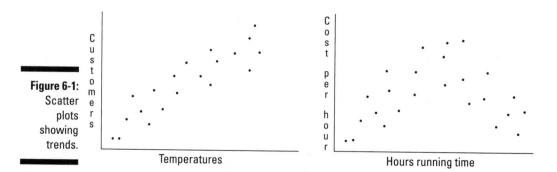

A scatter plot doesn't tell you the exact mathematical relationship between two values, but it does show you whether such a relationship in fact exists. If you think that a connection between two different entities does exist, you can then try to write an equation or formula linking the values. (In Chapter 18, you see how the behaviors of costs are related with a mathematical statement.)

To create a scatter plot, draw two intersecting perpendicular lines (called the *axes*) and assign units to the lines based on what's represented by the numbers you'll be plotting. You place a point or dot to represent each set of numbers by lining the points up with the units on the axes.

Draw a scatter plot showing the data collected on the number of inches of rain during the summer and the yield per acre of a crop. Table 6-1 shows the data collected over an 8-year period.

| Table 6-1 | Records of Rainfall and Crop Yield | |
|---|---|---|
| *Year* | *Inches of Rain* | *Yield per Acre* |
| 1 | 10 | 20 |
| 2 | 20 | 30 |
| 3 | 6 | 10 |
| 4 | 2 | 5 |
| 5 | 8 | 15 |
| 6 | 24 | 35 |
| 7 | 16 | 25 |
| 8 | 20 | 25 |

The scatter plot has the number of inches of rain along the *x*-axis (or bottom of the graph), and the yield per acre is shown along the *y*-axis (or side of the graph). If a relationship is present between the amount of rain and the yield, you should see a pattern emerging in the dots of the scatter plot. Just looking at the data tells you very little. The numbers are all over the place and difficult to decipher.

Figure 6-2 shows you the scatter plot constructed from the data in Table 6-1. The upward movement of the yield numbers as the rainfall increases seems to suggest that there's a relationship between the two values.

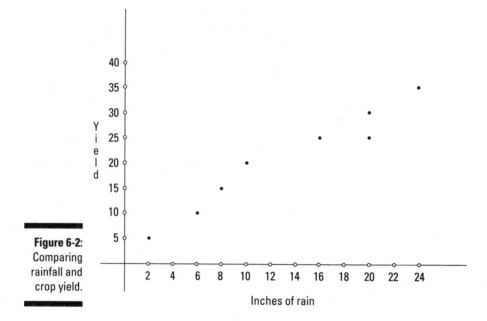

**Figure 6-2:**
Comparing
rainfall and
crop yield.

# Lining Up Data with Line Graphs

A *line graph* actually consists of a bunch of connected segments. A line graph connects data that occurs sequentially over a period of time. You use a line graph to show how values are connected when one point affects the next one.

For example, a line graph often is used to show how temperatures change during a day, because the temperature one hour affects the temperature the next hour. Another use for a line graph is to show the depreciation of a piece of machinery or other property. The value of an item one year has a bearing on the item's value the next year. The following sections explain everything you need to know about line graphs.

## Creating a line graph

The axes used when creating a line graph have numbers or values representing different aspects of the data. The two axes usually have different numbering systems, because the values being compared most likely don't even have the same units. But the numbering on each axis should be uniform — spaced equally and numbered consecutively.

When constructing a line graph, you start out somewhat like you do with a scatter plot: You draw your axes and then place points or dots to represent the numbers from your data. The difference is that you connect one dot to the next with a line segment. You do so in order to show that they're connected and changing.

Believe it or not, even a new Mercedes depreciates in value over the years. Table 6-2 shows the first few years of one car's value. Draw a line graph to illustrate the value.

| Table 6-2 | Depreciation of a Mercedes |
|---|---|
| *Year* | *Value* |
| 0 | $56,000 |
| 1 | $43,120 |
| 2 | $36,650 |
| 3 | $33,720 |
| 4 | $31,360 |
| 5 | $29,160 |
| 6 | $27,120 |
| 7 | $25,200 |

A line graph is used to display the information in Table 6-2, because one year's value of the car is tied to the previous year's value and the next year's value. The line graph shown in Figure 6-3 shows the number of years since the car's purchase along the horizontal ($x$) axis and the value of the car on the vertical ($y$) axis. The line graph helps you see how dramatically the value of the car drops at first and how the drops in the value taper off as the car gets older.

## Indicating gaps in graph values

You want your graphs and charts to do a lot of explaining without words. So you need to label the axes carefully, use a uniform scale on the axes, and plot the values carefully.

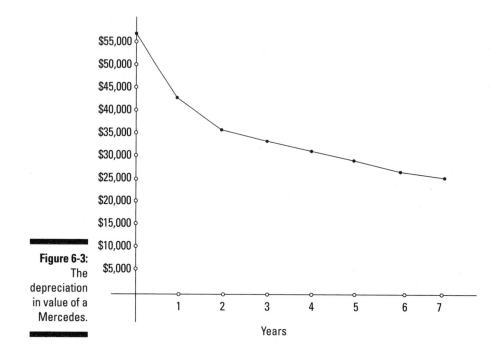

**Figure 6-3:**
The
depreciation
in value of a
Mercedes.

Some data sets don't cooperate very well when it comes to setting up axes that are constructed correctly and, at the same time, give good information. Sometimes, for instance, problems arise when you want the intersection of the axes to be 0 and the numbers to increase as you move to the right and upward. But you don't want your graph to be, say, 6 inches across and 90 inches high just to accommodate the numbering protocol. Instead, in this case, you should indicate a gap in the numbering with a zigzag on the axis.

For example, if you want to create a line graph of the total city budget from 1993 through 2000, you may have to enter numbers in the billions. Even if you knock off all the zeros and label your axis as being billions, the numbers go from $5,168 billion to $5,758 billion. So you wouldn't get much detail in the graph from the numbers. On the left-hand side of Figure 6-4, you see a graph of the total budget for 8 years. The scale is kept uniform and the numbers on the axis go from $0 to $6,000 billion. You don't see much detail or movement from year to year considering the breadth of the range.

An alternative is to put in a *break,* which looks like a zigzag, to show that values are missing in the numbering system on the axis. By doing so, you're still able to keep the remaining numbers uniformly distributed to give a better picture of what's going on. In the right-hand graph in Figure 6-4, you see how a break in the axis allows more information to be conveyed by spreading out the upper numbers for more detail. With this break, you see the detail on just how quickly the budget is changing during particular years.

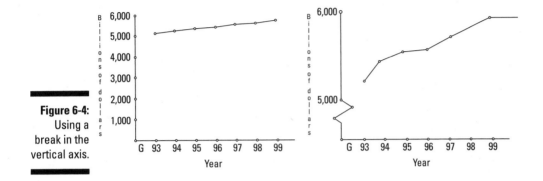

**Figure 6-4:**
Using a break in the vertical axis.

# Measuring Frequency with Histograms

A histogram isn't some cold remedy (though, doesn't it sure sound like it?). *Histogram* is another name for a bar graph. But, as you can probably tell, *bar graph* is a bit more descriptive; in fact, you probably already have a picture in your mind of what a bar graph is. However, because you'll often hear these graphs called histograms, I'll stick with that term throughout this section. Just know that you may see it referred to either way.

A *histogram* is a graph of frequencies — it shows how many of each. With a bar's height, a histogram shows the relative amount of each category. Unlike a line graph, a histogram doesn't have to have sequential numbers or dates along the horizontal axis. Why? Because the value of one category doesn't affect the next one.

For instance, you can list the states along the bottom of a histogram to show the area or population of each state. Or you can show the production of some product.

Another nice feature of a histogram is that you can compare two entries of each category — perhaps showing the difference between one year's production and the next. However, the vertical axis has to be uniform in labeling so that the amounts or frequencies are represented fairly.

To create a histogram, you start out with intersecting horizontal and vertical axes. You list your categories to be graphed along the horizontal axis, and you label the units along the vertical axis. Draw a thick line or rectangle above each category to its corresponding height. The rectangles will be parallel to one another, but they'll usually be different heights.

Create a histogram showing the number of hurricanes that hit the United States in each decade of the 20th century. The numbers are as follows:

1901–1910, 18

1911–1920, 21

1921–1930, 13

1931–1940, 19

1941–1950, 24

1951–1960, 17

1961–1970, 14

1971–1980, 12

1981–1990, 15

1991–2000, 14

The histogram you create should have the decades listed along the horizontal axis and the numbers 0 through 24 or 25 on the vertical axis. The bar representing each frequency should be shaded in. The bars can be touching, but they don't have to be. Figure 6-5 shows my version of the histogram. I chose not to have the bars touching, because only ten bars are needed.

A histogram makes comparing the relative number of hurricanes over the 20th century quite easy. With this graph, you can easily pick out when there were twice as many hurricanes during one decade than there were in another.

Try another example: Suppose a flag and decorating company sells five different types of products: flags, banners, table decorations, flag stands, and commemorative pins. The manager wants you to create a histogram showing the total sales of each category for two consecutive years. Table 6-3 shows the total sales for each category.

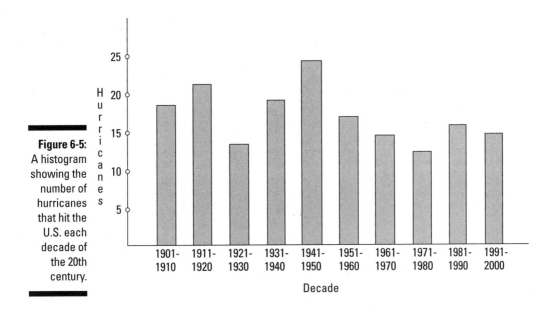

**Figure 6-5:** A histogram showing the number of hurricanes that hit the U.S. each decade of the 20th century.

| Table 6-3 | Total Revenue for a Flag and Decorating Company | |
|---|---|---|
| *Category* | *Year 1* | *Year 2* |
| Flags | $450,000 | $540,000 |
| Banners | $300,000 | $330,000 |
| Table decorations | $50,000 | $90,000 |
| Flag stands | $30,000 | $40,000 |
| Commemorative pins | $160,000 | $290,000 |

You see that each category had an increase in sales. By creating a histogram with the two years' sales side by side, you can see how the increases compare proportionately in each category. Also, the histogram allows you to better understand where the main emphasis is for the company — where most of the revenue comes from in the different sales. See Figure 6-6 to see a completed histogram for this scenario.

A picture can't tell you everything. You can see the relative changes and the comparable revenue amounts. But if you want to determine the actual percent changes and the proportionate amount that each product contributes to the total revenue, you have to figure the percentages. In Chapter 3, you find all the information you need on percent increases and decreases.

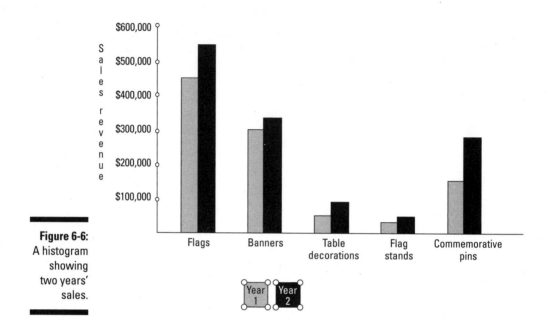

**Figure 6-6:**
A histogram
showing
two years'
sales.

# Taking a Piece of a Pie Chart

A *pie chart* is a circle divided into wedges where each wedge or piece of the pie is a proportionate amount of the total — based on the actual numerical figures. Pie charts are especially useful for showing budget items — where certain amounts of money are going. To create a pie chart, you divide the circle proportionately and draw in the *radii* (the edges of the pieces). The following sections explain pie charts in more detail.

## Dividing the circle with degrees and percents

Think of a circle as being divided into 360 separate little wedges. (A circle's angles all add up to 360 degrees.) Just one of the 360 little wedges is difficult to see — you may not even notice a drawing of one degree in a picture. Table 6-4 shows you many of the more useful fractional divisions of a circle. (I talk more about degrees in Chapter 7, if you need more information.)

| Table 6-4 | Fractional Portions of a Circle |
|-----------|----------------------------------|
| **Number of Degrees** | **Fraction and Percent of a Circle** |
| 30 | $\frac{30}{360} = \frac{1}{12} \approx 8.3\%$ |
| 45 | $\frac{45}{360} = \frac{1}{8} = 12.5\%$ |
| 60 | $\frac{60}{360} = \frac{1}{6} \approx 16.7\%$ |
| 90 | $\frac{90}{360} = \frac{1}{4} = 25\%$ |
| 120 | $\frac{120}{360} = \frac{1}{3} \approx 33.3\%$ |
| 135 | $\frac{135}{360} = \frac{3}{8} = 37.5\%$ |
| 150 | $\frac{150}{360} = \frac{5}{12} \approx 41.7\%$ |
| 180 | $\frac{180}{360} = \frac{1}{2} = 50\%$ |

To give you an idea of how much of a circle some of the angles account for, Figure 6-7 shows you two circles and some wedges drawn in with their respective fractions. You'll probably need fractions and degree measures other than those shown in the figure, but these circles show you some representative wedges that you can use to approximate other sizes.

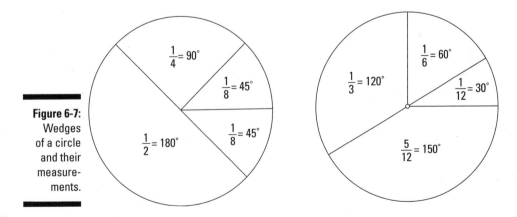

**Figure 6-7:** Wedges of a circle and their measurements.

# Chapter 7

# Measuring the World around You

· · · · · · · · · · · · · · · · · · · · · · · · · · · · · · · · · · · · · · · · · · · · ·

## In This Chapter

▶ Moving between different measurement units

▶ Using the metric system

▶ Understanding the secret of properly measuring lumber

▶ Matching angles and degrees

· · · · · · · · · · · · · · · · · · · · · · · · · · · · · · · · · · · · · · · · · · · · ·

*W*henever you use a ruler, a yardstick, a tape measure, a scale, or a thermometer, you venture into the world of units of measurement. Measures are pretty much standardized throughout the different countries of the world. And gone are the days of using hands (although horses are still referred to as so many hands tall), joints of the thumb, and the distance from your nose to the tip of your outstretched arm. You can now depend on the simple plastic ruler to give you the same measures as the next person's ruler.

When using measurements, you pick the most convenient unit for the situation. For instance, you aren't going to use inches when measuring the size of a parking lot, and you aren't going to use tons when weighing boxes of nails. You go with what makes sense. But having said that, you still may have opportunities to mix and match measures, so in this chapter I show you a quick and easy setup that works for all your conversion problems.

Also important in this chapter is my discussion on using the metric system. The world is roughly 96% metric and 4% English as far as measurement systems. The metric system seems to become popular periodically in the United States, but the interest comes and goes. In the mid-1970s, there was a huge push to go metric. (I still have my "Go Metric" bumper sticker.) The switch to metric in the U.S. will probably never happen, but as long as Americans continue to do business with other countries, you'll still occasionally (or often!) need to understand and work with metric measures.

Finally, I round off this chapter with a section on using angles and degrees. I show you how to break degrees into smaller units, and I also explain how to subdivide angles.

# Converting from Unit to Unit

You know that 12 inches equals a foot and that 4 cups are in a quart. But what about changing 16 yards to inches or 400 ounces to pounds? Are you still comfortable when faced with these measures? I hope so. But just in case you aren't, I offer a basic conversion formula in this section that works for all unit-swapping that you may have to do. You do need to start with the basic equivalents (such as 5,280 feet = 1 mile), but after those basics are established, the rest is straightforward and consistent.

## Using the conversion proportion

You can probably do the easier unit conversions in your head (such as 24 inches equals 2 feet), but sometimes you won't be sure whether you have to multiply or divide to change from one unit to another. Don't worry. In this case, you can fall back on a simple proportion. You put in the known value, create two fractions, cross-multiply, and pull out the needed values.

In general, you always start out with an equation of a basic, known measurement formula (such as 1 bushel equals 1.24 cu. ft.). Then you put the two sides of the formula into the two numerators (tops) of two fractions. The value to be converted goes under one side of the fraction (the side with the matching unit). You place an $x$ under the other side. Finally, you cross-multiply and solve for $x$. So in other words, the general setup for converting one unit of measure to another is

$$\frac{\text{Left side of known formula}}{\text{Same units as left side or } x} = \frac{\text{Right side of known formula}}{\text{Same units as right side or } x}$$

The proportion looks rather vague, so I offer several examples to help illustrate how to use the equation for conversions. (If you need help with proportions or basic algebraic equations, refer to those topics in Chapter 4.)

If 1 U.S. dollar is equal to 0.65 euro, how many euros is 45 dollars?

You may immediately say, "Piece of cake. I just multiply to find the euros." And you'd be correct. But please bear with me and use the conversion proportion. Getting used to setting up the formulas and values in a simpler problem helps you out when the situation gets more complex.

After plugging the numbers into the conversion proportion, you get this equation:

$$\frac{1 \text{ dollar}}{45 \text{ dollars}} = \frac{0.65 \text{ euro}}{x \text{ euros}}$$

Notice that you have dollars under dollars and euros under euros. Now cross-multiply to solve for $x$:

$$\frac{1 \text{ dollar}}{45 \text{ dollars}} = \frac{0.65 \text{ euro}}{x \text{ euros}}$$
$$1 \cdot x = 45 \cdot 0.65$$
$$x = 29.25$$

As you can see, 45 dollars equals 29.25 euros.

Here's another problem to practice: How many dollars is 39,600 euros? Do you multiply or divide to solve this problem? What do you divide by?

Using the conversion proportion, you start with the basic formula (which you discovered in the previous example problem and can find daily on the Internet or at a bank) and put the 39,600 euros under the part of the formula with euros:

$$\frac{1 \text{ dollar}}{x \text{ dollars}} = \frac{0.65 \text{ euro}}{39,600 \text{ euros}}$$

Now you cross-multiply. After doing so, you see that $x$ has a multiplier. Divide each side of the equation by that multiplier. Here's what your work should look like:

$$1 \cdot 39,600 = x \cdot 0.65$$
$$\frac{39,600}{0.65} = \frac{x \cdot \cancel{0.65}}{\cancel{0.65}}$$
$$60,923.07692 = x$$

You find that it takes almost $61,000 to equal 39,600 euros.

## Lining up the linear measures

A *linear measure* has one dimension — how long it is. Think of measuring along a straight line. The standard linear measures are inches, feet, yards, rods, and miles. (Metric linear measures are found in the later section, "Making Sense of the Metric System.")

The following are some of the most commonly used linear measure equivalences:

1 yard = 3 feet = 36 inches

1 rod = 16.5 feet

1 mile = 5,280 feet

1 furlong = 660 feet = 220 yards

Try your hand at some linear conversions with the following example.

Suppose you need to measure the length and width of a storage room but you forgot your tape measure. You have a pocket calendar that you know is 7 inches long, and you have your clipboard, which is 15 inches long. So you measure the length and width of the storage room with these items. You come up with a dimension of 18 clipboards and 12 pocket calendars long by 18 clipboards and 1 pocket calendar wide. What are the dimensions of the room in feet?

First, change the clipboard and calendar measures to inches. The length is $(18 \times 15 \text{ inches}) + (12 \times 7 \text{ inches}) = 270 + 84 = 354$ inches. The width is $(18 \times 15 \text{ inches}) + (1 \times 7 \text{ inches}) = 270 + 7 = 277$ inches.

Now change inches to feet using the conversion proportion. First, find the length:

$$\frac{1 \text{ foot}}{x \text{ feet}} = \frac{12 \text{ inches}}{354 \text{ inches}}$$
$$1 \cdot 354 = x \cdot 12$$
$$\frac{354}{12} = x$$
$$x = 29.5 \text{ feet} = 29 \text{ feet, 6 inches}$$

The width is

$$\frac{1 \text{ foot}}{x \text{ feet}} = \frac{12 \text{ inches}}{277 \text{ inches}}$$
$$1 \cdot 277 = x \cdot 12$$
$$\frac{277}{12} = x$$
$$x \approx 23.083 \text{ feet} = 23 \text{ feet, 1 inch}$$

So the garage is 29 feet 6 inches by 23 feet 1 inch.

## Spreading out with measures of area

An *area measurement* is really just the total number of a bunch of connected squares. When you say that you have a room that measures 36 square feet, you mean that 36 squares, each 1 foot by 1 foot, would fit in that room. Of course, most rooms aren't exactly 6 feet x 6 feet, 9 feet x 4 feet, or some other combination of whole numbers. More often, a room will be 8 feet, 3 inches by 4 feet, 4.5 inches. If you've ever had to lay square tiles in a room, you know that even if the room was meant to be square it doesn't always come out that way. Thank goodness for tile cutters.

Area is used for orders of carpeting or tile; area is used to determine how much land you have to build on; area is important when you plan your factory or storage area. To find the area of a region, you use an appropriate area formula. To find the area of a rectangle, for example, you use a different formula than you would to find the area of a triangle. In Figure 7-1, I show you several types of regions and the formulas you use for finding the area.

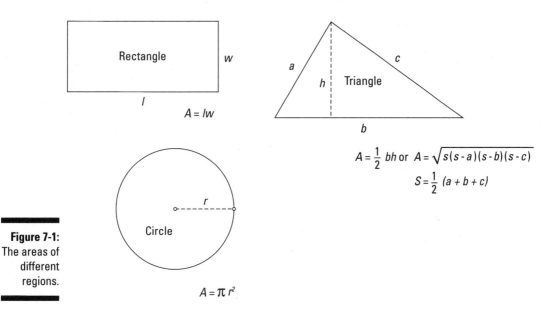

**Figure 7-1:**
The areas of different regions.

You need to identify which type of region you have before you can apply the correct formula. If your region doesn't match any of the types shown in Figure 7-1, you need to break the region into rectangles or triangles and find the area of each. You see examples of this method in Chapter 21.

The following examples give you a chance to practice calculating some areas with the formulas given in Figure 7-1.

Find the area of a rectangular room that measures 19 feet, 9 inches in length by 15 feet, 4 inches in width.

As you can see in Figure 7-1, the area of a rectangle is length × width. For the computations, you need the measures to be in either inches or feet, and feet makes the most sense when measuring rooms. So first you need to change the feet and inches to just feet. The length is 19 feet, 9 inches, which is 19¾ feet. The width of 15 feet, 4 inches is 15 ⅓ feet. Multiply the two measures, like so:

$$19\frac{3}{4} \times 15\frac{1}{3} = \frac{79}{4} \times \frac{46}{3} = \frac{79}{\underset{2}{\cancel{4}}} \times \frac{\overset{23}{\cancel{46}}}{3} = \frac{1817}{6} = 302\frac{5}{6}$$

The area is 302⅝ sq. ft. (If you don't remember how to deal with multiplying mixed numbers, head to Chapter 1 where I cover this math in detail.)

A triangular lawn measures 300 yards on one side, 400 yards on the second side, and 500 yards on the third side. Find the area of the triangle.

One method of finding the area of this triangle is to use this formula:

$$A = \frac{1}{2}bh$$

In this case, you multiply one side of the triangle, $b$, by the height drawn perpendicular to that side up to the opposite corner, $h$. Then you take one half of the product. Quite often, though, you don't have a way of measuring that perpendicular height. Your fallback in this situation is Heron's formula.

To find the area of a triangle whose sides measure $a$, $b$, and $c$ in length, use Heron's formula, which looks like this:

$$A = \sqrt{s(s-a)(s-b)(s-c)}$$

where $s$ is the semi-perimeter (half the perimeter). In other words, $s = ½ (a + b + c)$.

For example, to find the area of the triangular lawn whose sides measure 300, 400, and 500 yards, use Heron's formula. First, you have to find the semi-perimeter, which is half of the sum of the sides. 300 + 400 + 500 = 1,200. So, the semi-perimeter is found like this: ½ (1,200) = 600 yards. Now use Heron's formula to get

$$A = \sqrt{600(600-300)(600-400)(600-500)}$$
$$= \sqrt{600(300)(200)(100)}$$
$$= \sqrt{3,600,000,000} = 60,000$$

So the lawn measures 60,000 sq. yd. in area.

If you're a fan of Pythagoras (the guy who discovered the relationship between the squares of the sides of any right triangle), you probably noticed that the sides of the previously mentioned triangle make a right triangle. With a right triangle, the two shorter sides are perpendicular to one another, and you can use the quicker formula ($A = ½ bh$) for the area of a triangle. But the sides of a right triangle also make for a nice result using Heron's formula, and I preferred showing you a *nice* result.

You may have looked at the area of 60,000 sq. yd. and said, "Wow. That's a lot of yardage!" Or, maybe you're having difficulty imagining 60,000 sq. yd. and need more of a hint of how big that is. If so, read on.

# Who was Heron of Alexandria?

Heron of Alexandria, who lived around 67 A.D., discovered and wrote about many diverse mathematical topics. In fact, he's credited with several formulas, and he commented on the work of other mathematicians, such as Archimedes. Heron is also credited with the following:

✔ Determining the formula for the volume of the frustum of a cone

✔ Discovering a creative technique used to solve a quadratic equation

✔ Producing some good approximations of the square root of a number that isn't a perfect square

Much of Heron's math is found in his work, *Metrica,* which was discovered in the late 1800s in a 12th-century manuscript in Constantinople. Heron's real claim to fame, however, is his formula for finding the area of a triangle using the lengths of the sides.

The following are the most commonly used area equivalences:

1 square foot = 144 square inches

1 square yard = 9 square feet = 1,296 square inches

1 square mile = 3,097,600 square yards = 27,878,400 square feet = 640 acres

So 60,000 sq. yd. isn't all that big compared to 1 sq. mi.

How many acres are in 60,000 sq. yd.?

From the previous list of area equivalences, you see that 3,097,600 sq. yd. is equal to 640 acres. Set up the conversion proportion with the equation involving square yards and acres in the numerators, and with 60,000 sq. yd. under the 3,097,600 square yards. Solve for the unknown number of acres. Your math should look like this:

$$\frac{3,097,600 \text{ square yards}}{60,000 \text{ square yards}} = \frac{640 \text{ acres}}{x \text{ acres}}$$

$$3,097,600x = 640\,(60,000)$$

$$3,097,600x = 38,400,000$$

$$x = \frac{38,400,000}{3,097,600}$$

$$\approx 12.397$$

So 60,000 sq. yd. is a little over 12⅓ acres.

# Adding a third dimension: Volume

*Volume* is measured by how many cubes you can fit into a structure. (A cube is like a game die or a lump of sugar.) When you talk about the size of a factory or office building, you give measures in terms of square feet. But the building really has a third dimension: the height of each room. You don't know, from the square footage of a building, whether the rooms have ceilings that are 7 feet, 12 feet, or more above the floor. The amount of volume in a building plays a huge role when trying to heat and air-condition the space. It also makes a difference when it comes to painting, wallpapering, and decorating.

### Finding volume

The volume of any *right rectangular prism* (better known as a box) is found by multiplying the length by the width by the height of the prism. The volume then comes out in terms of how many cubes can fit in the prism. Not all structures are made completely of right angles, so pieces and slices of cubes get involved. But the different formulas you can use make up for any adjustments from the standard cube.

Here are the formulas for some of the more commonly used three-dimensional structures:

> **Right rectangular prism (box):** $V = lwh$, where $l$, $w$, and $h$ represent length, width, and height, respectively
>
> **Prism:** $V = Bh$, where $B$ is the area of the base and $h$ is the height
>
> **Pyramid:** $V = \frac{1}{3} Bh$, where $B$ is the area of the base of the pyramid and $h$ is the height perpendicular from the base to the tip of the pyramid
>
> **Cylinder:** $V = \pi r^2 h$, where $r$ is the radius and $h$ is the height
>
> **Sphere:** $V = \frac{4}{3} \pi r^3$, where $r$ is the radius of the sphere

Suppose that a room is 20 feet by 35 feet and has a vaulted ceiling that starts slanting at the 7-foot level and reaches a peak at 14 feet above the floor. The peak runs lengthwise down the room. What's the total air volume of the room?

Figure 7-2 shows a sketch of the room. The two ends of the vaulted ceiling form a triangle, so the area above the 7-foot level of the walls is determined by finding the volume of a triangular prism.

First, find the volume of the rectangular bottom part of the room with $V = lwh$:

> $V = (35 \text{ feet})(20 \text{ feet})(7 \text{ feet})$
>
> $= 4,900 \text{ cu. ft.}$

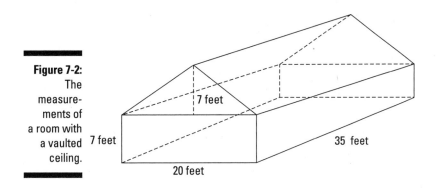

**Figure 7-2:**
The measurements of a room with a vaulted ceiling.

7 feet

7 feet

35 feet

20 feet

Now find the area of the triangular upper part of the room (which has a triangular base). You can find the area like this:

$$A = \tfrac{1}{2} bh = \tfrac{1}{2}(20)(7) = 70$$

The area of the triangle is 70 sq. ft. Multiply that area by the height of the triangular prism, 35 feet, and you get $70 \times 35 = 2{,}450$ cu. ft.

Don't confuse the *base* of the triangular prism that's used to find the volume of the three-dimensional figure with the *base* of the triangle that's used to find the area of the end piece of the room. The *Base* of the triangle is an area measure and the *base* of the triangle is a linear measure.

So the total air volume of the room is $4{,}900 + 2{,}450 = 7{,}350$ cu. ft. A lot of warm air goes up into the peak of the ceiling — which is good in the summer but bad in the winter.

---

# Heating and cooling the Great Pyramid

When you begin to think that your energy bills are over the top, think about the challenges of climate control in huge structures such as amphitheaters, factories, or malls. And what about the Great Pyramid? It isn't really hollow, but what if it were? Imagine the bill for air-conditioning a structure of that size and dimension.

The Great Pyramid was originally about 481 feet tall, and its square base had sides measuring about 756 feet. The volume of the pyramid was

$$V = \tfrac{1}{3} Bh = \tfrac{1}{3}(756 \cdot 756)(481) = 91{,}636{,}272$$

Imagine trying to air-condition *that* volume in the middle of the Egyptian desert.

### Understanding volume equivalences

If I declare that the volume of the Great Pyramid is over 91 million cu. ft., you may not quite see the magnitude of the landmark. Even 1 million cu. ft. is difficult to comprehend. So here are some equivalences to help put the Great Pyramid (and other large stuff) in perspective:

> 1 cubic foot = 1,728 cubic inches
>
> 1 cubic yard = 27 cubic feet = 46,656 cubic inches
>
> 1 cord = 128 cubic feet
>
> 1 cubic mile = 147,197,952,000 cubic feet

Some restaurants have wood-burning fireplaces to add ambience to the dining experience. So they have to order and store firewood. How many cords of firewood can Gloria, manager of Gloria's Fine Dining, fit into a storage shed that's 12 feet long by 8 feet wide by 6 feet high?

First find the volume in cubic feet by multiplying the length by the width by the height ($V = lwh$):

$$V = (12 \text{ feet})(8 \text{ feet})(6 \text{ feet})$$
$$= 576 \text{ cubic feet}$$

Now write a conversion proportion with the cord and cubic feet:

$$\frac{1 \text{ cord}}{x \text{ cords}} = \frac{128 \text{ cubic feet}}{576 \text{ cubic feet}}$$
$$1 \cdot 576 = x \cdot 128$$
$$\frac{576}{128} = x$$
$$x = 4.5 \text{ cords}$$

Gloria can fit 4.5 cords of firewood in the shed.

# Making Sense of the Metric System

The *metric system* of measurement is used by 96% of the world's population. The metric system is popular because of its simplicity. Each type of measure (length, volume, and mass) is based on a single unit and on powers of ten of that unit. For lengths or linear measures, the unit is the meter; for volume, it's the liter; and for mass or weight, it's the gram.

Another beauty of the system is the way the different types of measures interact with one another. The gram, for example, is equal to 1 cubic centimeter of water, which is equal to 1 milliliter. Add that to the fact that the metric units are interchangeable with one another and also that conversions use decimals rather than fractions, and the metric system seems to make the most sense for use universally. Even so, the United States is still holding out and chooses to stick to its traditional English system of measure.

No matter what the United States does, you're likely to run into metrics when you're dealing with numbers from other countries. When that happens, just read through this section, which provides the basics.

## *Moving from one metric unit to another*

Converting from one unit to another in the metric system involves nothing more than sliding the decimal point to the right or left. Changing 4.5 decameters to centimeters, for example, means moving the decimal point three points to the right. So 4.5 decameters = 4,500 centimeters. If you know your prefixes, you're in business to convert from one metric unit to another. The scale in Figure 7-3 shows the relative positions of the different prefixes used in the metric system.

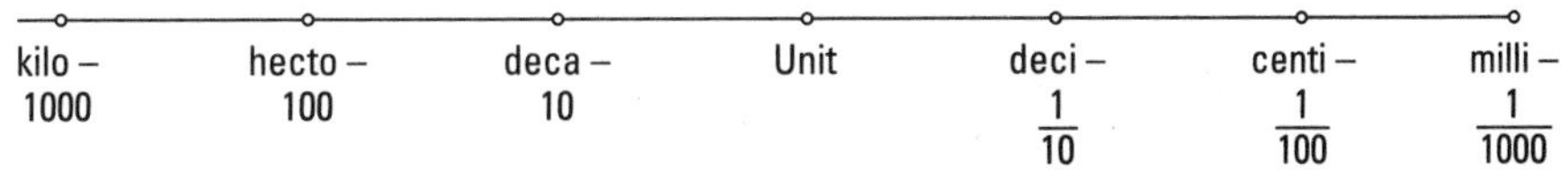

**Figure 7-3:**
A scale showing how Latin prefixes refer to numbers.

| kilo – | hecto – | deca – | Unit | deci – | centi – | milli – |
|---|---|---|---|---|---|---|
| 1000 | 100 | 10 | | $\frac{1}{10}$ | $\frac{1}{100}$ | $\frac{1}{1000}$ |

The prefixes tell you the size of a single metric measure. For example, a kilometer is 1,000 meters, and a milliliter is one-thousandth of a liter. You use the same prefixes for each of the types of measures. So 1 kilogram equals 1,000 grams. As you can imagine, this uniformity keeps the conversions much simpler.

Say that you buy 3.12 liters of olive oil and want to divide it into milliliter spoonfuls for your restaurant's signature dish. How many milliliters are in 3.12 liters?

# Blame it on the French: Discovering where the metric system came from

It was the French, in the late 1700s, who decided that enough was enough. The countries of the world used dozens of different measurement systems. The different standards of measurement caused a lot of confusion and difficulties, especially when the countries began trying to trade with one another. The French wanted to develop a system to be used throughout the world. They decided that a decimal-based system made the most sense; powers of ten allowed for simple conversions. Little by little, this same metric system has been adopted by countries around the world. As of 2007, only three countries in the world have chosen not to use the metric system: Liberia, Myanmar, and the United States. Even though the UK has officially adopted the metric system, it continues to use a mixture of metric and Imperial measures.

The meter, the gram, and the liter are the three different types of measures making up the metric system. An attempt was made to connect the basic measures to naturally occurring structures or physical quantities. The meter was originally defined as being one forty-millionth of the polar circumference of the world. And the kilogram was to be the mass of 1 liter (1 cubic decimeter) of water at 4 degrees Celsius. The meter was later defined as being the distance traveled by light in an absolute vacuum during about one three hundred millionth of a second. (Oh, yeah, I measure that every day.)

The metric system also has measures of temperature. In the metric or Celsius system, water freezes at 0 degrees and boils at 100 degrees (rather than the Fahrenheit scale of 32 degrees to 212 degrees).

The pros and cons abound regarding whether the United States should change or not change to the metric system. But I won't bore you with the details here. Suffice it to say that if the U.S. did decide to change, why would we ever have to teach students how to add fractions? Imagine generations of children not having to find common denominators — so sad.

Look at the scale shown in Figure 7-3. To go from liters (the main unit on the scale) to milliliters, you have to move the decimal point three places to the right. So 3.12 liters is 3,120 milliliters. Not too difficult, is it?

Now imagine that you measure along a stretch of road next to your bed-and-breakfast with a centimeter ruler (don't ask why you forgot your tape measure). You come up with 3,243 centimeters. When you order bedding plants to line that stretch of road from a European supplier, you have to use metrics. How many decameters is 3,243 centimeters?

By looking at Figure 7-3, you can see that to go from centimeters to decameters, you have to move three decimal places to the left. The number 3,243 has its decimal point at the right end, so you get 3.243 decameters. Nice work!

# Converting from metric to English and vice versa

One of the reasons that the metric system was developed was to eliminate the need for converting from one system to another. Two hundred years later, we're still doing the measurement conversions. At least, we're down to two basic systems and some pretty standard conversion values. However, in this section, you'll see that the numbers aren't exactly pretty. Converting from metric to English or English to metric isn't an exact science — and I mean that literally. The measures are approximate — or as close as three decimal places can get you.

Here are some of the more frequently used values when converting from metric to English or English to metric:

| | |
|---|---|
| 1 mile = 1.609 kilometers | 1 kilometer = 0.621 mile |
| 1 foot = 0.305 meters | 1 meter = 1.094 yards |
| 1 inch = 2.54 centimeters | 1 centimeter = 0.394 inch |
| 1 quart = 0.946 liter | 1 liter = 1.057 quarts |
| 1 gallon = 3.785 liters | 1 liter = 0.264 gallon |
| 1 pound = 453.592 grams | 1 kilogram = 2.205 pounds |

In practical, everyday computations, it's more common to use 1.6 for the number of kilometers in a mile and 2.2 for the number of pounds in a kilogram. You have to decide, depending on the application, just how precise you need to be.

While on a business trip in Europe, you read on a sign that it's 400 km to Hamburg. How far is that in miles?

You can use the following conversion proportion to determine the number of miles:

$$\frac{1 \text{ kilometer}}{400 \text{ kilometers}} = \frac{0.621 \text{ mile}}{x \text{ miles}}$$
$$1 \cdot x = 0.621(400)$$
$$x = 248.4$$

You have about 250 miles to go. If you can drive at 50 miles per hour, that's a 5-hour drive. How many hours is that in Germany? (Just kidding — no conversion needed here. But do watch out for the 24-hour clock!)

Suppose you're told that you need to buy 23 gallons of paint so that your maintenance worker can finish a project in the apartment complex that you own and rent out. The paint you want only comes in liter cans. How many liter cans will it take to finish the project?

Use the following conversion proportion involving gallons and liters to find out:

$$\frac{1 \text{ gallon}}{23 \text{ gallons}} = \frac{3.785 \text{ liters}}{x \text{ liters}}$$
$$1 \cdot x = 3.785\,(23)$$
$$x = 87.055$$

So you can see that it will take a few more than 87 liter cans of paint. Now the difficult decision is whether to buy that 88th can. You have to decide whether the maintenance guy can squeeze an extra 0.055 liter of paint out of the last can.

# Discovering How to Properly Measure Lumber

Have you ever been to a lumber yard? No aroma quite matches that of stacks and stacks of wood — all types of trees and all ages of planks. I have a friend whose business is carving out and constructing cellos. He creates cellos from hunks (very nice hunks) of wood. He buys the wood and then lets it sit for about five years to age before working on it. Such patience.

Your exposure to a lumber yard is probably more utilitarian: You need to do some repair to a rental property, you need more shelves in your shop, or you decide to subdivide the showroom portion of your business. In any case, you need to be aware of the pitfalls of measuring spaces and trying to get the lumber measurements to coincide.

You've undoubtedly heard of pieces of wood referred to as 2 x 4s or 4 x 6s. But did you also know that these measures are lies? A 2 x 4 is really a 1.5 x 3.5. This discrepancy isn't a conversion issue. It's basically a shrinkage and finishing issue.

Shocked? Yeah, I was too when I found out. But don't worry. You just need to take into account the actual size of the lumber you're buying if you need a particular thickness of a wall or deck area. If you don't add on enough lumber, you're apt to come up short! In Table 7-1, I give you some of the more common sizes of lumber pieces and their actual sizes.

| Table 7-1 | Lumber Dimensions |
|---|---|
| *Stated Size of Lumber* | *Actual Size* |
| 1 inch x 2 inches | ¾ inch x 1½ inches |
| 1 inch x 4 inches | ¾ inch x 3½ inches |
| 1 inch x 6 inches | ¾ inch x 5½ inches |
| 1 inch x 8 inches | ¾ inch x 7¼ inches |
| 2 inches x 4 inches | 1½ inches x 3½ inches |
| 2 inches x 6 inches | 1½ inches x 5½ inches |
| 4 inches x 4 inches | 3½ inches x 3½ inches |
| 4 inches x 6 inches | 3½ inches x 5½ inches |

Try your hand at this example: Suppose you're building a loading dock on the back of a floral business that's going to be 30 feet long and 12 feet wide. You plan to run the planks perpendicular to the building (and parallel to the 12-foot side) and allow for drainage between the planks. If you allow a ⅛-inch gap between each plank, how many rows of 1 x 6 planks do you need to build the deck?

You need 30 feet of 1 x 6 planks laid side by side. As I mention in Table 7-1, the actual size of a 1 x 6 is ¾ inch x 5½ inches. Don't worry about the thickness of ¾ inch. Instead, simply add ⅛ inch (the gap) to each plank width of 5½ inches to get 5⅝ inch units (plank plus gap). Here's the addition of the fractions:

$$5\frac{1}{2} + \frac{1}{8} = 5\frac{4}{8} + \frac{1}{8} = 5\frac{5}{8}$$

Need some guidance on adding fractions? I show you how in Chapter 1. Now you need to change 30 feet to inches. Do so by multiplying 30 feet by 12 inches to get 360 inches. Now divide 360 by 5⅝, like so:

$$360 \div 5\frac{5}{8} = 360 \div \frac{45}{8}$$

$$= 360 \times \frac{8}{45} = \frac{^8 360}{1} \times \frac{8}{45_1} = \frac{64}{1} = 64$$

This math tells you that it will take 64 rows of 1 x 6 planks to build your loading dock.

# Measuring Angles by Degrees

*Angle measurements* are important to carpenters, architects, pilots, and surveyors. The scientific measurement of angles is usually done in *radians,* because a radian is a more naturally occurring size (slightly more than 57 degrees). But most of us think of angles in terms of degrees.

A *degree* is $\frac{1}{360}$ of a circle — a very tiny wedge. Using the number 360 to divide a circle into equal pieces was rather clever. After all, 360 has plenty of divisors. You can divide a circle into many, many equal pieces, because 360 divides evenly by 2, 3, 4, 5, 6, 8, 9, 10, 12, 15, 18, 20, 24, 30, 36, 40, 45, 60, 72, 90, 120, and 180.

In this section, I show you the properties of angles in figures, and I explain how to subdivide an angle. In Chapter 6, I show you some sketches of the more commonly used angle measures. And in Chapter 21, you can find the different ways that angles are used in navigational and surveying directions.

## Breaking down a degree

Angle measurements are made with instruments specifically designed for the process. A *protractor* is one of these instruments. They're usually in the shape of a half circle and allow you to measure angles of up to 180 degrees. The protractor has marks on it for all the degree measures from 0 to 180.

It isn't practical to try to divide a degree into units that are smaller than the ones on these hand-held protractors. Most people wouldn't be able to see the divisions, anyway. When navigating the skies or the seas, though, a fraction of a degree one way or the other can make a huge difference in whether you reach your destination.

As I mention earlier, a degree is $\frac{1}{360}$ of a circle. Each degree is itself subdivided into 60 smaller units called *minutes.* Each minute is subdivided into 60 smaller units called *seconds.* So each degree — which is small to begin with — is divided into $60 \times 60 = 3,600$ smaller portions for increased accuracy. A single prime (') indicates minutes, and a double prime (") indicates seconds. So, the angle measure given as 55°45'15" is read as 55 degrees, 45 minutes, 15 seconds.

The degree-minute-second notation is just grand, but it isn't very helpful when combining with other numbers in operations. For instance, if you want to multiply the angle measuring 55°45'15" by 6, you have to multiply each subdivision by 6, rewrite the minutes and seconds so that they don't exceed 59 of each, and adjust the numbers accordingly. So usually you'll find it much easier to change the angle measure to a decimal before multiplying or otherwise combining. I show you how to do this in the following example. And, just in case you doubt my judgment on which method is easier, I include the math for both.

Multiply 55°45'15" by 6.

**Method 1:** Multiply each unit by 6 and adjust.

55° × 6 = 330°.

45' × 6 = 270'. To adjust, divide by 60. 270' ÷ 60 = 4 plus a remainder of 30. Because 60 minutes makes a degree, this result represents 4 degrees and 30 minutes.

15" × 6 = 90". Adjust by dividing 90 by 60. 90" ÷ 60 = 1 plus a remainder of 30. Sixty seconds makes a minute, so this result represents 1 minute and 30 seconds.

Now combine all the degrees, minutes, and seconds from the multiplications and adjustments. Degrees: 330 + 4 = 334°. Minutes: 30 + 1 = 31'.

So the final result is 334°31'30".

**Method 2:** Change the measure to a decimal and multiply by 6.

Divide the number of minutes by 60 and the number of seconds by 3,600 to get the decimal equivalent of each unit. Then add the decimals to the degree measure.

45' 60 = 0.75 and 15" ÷ 3,600 ≈ 0.0041667. If you add the decimals to the degree measure, you get 55 + 0.75 + 0.0041667 = 55.7541667.

Multiplying the sum by 6 gives you: 55.7541667 × 6 = 334.5250002.

Now to compare the answers from the two methods, change the 334°31'30" to a decimal by dividing the 31' by 60 and the 30" by 3,600. 31' ÷ 60 ≈ 0.5166667, and 30" ÷ 3,600 ≈ 0.0083333. So 334°31'30" = 334 + 0.5166667 + 0.0083333 = 334.5250000. With rounding, the two answers differ by 0.0000002. Not too bad.

## Fitting angles into polygons

A polygon is a dead parrot. Get it? Polly gone? Sorry. That's lame math teacher humor. Okay, the real scoop is that a *polygon* is a many-sided closed figure made up of segments. What I mean by *closed* is that each segment is connected by its endpoints to two other segments; in other words, you have an inside and an outside. You can determine the total number of degrees of the angles inside a polygon if you know how many sides there are.

REMEMBER

A polygon with $n$ sides has interior angles measuring a total of $180(n - 2)$ degrees. Here are some examples plugged into the formula:

A triangle has three sides, so the total of all the degrees inside it is $180(3 - 2) = 180(1) = 180$ degrees.

A rectangle has four sides, so the total of all the degrees inside it is $180(4 - 2) = 180(2) = 360$ degrees.

A hexagon has six sides, so the total of all the degrees inside it is $180(6 - 2) = 180(4) = 720$ degrees.

You know that not all areas in a building follow the nice geometric shapes — you have nooks and crannies and other odd outcroppings. So does this rule for the total interior angles still work for odd shapes? You betcha! Check out the following example, which proves my point.

GO FIGURE

Figure 7-4 shows a possible layout for a large work area. The angle measures are shown at each interior angle. What's the total of the interior angle measures of the area shown in the figure?

The area has eight sides. Using the formula, you get $180(8 - 2) = 180(6) = 1,080$ degrees for a total of the inside angles. Does this match with the degrees shown in the sketch? Add up the angles to find out: $77 + 90 + 90 + 270 + 270 + 90 + 90 + 103 = 1,080$ degrees. Yup! The formula works, even with odd-shaped areas.

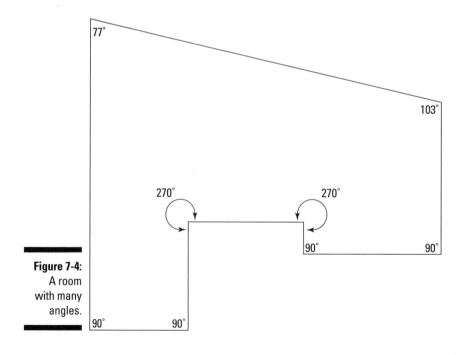

**Figure 7-4:**
A room
with many
angles.

# Chapter 8

# Analyzing Data and Statistics

. . . . . . . . . . . . . . . . . . . . . . . . . . . . . . . . . . . . . . . . . . . . . . . . . . . . . . . .

## In This Chapter

▶ Taking your raw information and turning it into something useful

▶ Choosing which of the three average values you should use

▶ Including standard deviation in your analysis

. . . . . . . . . . . . . . . . . . . . . . . . . . . . . . . . . . . . . . . . . . . . . . . . . . . . . . . .

**D**id you know that an author has published a book entitled *How to Lie with Statistics?* Actually, you could change the word *statistics* to just about any mathematical term and easily come up with some examples of how to mislead readers. But statistics especially seem to lend themselves to misinterpretation and abuse.

The world of statistics consists of figuring out how to collect data, organizing your data, and then drawing meaningful conclusions. The conclusions you draw will deal with the average (or middle value) and the spread around the average. You use computations with formulas or charts and graphs.

In business, it's important to present statistics as clearly and neutrally as possible. It's always tempting to put a positive spin on data, but if the numbers are trending down, the leaders in the company need to know. And if you're assessing the stats of another company, you need to understand how to determine whether the numbers have been slanted to present a pretty picture that doesn't match reality.

In this chapter, I don't tell you how to ferret out all the spin, but I can make you more aware of the potential for misuse. If you're a knowledgeable reader or consumer of statistics, you'll have less of a chance of being duped. So I offer information on how data is organized and how computations are made.

## Organizing Raw Data

Data is information. Data is facts and figures. Data is something that's known. One problem with raw numerical data is that it's often presented or available in a format that gives little useful information regarding what it's all about.

For instance, if you're given a list of the number of cars sold each day of the year at a dealership and it goes something like 4, 7, 6, 5, 8, 7, 6, 5, 7, 5, 4, and so on, you're likely to say, "Wait! This is meaningless!" And you'd be right. A list of numbers like this may tell you that no more than 8 cars were sold on any one day, but it doesn't tell you such things as

✔ How many cars are sold on each day of the week?

✔ How much revenue is attached to the number of cars sold?

✔ What number of cars sold in a day happens more frequently?

✔ What days of the week yield the most car sales?

✔ What's the average number of cars sold?

Many other useful bits of information could add meaning to these numbers. The key to using data successfully is to have it organized. When the data is organized, you can draw conclusions, do computations, and make predictions. In the following sections, I provide a few options for organizing your data.

## Creating a frequency distribution

A *frequency distribution* is one structure or process for organizing data. It tells you how often a particular score or value occurs. For example, a frequency distribution is a good way to present the following information:

✔ The number of customers who come into your shop each day.

✔ The number of cell phones activated during each hour that your business is open.

✔ The number of stops made by the delivery person during each shift.

With a frequency distribution, you gather all the scores or values that are the same and count how many of them you have. For example, say that you have a list of 75 numbers. Almost all the numbers repeat. Rather than add all 75 numbers one at a time, you can organize them in a frequency distribution. By doing so, you determine how many of each number there are (how many 5s, how many 7s, and so on), and then you use multiplication and addition to get the result. In other words, you multiply the number of times the number occurs by that number and add the products together.

This method gives you an easier and more accurate result. After all, when a list of numbers is too long, you may tend to lose track of where you are in your addition and make errors. How many times have you had to start all over again when dealing with long lists of numbers?

A frequency distribution is usually presented as a table in which all the possible scores or numbers are listed in order (from lowest to highest or highest to lowest). A tally is made in another column. A *tally* consists of hash marks (| | |) where you make a mark for each number and slice through four of the hash marks when you get to a group of five, like this: ‖‖. You probably used these tallies as a kid when you were in competition with your friends.

Table 8-1 shows a frequency distribution of the following numbers: 4, 7, 6, 5, 8, 7, 6, 5, 7, 5, 4, 5, 6, 8, 8, 6, 4, 3, 5, 6, 7, 6, 5, 4, 3, 4, 5, 6, 7, 8, 8, 8, 6, 6, 3, 4, 3, 6, 4, 5, 6, 0, 6, 5, 3, 6, 3, 5, 6, 7, 7, 6, 5, 4, 3, 2, 2, 4, 5, 6, 7, 6, 7, 7, 8, 5, 6, 5, 4, 6, 6, 6, 4, 3, 3. The numbers in the table go from a low of 0 to a high of 8. The third column gives the frequency of each number, which is simply the result of the tallies in the second column. The total is obtained from adding the numbers in the frequency column.

| Table 8-1 | A Frequency Distribution of the 75 Numbers | |
|---|---|---|
| **Number** | **Tally** | **Frequency** |
| 0 | | | 1 |
| 1 | | 0 |
| 2 | || | 2 |
| 3 | ‖‖ |||| | 9 |
| 4 | ‖‖ ‖‖ | | 11 |
| 5 | ‖‖ ‖‖ |||| | 14 |
| 6 | ‖‖ ‖‖ ‖‖ ‖‖ | | 21 |
| 7 | ‖‖ ‖‖ | 10 |
| 8 | ‖‖ || | 7 |
| **Total** | | 75 |

# Grouping values together in a frequency distribution

Sometimes a tally involves many items with a wide range of values. For instance, you may have the length of time (in minutes) that customers walk through your company's showroom. Listing all the possible values may take a page or two of paper — especially when the range is from 0 to 100 or from

1,000 to 10,000 (that would be a long time to wander the showroom). One method that eliminates the need for so many numbers is to do groupings of values.

For example, you can make groups that consist of all the numbers from 1 to 5, 6 to 10, 11 to 15, and so on. The only stipulation is that you keep the groupings equivalent in terms of size. A good rule of thumb is to plan on having about ten different groupings of numbers (give or take a few).

Groupings are a good way to present information such as

- ✔ The salary amounts of your office's employees.
- ✔ The number of light bulbs sold in your store on any given day.
- ✔ The number of plates of spaghetti sold in an evening at your restaurant.

To determine how to divide a set of numbers into groups, look at the range of numbers or values that you have collected. What's the range from the highest number to the lowest? If your listing of numbers goes from a low of 3 to a high of 97, your range is 95 numbers (97 – 3 + 1; you add the 1 so that both 3 and 97 are included in the count).

After you find your range, divide that number by 10 (which, as I mention earlier, is a good number of groupings to have). If you divide 95 by 10, you get 9.5. In this case, the best arrangement probably would be to have ten groupings of ten numbers.

So now you can let the first grouping be the numbers 1 through 10; the second grouping would go from 11 through 20. From there you can calculate the groupings all the way up to the last grouping going from 91 through 100. I know that the numbers you've collected don't go down to 1 or up to 100, but this way the intervals are all the same size, and they cover all the numbers pretty symmetrically.

Ready for some practice in setting up a frequency distribution with groupings? Great! Try this one out: Say that you've been keeping track of the number of miles driven each week by your sales associates. You want to organize them in a frequency distribution. What does your distribution table look like?

To begin, take a look at Table 8-2, which shows the mileage records of your eight associates.

**Table 8-2**    **The Mileage Records of the Company's Sales Associates**

|         | Associate 1 | Associate 2 | Associate 3 | Associate 4 | Associate 5 | Associate 6 | Associate 7 | Associate 8 |
|---------|-------------|-------------|-------------|-------------|-------------|-------------|-------------|-------------|
| Week 1  | 123         | 444         | 574         | 893         | 938         | 722         | 422         | 349         |
| Week 2  | 285         | 483         | 533         | 563         | 234         | 442         | 433         | 532         |
| Week 3  | 439         | 842         | 576         | 488         | 574         | 267         | 55          | 466         |
| Week 4  | 456         | 599         | 444         | 113         | 634         | 331         | 165         | 986         |
| Week 5  | 898         | 456         | 522         | 464         | 533         | 222         | 453         | 886         |
| Week 6  | 453         | 228         | 866         | 856         | 266         | 234         | 155         | 442         |
| Week 7  | 435         | 868         | 456         | 456         | 235         | 956         | 762         | 544         |

Looking through the list, you can see that the numbers range from a low of 55 miles to a high of 986 miles. The range is 932 miles (986 − 55 + 1), so your best bet is to have 10 groupings starting with 0 to 99 and going up to 900 to 999. Table 8-3 shows you the tally created and the frequency in each grouping.

| Table 8-3 | The Frequency Distribution of Sales-Associate Mileage | |
|---|---|---|
| *Range* | *Tally* | *Frequency* |
| 0 – 99 | \| | 1 |
| 100 – 199 | \|\|\|\| | 4 |
| 200 – 299 | ⧸⧸⧸⧸ \|\|\| | 8 |
| 300 – 399 | \|\| | 2 |
| 400 – 499 | ⧸⧸⧸⧸ ⧸⧸⧸⧸ ⧸⧸⧸⧸ \|\|\| | 18 |
| 500 – 599 | ⧸⧸⧸⧸ ⧸⧸⧸⧸ | 10 |
| 600 – 699 | \| | 1 |
| 700 – 799 | \|\| | 2 |
| 800 – 899 | ⧸⧸⧸⧸ \|\| | 7 |
| 900 – 999 | \|\|\| | 3 |
| **Total** | | 56 |

This frequency table is much more informative than the original listing of numbers in Table 8-2. In fact, this table is probably even more informative (and easier to set up) than listing all the numbers in order from smallest to largest or vice versa. Why? Well, for starters, the manager gets a feel for the expected or more frequent mileage numbers. Also, the highest and lowest numbers stand out more and may trigger some inquiries. Finally, the mileage information helps in planning the salesperson's time and reimbursement projections.

# Finding the Average

When you read about the average income across the country, you probably think in terms of a number that falls in the middle of the entire spectrum of incomes. In general, that's a correct thought — the average usually is the middle or most used value. But, in fact, the average value can be determined one of three ways. In other words, the "middle" average is just one way. The average can be the *mean,* the *median,* or the *mode.*

The purpose of having three different ways of measuring the average is to give the best and most descriptive number for the average of a particular collection of numbers. Unfortunately, having the three choices can lead to some abuse of numbers and some misrepresentation of what's in the collection. So, in the following sections, I show you all three methods of finding the average, and I let you decide what works best for describing the average value in your situation.

# Adding and dividing to find the mean

The *mean average* is the number determined by adding all the numbers in a collection and dividing by the number of numbers that you've added together.

Want to see the actual formula? The mean average of the numbers $a_1$, $a_2$, $a_3$, . . . $a_n$ is equal to the sum of the $n$ numbers divided by $n$:

$$\frac{a_1 + a_2 + a_3 + \cdots + a_n}{n}$$

Find the mean average of these numbers: 8, 6, 4, 9, 2, 2, 5, 4, 7, 3, 9, 6.

To find the mean, add the numbers and divide by 12. You divide by 12 because you have 12 numbers in the list. Here's what the math looks like:

$$\frac{8 + 6 + 4 + 9 + 2 + 2 + 5 + 4 + 7 + 3 + 9 + 6}{12}$$

$$= \frac{65}{12} = 5\frac{5}{12} \approx 5.417$$

The mean average, 5.417, isn't on the list of numbers, but it somewhat describes the middle of the collection.

Suppose a not-so-ethical factory owner objects to the bad press that he has been receiving from his employees, who are complaining that his pay scale is much too low. When interviewed, he expresses incredulity that his employees would complain. After all, the average salary for everyone at the factory is $53,500. The nine employees who are complaining claim that they're making only $15,000. How can this be?

If the nine employees are each making $15,000, and if the average salary of the nine employees and the owner is $53,500, you can easily find the owner's salary by solving for $x$, like this:

$$\frac{9(15,000) + x}{10} = 53,500$$

$$\frac{135,000 + x}{10} = 53,500$$

$$135,000 + x = 535,000$$

$$x = 535,000 - 135,000$$

$$= 400,000$$

If the owner is taking home $400,000, the mean average salary of the ten people is, in fact, $53,500. So the owner isn't exactly lying, but he's definitely using a method that doesn't paint a true picture. This is a case of the mean average not really representing the true average or middle value. A better measure of the average salary in this case is the median or the mode, which I explain later in this chapter.

## Locating the middle with the median

The *median* is the middle of a specifically ordered list of numbers (to remember this one, just think about a median that runs down the middle of the highway). The median as an average or middle of a set of numbers is a great representation of the average when you have one or more *outliers*. An outlier is a number that's much smaller or much larger than all the other values.

For instance, you may remember in school when that one classmate always scored 100 or 99 even when the rest of you struggled for a 70. Those curve-breakers were so annoying. A curve-breaker is sort of like an outlier. The higher-than-usual score hiked up the average for everyone. The median average is designed to lessen the effect of that unusual score.

The median is the middle number in an ordered (highest to lowest or lowest to highest) list of numbers. If the list has an even number of entries (meaning there is no middle term), the median is the mean average of the middle two numbers. (Refer to the previous section if you need guidance on finding the mean average.)

Find the median of the following numbers: 8, 6, 4, 9, 2, 2, 5, 4, 7, 3, 9, 6.

First put the numbers in order from least to greatest: 2, 2, 3, 4, 4, 5, 6, 6, 7, 8, 9, 9. Because the list has 12 numbers, the 6th and 7th numbers in the list are in the middle. The two middle numbers are 5 and 6. So to find the median, you have to find the mean average of 5 and 6. You do so by adding the numbers together and dividing by 2. You get 5.5 as the mean.

As it turns out, the mean average of the entire list is 5.417 (see the example in the previous section), so the mean and median are pretty close in value in this situation. Having two averages of the same list that are close together in value gives you more assurance that you have a decent measure of the middle.

The ages of the 15 employees at Methuselah Manufacturing are 18, 25, 37, 28, 23, 29, 31, 87, 20, 24, 30, 21, 22, 93, 19. Find both the mean and median ages and determine which is a better representation of the average age.

The mean average is the sum of the ages divided by 15:

$$\frac{18 + 25 + 37 + \cdots + 19}{15} = \frac{507}{15} = 33.8$$

As you can see, the mean average age is 33.8, which is a higher number than all but three of the actual ages.

Now find the median by putting the ages in order from lowest to highest:

18, 19, 20, 21, 22, 23, 24, 25, 28, 29, 30, 31, 37, 87, 93

You have an odd number of values, so your middle age is the eighth number in the list: 25. An average age of 25 (rather than 33.8 with the mean average) better represents most of the actual ages. The two oldest employees are so much older than the others that their ages skew the middle or average age.

## Understanding how frequency affects the mode

The *mode* of a set of numbers is the most-frequently occurring number — it's the one that's listed most often. The mode is a good average value when the number occurs overwhelmingly frequently in the list. The following three sets of numbers all have a mode of 5:

A) 1, 1, 2, 2, 2, 3, 3, 4, 4, 5, 5, 5, 5, 5, 5, 6, 6, 6, 7, 7, 8, 10

B) 1, 2, 3, 4, 5, 5, 6, 7, 8, 9, 10, 11, 12, 13, 14, 15, 16, 17, 18

C) 5, 5, 5, 5, 5, 5, 5, 5, 5, 5, 5, 5, 5, 5, 5, 5, 5, 5, 999, 999

The mode of 5 in the first and third lists seems to make the most sense as representing the average of the list. In the middle list, however, even though 5 is technically the mode, it doesn't seem to represent what the list of numbers is. In this case, you might want to find the average another way.

Suppose you want to show that your nonprofit foundation is kid-friendly, but you don't want to reveal how many children are in specific families. Determine which average is the best representation of the number of children of the 43 families served by your foundation: 0, 5, 6, 0, 3, 4, 0, 4, 10, 14, 7, 0, 6, 2, 3, 0, 4, 5, 4, 8, 7, 3, 5, 3, 2, 6, 3, 3, 0, 2, 2, 6, 1, 2, 3, 2, 2, 4, 4, 4, 6, 1, 2.

First put the numbers in order. You must do this when finding the median, but it's also helpful to have an ordered list when figuring the mean and mode. In order, the numbers are 0, 0, 0, 0, 0, 0, 1, 1, 2, 2, 2, 2, 2, 2, 2, 2, 3, 3, 3, 3, 3, 3, 3, 4, 4, 4, 4, 4, 4, 4, 5, 5, 5, 6, 6, 6, 6, 6, 7, 7, 8, 10, 14.

The mean is the sum of the numbers divided by 43. The sum is 158, and $158 \div 43 \approx 3.674$.

The median is the middle number, which in this case is the 22nd number. So, the median is 3.

The mode is the most frequent number, which after counting is 2 (there are eight 2s).

So, which outcome —3.674, 3, or 2 — do you think is the best representation of the list? The two outliers (10 and 14) pull the average up a bit. I'd probably go with the middle (median) of the averages — but it's really your call on this. You can correctly say that any of the values is the average.

# Factoring in Standard Deviation

The average of a listing of numbers tells you something about the numbers, but another important bit of information is the variance or *standard deviation* — how much the numbers in the listing deviate from the mean average (see the earlier section, "Adding and dividing to find the mean," for more on mean averages). The standard deviation is a measure of variation and can be a decent comparison between two sets of numbers — as long as the numbers have some relation to one another. After all, as with other facets of statistics, you have to be careful not to misrepresent what's going on just because you can.

## Computing the standard deviation

Standard deviation is a measure of *spread*. This spread has to do with how far most of the numbers are from the average. For example, you may be interested not only in the average number of sales of your staff members, but also whether the sales cluster closely around that average or are much higher and much lower than the average. Is your staff pretty predictable and steady, or are they all over the place with their sales endeavors? Standard deviation is a number representing the deviation from the mean. So the greater the standard deviation, the more numbers you'll find that are farther from the mean average.

To find the standard deviation of a list of numbers, use the following formula:

$$s = \sqrt{\frac{\sum x^2 - n(\overline{x})^2}{n-1}}$$

where $\Sigma x^2$ represents the sum of all the squares of the numbers in the list, $n$ is the number of numbers, and $\bar{x}$ represents the mean average of the numbers (which is squared in the formula before multiplying it by $n$).

Each of the following three lists of numbers has a mean average of 5:

A) 1, 2, 3, 4, 5, 5, 5, 5, 5, 5, 5, 5, 5, 5, 5, 5, 5, 5, 5, 6, 7, 8, 9

B) 1, 2, 2, 3, 3, 3, 4, 4, 4, 4, 5, 5, 5, 6, 6, 6, 6, 7, 7, 7, 8, 8, 9

C) 1, 1, 1, 1, 1, 1, 1, 1, 1, 1, 1, 5, 9, 9, 9, 9, 9, 9, 9, 9, 9, 9, 9

To find the standard deviation of list A, you square each number and then add up all the squares. Because the number 5 repeats 15 times, I show that as a multiple of 12:

$$1^2 + 2^2 + 3^2 + 4^2 + 15(5^2) + 6^2 + 7^2 + 8^2 + 9^2 = 635$$

You know that the mean average of the list is 5, and you know that there are 23 numbers in the list. Now all you have to do is substitute the numbers into the formula:

$$s = \sqrt{\frac{\sum x^2 - n(\bar{x})^2}{n-1}}$$

$$= \sqrt{\frac{635 - 23(5)^2}{23-1}} = \sqrt{\frac{635 - 575}{22}}$$

$$= \sqrt{\frac{60}{22}} \approx 1.65$$

Using the same process on lists B and C, you get standard deviations of about 2.132 for list B and exactly 4 for list C. So the first list of numbers has the smallest deviation, and the last list of numbers has the greatest deviation.

 GO FIGURE

At a local ice cream factory, the hourly wages of the employees are as follows:

4 employees earn $8.00 per hour

7 employees earn $8.50 per hour

20 employees earn $9.25 per hour

25 employees earn $10.00 per hour

4 employees earn $12.00 per hour

What's the (mean) average wage of the employees, and what's the standard deviation?

TIP

This type of problem is solved more easily with a chart or table. So take a look at Table 8-4, which I created for this scenario.

| Table 8-4 | | Hourly Wages of Employees | | |
|---|---|---|---|---|
| Hourly Rate | # Employees | Rate × # of Employees | Square of Rate | Square × # of Employees |
| 8.00 | 4 | 32 | 64 | 256 |
| 8.50 | 7 | 59.50 | 72.25 | 505.75 |
| 9.25 | 20 | 185 | 85.5625 | 1,711.25 |
| 10.00 | 25 | 250 | 100 | 2,500 |
| 12.00 | 4 | 48 | 144 | 576 |
| Totals | 60 | 574.5 | | 5,549 |

First, to compute the mean, you need to find the sum of the hourly rate of all 60 employees. Look at the sum of the third column, 574.5, where each rate has been multiplied by the number of employees earning that rate. Divide 574.5 by 60 (the number of employees) and you get 9.575. The (mean) average hourly rate is about $9.58.

To compute the standard deviation, you need the sum of the squares of all 60 wage rates. The last column has the square of each rate multiplied by the number of employees involved. The sum of the squares is 5,549. You have to multiply the number of employees, 60, by the square of the mean. $9.575^2 = 91.680625$. Now you're all set to put the numbers into the formula for the standard deviation:

$$s = \sqrt{\frac{\sum x^2 - n(\bar{x})^2}{n-1}} = \sqrt{\frac{5549 - 60(91.680625)}{60-1}}$$

$$= \sqrt{\frac{5549 - 5500.8375}{59}} = \sqrt{\frac{48.1625}{59}} \approx \sqrt{0.816314} \approx 0.904$$

The standard deviation is about $0.90, or 90 cents.

# Part III

# Discovering the Math of Finance and Investments

The 5th Wave                    By Rich Tennant

"I was looking for an annuity that came with a rate that was fixed, not neutered."

## In this part . . .

Nothing is more common to different peoples' interest than the mathematics of finance. Whether you're earning interest compounded quarterly or paying off an amortized loan, you're involved with the intricacies of interest, principals, and percentage rates. With the chapters in this part, you can become better prepared to compare the many different financial products out there. By the end of this, you'll be ready to stand as your own financial advocate. But it never hurts to get some professional advice either.

# Chapter 9

# Computing Simple and Compound Interest

*I*nterest is basically a percentage of some amount of money. It's either the amount of money that you pay someone for the use of their money (such as for a loan or credit card) or the amount that you're paid for the use of *your* money (such as with a savings account at a bank). The concept and practice of determining interest has been around since ancient Babylonian mathematicians created tables of numbers to determine how long it takes to double one's money at a particular interest rate.

Interest can be simple — as is the case with simple interest — but it can also be complex — as with compound interest. (I explain each of these in this chapter.) Extensive tables that provide numbers that you can use in calculations of interest are available, but the computations are so easily achieved with a standard scientific calculator that you don't need to tote tomes of numbers around with you. In this chapter, you discover the hows and whys of computing interest. You also see the effect of time and rate.

It's best if you have a calculator at hand to confirm or deny the computations of the example problems. However, even though the calculations aren't difficult, you need to be careful when entering the entries into your calculator. Why? What you type in may not be what you mean. For example, if you want to raise a number to a fractional power, you need to put the fraction in parentheses. Typing 8^2/3 tells your calculator to raise 8 to the second power and then divide by 3. If you really meant that you want to raise 8 to the ⅔ power, you have to put the ⅔ in parentheses. Calculators use the *order of operations* and perform powers before multiplication or subtraction (see Chapter 5 for more on the order of operations).

# Understanding the Basics of Interest

Interest is money. And if you're the *borrower,* you pay for the privilege of using the money. If you're the *lender,* you're paid the interest for your service of providing the money. The amount of money being borrowed or loaned is the *principal,* or the initial amount. The *rate* of interest is the percentage of the principal that it costs to borrow the money (or that you're paid for using the money). And the *time* is the period — in years, months, or days — that the transaction is taking place. Yes, I know, that's a lot of terms to keep straight, but they're all important when it comes to understanding how to compute interest.

Here are the two basic types of interest that you're likely to come across:

- ✔ **Simple interest:** This type of interest is computed on the principal — the amount borrowed — for the entire length of time of the transaction.

- ✔ **Compound interest:** This interest builds on itself. Money earned in interest for part of the time period is reinvested and used in the computation of interest for the rest of the time period.

I explain both of these types of interest later in the chapter.

# Simply Delightful: Working with Simple Interest

Interest on money is the cost of buying things on credit. The fee you pay for the privilege of using someone else's money is the *interest,* and the amount you get for lending someone money is also *interest.* The simplest computation of interest is *simple interest* (now, isn't that handy).

In this section, you see how simple interest works and how the different components, such as principal, rate, and time, interact with one another. *Note:* Many of the properties of simple interest are closely related to those of compound interest, so you see properties that are introduced here repeated — with an interesting twist — in the sections on compound interest.

# Computing simple interest amounts the basic way

The formula for computing the amount of simple interest earned on a particular amount of money is

$I = Prt$

where $P$ is the principal (the amount of money involved in the transaction), $r$ is the rate of interest (written as a decimal number), and $t$ is the amount of time (usually in years, if $r$ is the yearly rate).

To compute, you simply fill in the numbers you have and solve for the missing ones. It can't get much easier than that! (If you need guidance on decimal equivalents, check out Chapter 2.)

Check out this example for practice: Suppose Jake borrows $4,000 for a new piece of equipment. He borrows the money for 2 years at 11.5% interest. How much does he pay in interest, and what's the total amount he has to repay?

To solve this problem, simply plug your numbers into the formula ($I = Prt$), like this: $4,000(11.5\%)(2 \text{ years}) = \$4,000(0.115)(2) = \$920$. So Jake owes $920 in interest plus what he borrowed, which is $920 + $4,000 = $4,920.

### Seeing how small businesses use simple interest

Simple interest is frequently used when small businesses act as lenders in order to sell products. For instance, a local hardware store may sell you a mechanized posthole digger and arrange for payments over the next two years. In this situation, an interest rate is set and a time period is agreed on. The amount loaned (plus the interest) is repaid periodically rather than at the end of the time period.

Here's an example that shows what I mean: Say that Delores purchases a new bedroom set from a local furniture store. She makes arrangements with the store to pay for the $3,995 bedroom set over the next 4 years at 12% interest (simple interest). If she is to make equal monthly payments, how much are those payments?

First determine the amount of interest that she's paying by using $I = Prt$: $3,995(0.12)(4) = \$1,917.60$. Add the interest to the cost of the bedroom set to get the total amount: $3,995 + $1,917.60 = $5,912.60. Now divide the total amount by 48 (4 years × 12 months per year) to get $5,912.60 ÷ 48 = $123.17916.

The division doesn't come out evenly, so Delores will pay $123.18 each month for the first 47 months, and then she'll pay $123.14 for her last payment. How did I figure the last payment? Well, if you multiply $123.18 by 47, you get $5,789.46 in payments. That leaves $5,912.60 − $5,789.46 = $123.14 for the 48th payment. Sure, it's only 4 cents, but Delores is a stickler for detail.

### Solving for different values using the simple interest formula

The simple interest formula allows you to solve for more than just the amount of interest accumulated. You also can use the formula to find the value of any of the variables — if you have the other three. For instance, you can determine the interest rate from the amount of interest paid. See how in the following example.

Freddy agreed to make quarterly payments of $781.25 for 4 years on a loan of $10,000. What simple interest rate is he paying?

You first need to determine the total amount of money being repaid. If Freddy is making 4 payments a year for 4 years, you know that he's making 16 payments of $781.25. So his total repayment is $16 \times \$781.25 = \$12,500$. You also know that he borrowed $10,000, so the interest he's paying is: $12,500 − $10,000 = $2,500. The interest amount, $2,500, is the answer to the interest formula, $I = Prt$. So replace the $I$ with $2,500, replace $P$ with $10,000, replace $t$ with 4, and then solve for $r$. Your math should look like this:

$$2,500 = 10,000(r)(4)$$
$$2,500 = 40,000r$$
$$\frac{2,500}{40,000} = \frac{40,000r}{40,000}$$
$$0.0625 = r$$

So, as you can see, Freddy is paying 6.25% interest on his loan.

Now try out this example, which asks you to use $I = Prt$ to solve for the principal amount. Say Hank was helping out his aunt with her finances. Auntie Em took out a loan for a new television set 2 years ago and agreed to pay $121.88 each month for 5 years at 12½% interest. How much did the TV cost (before interest is added in)?

The repayment amount is the principal (cost of the TV) plus the interest on the loan. If she's paying $121.88 per month, that's 60 payments (5 years $\times$ 12 months). So the total she'll pay back over the 5-year period is $121.88 $\times$ 60 = $7,312.80.

The total, $7,312.80, represents the principal plus the interest ($P + I = \$7,312.80$). You replace the $I$ in this equation with $Prt$ (so you can solve for the cost of the

TV), which gives you $P + Prt = \$7,312.80$. Now factor out the $P$, fill in the interest rate and time, and then solve for $P$, like so:

$$P + Prt = \$7,312.80$$
$$P(1 + rt) = \$7,312.80$$
$$P = \frac{\$7,312.80}{1 + rt}$$
$$= \frac{\$7,312.80}{1 + 0.125(5)}$$
$$= \frac{\$7,312.80}{1.625}$$
$$= \$4,500.18$$

The principal comes out to be $4,500.18. That's some jazzy TV set that Auntie Em bought.

## *Stepping it up a notch: Computing it all with one formula*

The formula for simple interest, $I = Prt$, gives you the amount of interest earned (or to be paid) given an amount of money, an interest rate, and a period of time. You want the amount of interest as a separate number when you're figuring your expenses or taxes as a part of doing business. But, as I show in the previous section, you frequently want the total amount of money available (or to be repaid) at the end of the time period. You add the principal to the interest to get the total. But guess what? There's a better way! Here's a formula that allows you to find the total amount all in one computation:

$$A = P(1 + rt)$$

where $A$ is the total amount obtained from adding the principal to the interest, $P$ is the principal, $r$ is the annual rate written as a decimal, and $t$ is the amount of time in years.

Try your hand at using this formula: Imagine that Casey loaned her business partner $20,000 at 7% simple interest for 3 years. How much will Casey be repaid by the end of the time period?

Using $A = P(1 + rt)$, Casey gets $A = \$20,000(1 + 0.07 \times 3) = \$20,000(1.21) = \$24,200$. From this total, you can determine the interest paid by subtracting the principal amount from the total: $24,200 – $20,000 = $4,200.

# Taking time into account with simple interest

Loans don't have to be for whole years. You can borrow money for part of a year or for multiple years plus a fraction of a year. If the time period is half of a year or a quarter of a year or a certain number of months, the computation is pretty clear. You make a fraction of the time and use it in place of the $t$ in $I = Prt$.

When you get into a number of days, however, the math can get pretty interesting. For instance, when dealing with a number of days, the discussion then involves *ordinary* or *exact* simple interest. I explain each of the scenarios in the following sections.

### Calculating simple interest for parts of years

The simple interest formula (which I explain earlier in the chapter) uses the annual interest rate and $t$ as the number of years. When you use a fraction of a year, you essentially reduce the interest rate. For example, to compute the interest on $1 for 1 year at 4% interest, your math looks like this: $I = Prt = \$1(0.04)(1) = 0.04$. Now, if you take that same $1 at 4% for half a year, you get $I = Prt = \$1(0.04)(0.5) = 0.02$.

It doesn't matter whether you take half the interest rate or use half a year; either way you get the same number. You just choose the one that makes the computation easier.

Say Abigail is borrowing $6,000 for 3 months at 8% simple interest. How much does she repay at the end of 3 months?

Using $I = Prt$, you let the time be ¼ of a year (3 months ÷ 12 months = ¼ ). So you get $I = \$6,000(0.08)(0.25) = \$120$. In this case, I replaced the ¼ with 0.25 to enter the numbers in the calculator. I also could have taken advantage of the fact that 8% divides evenly by 4 and written $I = \$6,000(0.02) = \$120$. In any case, Abigail owes $6,000 + $120 = $6,120 at the end of the 3-month loan period.

Try out another example that stirs things up a bit: Say that Gustav repaid a loan in 30 months by making bimonthly payments of $880. How much did Gustav borrow, if the simple interest rate was 4%?

*Bimonthly* means every other month, so a 30-month loan period has 15 payments. So you multiply the payment amount by the number of payments to get the total payment amount: $880 × 15 = $13,200. The total of all the payments includes both interest and principal, so use this formula to solve for the principal amount:

$$P(1 + rt) = \text{total repaid}$$

I show you how to use this formula in the earlier section, "Computing simple interest amounts the basic way." There I also show you how the formula is derived.

Solving for the principal, you let 30 months be 2½ years, and then you plug in all the numbers. Your math should look like this:

$$P(1 + rt) = \$13,200$$
$$P = \frac{\$13,200}{1 + rt}$$
$$= \frac{\$13,200}{1 + 0.04(2.5)}$$
$$= \frac{\$13,200}{1.1}$$
$$= \$12,000$$

So you can see that Gustav borrowed $12,000 2½ years ago.

### Distinguishing between ordinary and exact simple interest

To liven things up a bit, here in this section I introduce two ways of looking at computing interest: using *ordinary* simple interest and using *exact* simple interest. You don't need to be concerned about distinguishing between *ordinary* and *exact* interest unless you're dealing with small periods of time — usually a number of days instead of a number of years.

When computing simple interest in terms of a number of years or parts of years (half a year or a quarter of a year), you use whole numbers or fractions in your calculations. Short-term transactions, on the other hand, may be measured in days. In that case, you have to decide whether to use ordinary or exact interest. The difference between these two is that *ordinary interest* is calculated based on a 360-day year, and *exact interest* uses *exactly* the number of days in the year — either 365 or 366, depending on whether it's a leap year. (See the nearby sidebar, "Don't forget about the leap years!" for more information.) The following examples will help you understand how to use both ordinary and exact interest.

What's the difference between the ordinary interest and the exact interest on a loan of $5,000 at 6% interest for 100 days if the year is 2008?

First, to compute ordinary interest, you divide the number of days by 360 (the number of days in an "ordinary" year), which gives you the fraction you use for the amount of time in the formula. In this case, you divide 100 by 360. Now you plug all your numbers into $I = Prt$, like this, to get your answer:

$$I = \$5,000(0.06)\left(\frac{100}{360}\right) = \$300\left(\frac{5}{18}\right) \approx \$83.33$$

---

# Don't forget about the leap years!

Leap year days were added to the calendar in the 1700s when Pope Gregory determined that our seasons weren't jibing with the actual weather. Our calendar had become out of sync with the sun. As it turns out, the lack of coordination of the calendar with the seasons came about because the Earth's years are actually 365¼ days long. The Pope's new Gregorian calendar added leap year days to fix this problem.

Here's how to tell whether a year is a leap year: If a year is evenly divisible by 4, the year is a leap year and has 366 days instead of 365 (for example, 2008 and 2012 are leap years, while 2009, 2010, and 2011 are not). The added day is February 29. But because this addition causes a bit of an error (the year isn't *quite* 365¼ days long), another correction in the calendar is applied to *century* years, such as 1700, 1800, 1900, 2000, 2100, and so on. A century year is a leap year only if it's divisible by 400. So the year 2000 was a leap year, but the year 2100 will *not* be. (Hope you're planning on sticking around to remind the calendar makers of all this!)

---

Next, you compute the exact interest. To do so, you divide 100 by 366 (because 2008 is a leap year), and then you use that fraction for the amount of time. Now fill in your formula:

$$I = \$5,000\,(0.06)\left(\frac{100}{366}\right) = \$300\left(\frac{50}{183}\right) \approx \$81.97$$

The difference between the two interest methods is just over $1. The ordinary interest number is more convenient, but the number 360 in the denominator of the fraction *does* make the fraction value larger than with a denominator of 365 or 366. And the difference will become more significant as the amount borrowed increases.

Exact interest also is used when the starting and ending dates of a loan are given and you have to figure out the number of days involved. Most desk calendars include numbers on each date indicating which day of the year it is. For example, in the year 2007, November 4 was the 308th day of the year. I include the year because in 2008, November 4 is the 309th day of the year due to adding the leap year day.

Say, for example, that Timothy borrowed $10,000 on April 17 and paid it back, with exact simple interest of 5¼%, on November 22 in the year 2007. What interest was charged on the $10,000 loan?

First refer to a calendar and verify that April 17 was the 107th day in 2007 and November 22 was the 326th day. Find out the total number of days of the loan by subtracting: 326 − 107 = 219 days. Now, using the simple interest formula and 365 days, plug in your numbers:

$$I = \$10,000\,(0.0525)\left(\frac{219}{365}\right) = \$525\left(\frac{219}{365}\right) = \$315.00$$

Timothy will repay a total of $10,315 (which I found by adding $10,000 + $315) after 219 days.

## Surveying some special rules for simple interest

Simple interest is relatively easy to compute, but leave it to some folks to come up with shortcuts or rules to make the computation — or estimations of the computation — even easier. The rules I'm referring to are the Banker's Rule, the 60-day 6% method, and the 90-day 4% method. I explain each in the following sections.

### The Banker's Rule

*The Banker's Rule* is a common method of computing interest that combines ordinary interest and exact time. This definition seems a bit of a contradiction, but here's how it works: The ordinary interest rule of using 360 days is applied, and the exact number of days is used. So instead of computing interest on three months and calling it ¼ of a year, you determine exactly how many days are in those three months and divide by 360.

You may wonder just how different the figures might be when using the exact days in three months versus ¼ of a year. Consider some of the months in the year 2007 in groups of three:

>    January through March: 31 + 28 + 31 = 90 days

>    February through April: 28 + 31 + 30 = 89 days

>    March through May: 31 + 30 + 31 = 92 days

The fractions formed by dividing the number of days by 360 are:

$$\frac{90}{360} = \frac{1}{4} = 0.25, \ \frac{89}{360} \approx 0.247222\ldots, \ \frac{92}{360} \approx 0.25555\ldots$$

The decimal equivalents don't seem to be too different, but when you multiply these decimals by big numbers, the slight variations make a significant difference!

How about an example? Using the Banker's Rule, what's the interest on a loan of $10,000 at 7% from April 1 to October 1?

Determine the number of days involved. You can use a calendar or table that provides the number of each day and get 274 – 91 = 183 days. Or you can add up the days in each month: 30 + 31 + 30 + 31 + 31 + 30 = 183. (***Note:*** You don't count October 1st). Now, using the simple interest formula, calculate the interest:

$$I = \$10{,}000\,(0.07)\left(\frac{183}{360}\right) = \$700\left(\frac{183}{360}\right) \approx \$355.83$$

### The 60-day 6% or 90-day 4% rules

The 60-day 6% and 90-day 4% rules are two shortcut estimations of interest. These rules are great because both 60 and 90 divide 360 evenly and form nice fractions when you divide them by 360. For instance, take a look:

$$\frac{60}{360} = \frac{1}{6} \text{ and } \frac{90}{360} = \frac{1}{4}$$

When you couple the fractions formed by dividing 60 or 90 by 360 with interest rates of 6% or 4% or 12%, you can do quick calculations of the interest (as long as the combination of the fraction and the percentage comes out to be a nice number).

For example, if you have a 60-day loan at 6%, you can multiply ⅙ × 6% = 1%. A 90-day loan at 8% gives you ¼ × 8% = 2%. The rest of the computations involving multiplying fractions and interest rates work the same, and, hopefully, you can do them in your head.

As an example, use the 90-day 4% method to estimate the interest on a loan of $25,000.

This one is so easy. First, multiplying 4% by ¼ gives you 1%. Then multiplying $25,000 by 0.01 gives you $250. Not too bad.

# Looking into the future with present value

When you're talking money, the expression *present value* describes itself fairly well, but it's also a bit confusing. Just remember this: What you're really looking forward to is having a certain amount of money in the *future* when you talk about present value. In other words, if you have some goal for a particular amount of money in the future, you want to know how much money to deposit *now* (in the present) so that the addition of simple interest will create the sum of money that you need.

For example, if you want $10,000 to buy a new piece of machinery 3 years from now, do you invest $8,000, $9,000, or $9,500 today? When these amounts of money earn interest, will the added interest bring you up to $10,000? Well, you can't really answer this question yet, because I haven't given you any interest rate. So this type of question will be a two-pronged one. One plan of attack will be to determine how much money to deposit if you know the interest rate, and the other approach will be to find an interest rate if you know the amount of money available for deposit.

In general, you can solve for the principal needed or the rate needed (or even the amount of time needed) after you know what your money goal is. The formulas used in the computations are all derived from the basic $A = P(1 + rt)$ formula for the total amount of money resulting from simple interest. (You can find more on this basic formula in the earlier section, "Stepping it up a notch: Computing it all with one formula.") Here are all the formulas that you'll use when dealing with present value:

$$A = P(1 + rt) \qquad P = \frac{A}{1 + rt}$$

$$r = \frac{A - P}{Pt} \qquad t = \frac{A - P}{Pr}$$

where $A$ is the total amount after adding simple interest, $P$ is the principal, $r$ is the interest rate as a decimal, and $t$ is the amount of time in years.

Now it's time to try your hand at these formulas with a few examples.

Say that 3 years from now, you want to buy a new storage shed — and you need to have $27,000 set aside at that time. You intend to deposit as much money as is necessary right now so that when it earns simple interest at 5¾% it will be worth $27,000 in 3 years. How much do you have to deposit?

You know the total amount needed, $A$, the interest rate, $r$, and the amount of time, $t$. So, use the formula to solve for the principal needed:

$$P = \frac{A}{1 + rt} = \frac{\$27,000}{1 + 0.0575(3)} = \frac{\$27,000}{1.1725} \approx \$23,028$$

As you can see, you need to set aside about $23,000 right now.

Here's the same problem with a twist: Three years from now, you want to buy a new storage shed — and you need to have $27,000 set aside by that time. You have $20,000 at hand, which means that you need to find someone who will give you a high enough simple interest rate so that the $20,000 will grow to $27,000 (by adding the interest) in 3 years' time. What interest rate do you need?

This time you know the total amount, *A*, the principal, *P*, and the time, *t*. Use the formula to solve for the rate needed, like this:

$$r = \frac{A-P}{Pt} = \frac{\$27{,}000 - \$20{,}000}{\$20{,}000\,(3)} = \frac{\$7{,}000}{\$60{,}000} \approx 0.1167$$

You need an 11.67% interest rate. Lots of luck finding that rate.

# Getting to Know Compound Interest

The biggest difference between simple interest and compound interest is that simple interest is computed on the original principal only (see the earlier section, "Simply Delightful: Working with Simple Interest," for details). In other words, no matter how long the transaction lasts, the interest rate multiplies only the initial principal amount.

*Compound interest,* on the other hand, multiplies the interest rate by the original principal plus any interest that's accumulated during the time period of the transaction. As you can imagine, compound interest is pretty powerful. After all, the principal keeps growing, making the interest amount increase as time passes. The bigger interest amount is added to the principal to make the interest amount even bigger.

Compound interest is the type of interest that financial institutions use. In this case, your money grows exponentially, so exponents (those cute little superscripted powers) are a part of the formula for determining how much money you have as a result of compound interest. Of course, for compound interest to have its full effect, you can't remove the interest earned; compound interest is based on the premise that you leave the money alone and let it grow.

## Figuring the amount of compound interest you've earned

Computing the total amount of money that results from applying compound interest takes a jazzy formula. Besides involving multiplication, addition, and division, this formula also requires you to work with exponents (and who but mathematicians like to do that?). The hardest parts of working with the formula are entering the values correctly into the formula and performing the operations in the right order.

The formula for computing compound interest is

$$A = P\left(1 + \frac{r}{n}\right)^{nt}$$

where $A$ is the total amount of money accumulated (principal plus interest), $P$ is the principal (the amount invested), $r$ is the rate of interest (written in decimal form), $n$ is the number of times each year that the compounding occurs, and $t$ is the number of years.

Some scientific and graphing calculators allow you to type the numbers into the formula pretty much the same way you see them. But if you don't have a calculator with all the bells and whistles, you still can do this problem correctly — as long as you perform the steps of the equation in the correct order. You need to apply the *order of operations*. (If you don't know how the order of operations works, refer to Chapter 5, where I discuss it in detail.)

For the compound interest formula, you need to perform the operations in the following order:

1. **Determine the value of the exponent by multiplying $n \times t$ (the number of times compounded each year times the number of years).**

2. **Inside the parentheses, divide the interest rate, $r$, by the number of compoundings each year, $n$.**

3. **Add 1 to the answer in Step 2.**

4. **Raise the result from Step 3 to the power that you got in Step 1.**

5. **Multiply your answer from Step 4 by the principal, $P$.**

Put these steps to use in the following examples.

How much money has accumulated in an account that's earning interest at the rate of 4% compounded monthly if $10,000 was deposited 7 years ago?

Fill in the formula letting $P$ = $10,000, $r$ = 0.04, $n$ = 12, and $t$ = 7:

$$A = \$10,000\left(1 + \frac{0.04}{12}\right)^{12(7)}$$

Now go through these steps (the order of operations) to find your answer:

1. **Multiply $12 \times 7 = 84$.**

   This is the value of the exponent.

2. **Divide $0.04 \div 12 = 0.0033333.\ldots$**

   Round this answer to five decimal places to get 0.00333.

---

## The power of compound interest

What if a member of Christopher Columbus' crew went into the Bank of the West Indies in 1492 and deposited $1 with the agreement that it would earn 3% interest compounded annually? And what if that crew member never returned (maybe he was on the ship that sank)? Now come back to the year 2007, and say that you were contacted by the Bank of the West Indies as the only living heir of this sailor. The bank now wants to close out this "nuisance account." What if it was willing to give you the money accumulated in this account for over 500 years — as long as you pay a $50 per year maintenance fee retroactively? Does this sound like one of those spam e-mails that you've been getting recently? Well, before hitting the delete button, take a gander at what the numbers say.

A deposit of $1 earning interest at 3% compounded annually for 515 years (computing this to 2007) amounts to $4,084,799.30. Even if you have to pay 515 years of maintenance fees at $50 per year, that's only $25,750 out of your $4 million. Of course, you'll ask the bank to deduct the maintenance fee and send you the difference. If they insist on having the maintenance fee up front, it's time to delete the message.

This situation just goes to show you the power of compounding — how just one dollar can grow to millions of dollars — in a mere 500 years.

---

3. **Add 1 to the answer in Step 2.**

   After adding, you get 1.00333.

4. **Raise 1.00333 to the 84th power.**

   So $1.00333^{84} \approx 1.32214$.

5. **Multiply 1.32214 × $10,000 = $13,221.40.**

   Now you have the total amount accumulated in 10 years.

 Curious to know how this compares to investing the same amount of money for the same amount of time at the same rate using simple interest? Well, use the formula for the total amount, $A = P(1 + rt)$, to find out. $10,000(1 + 0.04 \times 7) = $12,800. By using compound interest, you earn about $400 more than you would have with simple interest.

## *Noting the difference between effective and nominal rates*

When you walk into a bank, you're faced with an easel or other display that's covered with decimal numbers. You see the latest on rates for car loans and home equity loans and the update on the effective interest rates for various products. So what's this effective interest rate business? The *effective interest rate* is what you actually get as a result of compounding.

The *nominal rate* is the named rate, or what you'd get without compounding. For instance, you may read that the nominal rate is 4.5% applied to certain deposits for 6 months. The effective interest rate is what you get as a result of compounding that nominal rate.

To determine the effective rate of interest, use this formula:

$$\left(1 + \frac{r}{n}\right)^n - 1$$

where *r* is the nominal interest rate (expressed as a decimal) and *n* is the number of compounding periods in a year.

When doing the computation for the effective interest rate, be sure to use the order of operations correctly. Otherwise your calculations will be off. Here are the steps you need to follow:

1. **Simplify the fraction inside the parentheses.**

2. **Add 1 to the result of Step 1.**

3. **Raise the sum from Step 2 to the power outside the parentheses.**

4. **Subtract 1 from the number you get in Step 3.**

What's the effective rate of 4.5% compounded quarterly?

Using the formula, the effective rate is computed like this:

$$\left(1 + \frac{0.045}{4}\right)^4 - 1 = (1.01125)^4 - 1 \approx 1.045765086 - 1 = 0.045765086$$

So, while the nominal rate is 4.5%, the effective rate is about 4.5765%.

Just how much difference does using the effective interest rate (versus the nominal rate) make? In other words, why is 4.5765% so much better than 4.5%? It's all in the degree of involvement — or, to be blunt, in how much money you have. If you deposit $1,000 and use 4.5% interest, you'll earn $1,000 × 0.045 = $45 in a year. The same $1,000 at 4.5765% earns about $45.76; you get a whole 76 cents more in a year. But if you have $10 million to invest, the difference between what's earned at 4.5% and 4.5765% is $7,650 ($457,650 − $450,000 = $7,650). Guess the saying "Money begets money" is true after all.

## Finding present value when interest is compounding

The present value of your money is what you deposit in your account today in order to have a certain target amount in your account in the future. Present value using simple interest is covered in the earlier section, "Looking into the future with present value." In this section, however, you get to use the power of compounding to make the present value smaller — meaning that you don't have to commit as much money right now in order to get your target amount later.

To find the present value, $P$, of an amount of money, $A$, in $t$ years when the money grows at $r$ interest (expressed as a decimal) compounded $n$ times each year, use the following formula:

$$P = \frac{A}{\left(1 + \frac{r}{n}\right)^{nt}}$$

To use this formula, use this order of operations:

1. **Find the value of the exponent $nt$ by multiplying the number of times you compound by the number of years.**

2. **Divide the rate, $r$, by $n$ and add the result to 1.**

3. **Raise the sum (after adding the 1) to the $nt$ power to get the value in the denominator of the fraction.**

4. **Divide the amount of money, $A$, by the denominator.**

Say, for example, that you want to have $400,000 in 5 years to purchase a new piece of equipment. How much do you need to deposit today at 6¾% interest, compounded monthly, so there's $400,000 in your account when you need it?

To solve, you just need to plug your numbers into the formula for present value, like so:

$$P = \frac{400,000}{\left(1 + \frac{0.0675}{12}\right)^{12\,(5)}} = \frac{400,000}{(1.005625)^{60}} = \$285,690.83$$

After doing the math, you can see that you need to deposit about $286,000 right now. But look how much it grows in just 5 years!

# Determining How Variable Changes Affect Money Accumulation

Many different banks are available these days, and there are just as many options, offers, and arrangements that can be made with your money at the different institutions. For instance, do you choose the free checking or the unlimited account transfers? Decisions, decisions. I can't help you in choosing one bank over another, but I can show you just how much the interest rate, the number of compoundings, or the time invested actually affect the end result. The rest is for you to sort through and decide.

As you may know from reading this chapter, the compound interest formula has four different *variables,* or things that can change. The variables are

- ✔ *P*, the principal or amount of money invested

- ✔ *r*, the interest rate, entered as a decimal

- ✔ *n*, the number of times each year that the interest is compounded

- ✔ *t*, the time in number of years

If you adjust any of the four variables, you change the output or end result. In general, increasing any of the variable numbers increases the output. But how much is the increase, and which variable has the greatest impact on an increase? I show you in the following sections.

## Comparing rate increases to increased compounding

Say that you have $10,000 and need to determine where to invest that money to earn the greatest amount of interest. If you invest your money at 4% interest, compounded annually, you'd have $10,000(1 + 0.04) = $10,400 at the end of one year. (Refer to the earlier section, "Getting to Know Compound Interest," for the details on how to do this computation.) Knowing that, your next question is this: Will you do better to invest your $10,000 with an institution that increases the interest rate by one-quarter of a percent, or should you stick with the same interest rate and go someplace that compounds interest quarterly?

To answer this question, you need to determine the total amount of money in an account if $10,000 is deposited for 1 year at 4¼% compounded annually. Then you simply compare that total amount with the result of $10,000 being deposited in an account that earns interest at 4% compounded quarterly. Here's what the computations look like:

One year at 4¼%, compounded annually: $A = \$10,000\left(1 + \dfrac{0.0425}{1}\right)^{1(1)} = \$10,425$

One year at 4%, compounded quarterly: $A = \$10,000\left(1 + \dfrac{0.04}{4}\right)^{4(1)} = \$10,406.04$

Clearly, the increase in the interest rate has the greater impact on the total amount after a year. A ¼% increase may not be realistic, however. In banking circles, a quarter of a percent is big money.

Imagine that the total amount in an account where a deposit of $10,000 is earning 4% interest compounded quarterly is $10,406.04 at the end of 1 year. What are the effects of increasing the interest rate by one-hundredth of a percent at a time and applying annual compounding?

The following list shows the end results of interest rates of 4.01%, 4.02%, and so on, with annual compounding:

$A = \$10,000(1 + 0.0401)^{1(1)} = \$10,401$

$A = \$10,000(1 + 0.0402)^{1(1)} = \$10,402$

$A = \$10,000(1 + 0.0403)^{1(1)} = \$10,403$

$A = \$10,000(1 + 0.0404)^{1(1)} = \$10,404$

$A = \$10,000(1 + 0.0405)^{1(1)} = \$10,405$

$A = \$10,000(1 + 0.0406)^{1(1)} = \$10,406$

It only takes a six-hundredths of a percent increase in the interest rate for the end result to equal the effect of compounding quarterly instead of annually.

## Comparing rate increases to increases in time

Maybe you're stuck with one type of interest compounding. For instance, maybe you only deal with institutions that compound quarterly. Now you want to compare the effect of increasing the interest rate with the effect of increasing the amount of time you leave the money in the account. If you have your money invested at 4% compounded quarterly, would you be better off increasing the rate of interest to 4¼% or leaving it in the original account for an extra quarter of a year?

Your math for solving this problem looks like this:

One year at 4¼% compounded quarterly:

$$A = \$10{,}000\left(1 + \frac{0.0425}{4}\right)^{4\,(1)} = \$10{,}431.82$$

One and a quarter years at 4% compounded quarterly:

$$A = \$10{,}000\left(1 + \frac{0.04}{4}\right)^{4\,(1.25)} = \$10{,}510.10$$

As you can see, the increase in interest rate resulted in a smaller total than the increase in time.

**GO FIGURE**

Using the information from the previous example, determine what rate of interest, compounded quarterly, earns the same amount of interest on $10,000 invested for one and a quarter years.

Here's how to get started:

$$4\%,\ \text{comp. quarterly}, 1\tfrac{1}{4}\ \text{year} = x\%,\ \text{comp. quarterly}, 1\ \text{year}$$

$$\$10{,}000\left(1 + \frac{0.04}{4}\right)^{4\,(1.25)} = \$10{,}000\left(1 + \frac{x}{4}\right)^{4\,(1)}$$

Next, divide each side by $10,000, and then simplify the exponents and the terms that are inside the parentheses, like so:

$$(1.01)^{5} = \left(1 + \frac{x}{4}\right)^{4}$$

Now finish by computing the power on the left, taking the fourth root of each side, and then solving for the value of $x$:

$$1.05101005 = \left(1 + \frac{x}{4}\right)^{4}$$

$$\sqrt[4]{1.05101005} = \sqrt[4]{\left(1 + \frac{x}{4}\right)^{4}}$$

$$1.012515586 = 1 + \frac{x}{4}$$

$$0.012515586 = \frac{x}{4}$$

$$0.050062344 = x$$

So, as you can see, you need an interest rate of about 5% to equal the effect of leaving the money in for 3 extra months.

The problems involving comparing interest rates and compounding and time show you that many variations are available when investing your money or borrowing money. Find some interest charts and look to see how much your money earns at each rate and amount of time. Sometimes, though, it's more a matter of convenience and service (rather than a few extra dollars here and there) that draws you to a particular institution — and that's okay.

# Chapter 10

# Investing in the Future

**In This Chapter**

▶ Investing with a lump sum

▶ Dealing with annuity calculations

▶ Understanding annuity payouts

*Y*ou earn money in many ways. First and foremost, you work hard at your job. And maybe you also buy and manage apartment buildings, buy stocks or bonds, or invest your money with some financial institution where you have the understanding that the institution will put your money to work to earn interest or dividends.

This chapter deals with the different opportunities available for investing your money. For instance, you can deposit a lump sum and let it grow by compounding. Or you can deposit that same lump sum and withdraw regular amounts until it's all gone. You may even opt to make regular deposits of money and allow them to grow over the years. The choices (and variations on the choices) are many.

When investing money, many folks use annuities. And sinking funds are often a form of annuity. The present value and future value of annuities translate into the various uses for annuities. In this chapter, by covering some examples, emphasizing the financial-speak vocabulary, and showing you the step-by-step process of the mathematics of investing, I show you how to take more control of your financial plans.

# Calculating Investments Made with Lump Sums

Businesses are created with the ultimate goal of making a profit. And profit occurs when the revenue is greater than the cost. When a business produces items to sell, revenue is generated by that sale. But businesses have other options for creating revenue, in the form of investments of capital. When situations arise and a business can take advantage of the opportunity to make a greater profit through an investment, a firm understanding of the math involved is a must.

Your investment opportunities are abundant, so care must be taken to ensure a safe investment as well as the best return possible. To determine the return or interest on deposited money, you can use either of the following methods:

- ✔ **A formula and a calculator:** This method has the advantage of compactness and flexibility.

- ✔ **A table (and, frequently, a calculator):** A table of values has the advantage of quick review of the results of different rates.

Using a formula and calculator is okay, but for the purposes of this chapter, I show you how to calculate investment info for a lump sum using the tables — mostly because I've found them to be the quickest and easiest route.

## Reading interest earnings from a table

Investments earn interest based on four elements:

- ✔ The total amount of money invested
- ✔ The rate of interest
- ✔ The number of times the interest is compounded per year
- ✔ The number of years the money is invested

The interest rate and number of times an amount is compounded are closely related. How so? Well, you divide the annual interest rate by the number of times it's compounded each year. (Refer to Chapter 9 and the later sections in this chapter for more on how interest is computed.)

One way to simplify all the possible variations of interest rates and the number of times the interest is compounded each year is to use a *periodic interest rate* in your computations. A periodic interest rate is what the percentage is for that short period of time; you divide the yearly rate by the part of the year being considered. For example, if the yearly interest rate is 8%, and the interest is compounded quarterly, you want the rate for one-fourth of a year; use this calculation each period: 8% ÷ 4 = 2% per time period.

You can obtain a particular periodic interest rate in several different ways — dividing a yearly rate by a carefully selected number.

For instance, to have a periodic interest rate of 1% per period

> 1% per month comes from a 12% annual rate.
>
> 1% per quarter comes from a 4% annual rate.
>
> 1% per half-year (semiannual) comes from a 2% annual rate.

To have a periodic interest rate of 1.5% per period

> 1.5% per month comes from an 18% annual rate.
>
> 1.5% per quarter comes from a 6% annual rate.
>
> 1.5% per half-year (semiannual) comes from a 3% annual rate.

### Using a table to determine the future value of an investment

Table 10-1 is a table that you can use to quickly determine the future value of your investment. The table is relatively limited, however. You need a complete book of compound interest tables if you intend to compute values of your investments using tables. In the meantime, follow my instructions and give the table a try; you may find that you like using interest tables.

To determine the future value of an investment from a table, multiply the amount of your investment by the entry corresponding to the number of periods at which the investment has been compounded at the particular interest rate per period.

Table 10-1 lists several interest rates and the number of periods that $1 is subject to compounding. You intend to invest more than $1? I hope so! Just multiply your investment amount by the corresponding value from the table.

| Table 10-1 | Future Value of $1 at Given Interest Rates | | | |
|---|---|---|---|---|
| Periods | 1% | 1.5% | 2% | 2.5% |
| 4 | 1.040604 | 1.061364 | 1.082432 | 1.103813 |
| 8 | 1.082857 | 1.126493 | 1.171659 | 1.218403 |
| 12 | 1.126825 | 1.195618 | 1.268242 | 1.344889 |
| 16 | 1.172579 | 1.268986 | 1.372789 | 1.484506 |
| 20 | 1.220190 | 1.346855 | 1.485947 | 1.638616 |
| 24 | 1.269735 | 1.429503 | 1.608437 | 1.808726 |
| 28 | 1.321291 | 1.517222 | 1.741024 | 1.996495 |
| 32 | 1.374941 | 1.610324 | 1.884541 | 2.203757 |
| 36 | 1.430769 | 1.709140 | 2.039887 | 2.432535 |
| 40 | 1.488864 | 1.814018 | 2.208040 | 2.685064 |

Here's an example to help you understand how to determine the future value of an investment using Table 10-1: Say that you invest $40,000 for 10 years at 6% annual interest compounded quarterly. How much is your money worth at the end of the 10 years?

After only a first glance, you may be wondering how you can use Table 10-1 to determine the amount of your investment. After all, 10 doesn't appear as a period number, and 6% doesn't appear as an interest rate. But that was just your *first* glance. Take another look and note that you can count how many times interest is compounded in 10 years if it's compounded quarterly. You simply multiply 10 years × 4 quarters = 40 compounding periods.

The rate of 6% annually becomes 6% ÷ 4 = 1.5% each quarter. So, in Table 10-1, reading across the row for 40 periods and down the 1.5% column, you find the number 1.814018. Now multiply your money like this: $40,000 × 1.814018 = $72,560.72. Your investment hasn't quite doubled in that 10 years. That wasn't so difficult, was it?

### Interpolating for investment value

Most compound interest tables are more complete than the example table that you find in Table 10-1. This sample is just a small excerpt of what usually are pages and pages of entries. Nonetheless, Table 10-1 still shows you how to use one of these tables. And now, with the same table, I want to show you how to find values that lie in between two rows or two columns (by the way, this type of estimation involving averaging is called *interpolation*).

For instance, you may want to use the table to find the value of an investment at the end of, say, 30 periods. To do so, you need to use the entries for 28 periods and 32 periods and find the halfway point. To find the halfway point, just average the two values (add them together and divide by 2).

What's the value of an investment of $100,000 after 2.5 years if it's earning 12% annual interest compounded monthly?

First you have to determine how many periods 2.5 years of monthly compounding means. Here's what your equation would look like: 2.5 × 12 = 30 periods. An annual interest rate of 12% divided by 12 is 1% per period.

So, in Table 10-1, look down the first column to the entries for 28 periods and 32 periods. Now you have to average these entries. To do so, first add the entries together. 1.321291 + 1.374941 = 2.696232. Divide that sum by 2, and the average of the two numbers is 1.348116.

Now all you have to do is multiply the amount of your investment by the average you just figured to get the value of your investment after 2½ years: $100,000 × 1.348116 = $134,811.60.

### Taking advantage of tables to determine growth time

Tables of interest rates for compound interest are most useful for determining future values of an investment. But you also can use these tables to determine how long it would take a lump sum investment to grow to a particular level.

Take another look at Table 10-1, and you see that the numbers in the rows and columns are all bigger than 1. The entry 1.429503 represents a resulting amount of money that's about 1.4 times as much as what you started with. So, if you invested $100, you have $100 × 1.4, or about $140. Now you're probably looking at the numbers starting with 2. That's where I look, too; after all, doubling my money sounds rather attractive.

The key to using a table to determine the amount of time needed to create a particular amount of money is to pick the multiplier that represents how much money you want. Then you work backward to determine interest rates.

Say you have $1,000,000 and decide to invest it until it earns $500,000 in interest. You're in a hurry, so you find an investment opportunity that promises an annual return of 18% compounded monthly. How long will you have to leave your money invested at this high (and scary) rate in order for it to earn what you need?

You can solve this problem using Table 10-1. Well, you can get a pretty good estimate anyway. First, add the principal to the interest to get a total amount in the account: $1,000,000 + $500,000 = $1,500,000. Then divide that total by the initial amount: $1,500,000 ÷ $1,000,000 = 1.5. So this means that you want every $1 invested to grow to $1.50.

To determine which column in Table 10-1 contains the 1.5 that you want, you need the interest rate. A rate of 18% annually when divided by 12 for the monthly compounding gives you 18% ÷ 12 = 1.5%.

Use the 1.5% column and read down until you find the number closest to 1.5. The row for 28 periods has the number 1.517222. That's pretty close to 1.5. So you'd have to have your money invested for about 28 periods. If each period is one month, you can figure out the amount of time that is by dividing the number of periods by 12, which would give you about 2 years and 4 months.

You can get a more exact answer by either interpolating (which I discuss in the previous section) or using the formula for compound interest. In the upcoming section, "Doubling your money, doubling your fun," I show you how to use the formula to solve for a time period.

## *Doubling your money, doubling your fun*

How long does it take to double your money? Well, that depends. If you have your money invested in an institution that pays 100% interest, compounded annually, your money doubles in one year. Okay, 100% interest that's compounded annually isn't reasonable — unless you have some interesting connections.

Instead, how about 10% compounded quarterly? Look at Table 10-1. Under the 2.5% interest rate column (10% compounded quarterly is 10% ÷ 4 = 2.5% each period) you find the entry 1.996495 opposite 28 periods. The number 1.996495 is pretty close to 2, so a good estimate of the amount of time needed is 7 years (because 28 periods ÷ 4 = 7).

You may have noticed that I carefully chose my examples for doubling using the table. One other entry is pretty close to 2. For instance, 2% interest with 36 periods also gives you twice your investment. But 2% interest might be 8% compounded quarterly or 12% compounded bimonthly or 24% compounded monthly. In other words, you can find many scenarios that accomplish the doubling. Likewise, with 2.5% and 28 periods, you could be considering 5% compounded biannually or 15% compounded every other month. Table 10-2 shows you how many years it takes to double your money, using the convenient 2% and 2.5% interest rates.

Of course, the higher the interest rate, the less time it takes for the investment to double. But another consideration in the computation is the number of times each year that the interest is compounded. The first four lines of Table 10-2 show you situations where it takes a total of 36 periods to double the money. The last four lines of the table show you how it can take only 28 periods.

| Table 10-2 | Years It Takes to Double Your Money | | |
|---|---|---|---|
| **Annual Interest Rate** | **Times Compounded Each Year** | **Percent** | **Years to Double** |
| 4% | 2 (biannually) | 2% | 18 |
| 8% | 4 (quarterly) | 2% | 9 |
| 12% | 6 (every other month) | 2% | 6 |
| 24% | 12 (monthly) | 2% | 3 |
| 5% | 2 (biannually) | 2.5% | 14 |
| 10% | 4 (quarterly) | 2.5% | 7 |
| 15% | 6 (every other month) | 2.5% | 4⅔ |
| 30% | 12 (monthly) | 2.5% | 2⅓ |

Using a table is a fine way to determine how long it takes for your money to double, but there are other ways that are also fabulous. The following sections explain these methods.

### Taking the easy way out with a doubling formula

You can use the following formula for finding the time it takes to double your investment:

$$n = \frac{\ln 2}{\ln (1 + i)}$$

where *n* is the number of periods needed and *i* is the interest rate per period.

This formula for doubling your investment is used for doubling and doubling alone. Just be careful when using the quick, slick formula that you've entered the correct value for *i*.

Using the slick formula that I just introduced, figure the following example: How long does it take to double your money if you can invest at 4.5% compounded monthly?

To begin, obtain the interest rate per period by dividing the interest rate by 12. 4.5% ÷ 12 = 0.375% = 0.00375. After that, plug 0.00375 into the formula, which gives you:

$$n = \frac{\ln 2}{\ln(1 + 0.00375)} = \frac{\ln 2}{\ln 1.00375} \approx 185.1856$$

Because $n$ is the number of periods, you divide the interest rate per period by 12 to get 185.1856 ÷ 12 ≈ 15.4321. So, as you can see, it will take almost 15½ years to double the value at this interest rate.

## Working with the Rule of 72

The *Rule of 72* is a quick and fairly accurate method of determining how long it takes to double your money at a particular interest rate. You don't get exact answers, but that's okay because once you're armed with an approximation, you can then decide if you want to pursue the issue further by hauling out your tables or calculator.

Here's how the Rule of 72 works: If you simply divide 72 by the annual interest rate, you'll get an estimate on how long it will take to double your money. Check out these examples:

72 ÷ 7.2% ≈ 10 years to double your money

72 ÷ 10% ≈ 7.2 years to double your money

72 ÷ 6% ≈ 12 years to double your money

72 ÷ 9% ≈ 8 years to double your money

One reason that this method is so popular is that it allows you to do quick calculations in your head. The number 72 is evenly divisible by so many numbers that you can either use exactly the interest rate you have in mind or you can do an estimate of an estimate by averaging for rates in between.

For instance, the Rule of 72 says that it takes 12 years to double your money at 6% interest and 9 years to double your money at 8% interest, so you can guess that it will take between 9 and 12 years to double your money at 7% interest.

You can also use the Rule of 72 to determine your rate of return when your money doubles. If you invested $4,000 8 years ago and now have $8,000, simply divide 72 by 8 and you see that the rate of return was 9%.

# Going the Annuity Route

Sometimes you don't want to invest a lump sum all at once (see the earlier section, "Calculating Investments Made with Lump Sums," if you do). In that case, you might choose to invest with an annuity. An *annuity* is a series of equal payments made at equal periods of time. You may, for example, sign up to contribute to an annuity in which you deposit $400 each month for the next 10 years. With annuities, your deposits or contributions earn interest, with the first payment or deposit earning interest the longest. And each payment's interest earns interest — it's a compounding of a piece at a time.

For example, if you deposit your $400 each month in an account earning 6% interest compounded monthly, each month the interest is 6% ÷ 12 = 0.5%, which is 0.005 in decimal form. The following shows the first few payments and how each payment grows due to compound interest:

**Month 1:** $400

**Month 2:** $400 + $400(1 + 0.005)

**Month 3:** $400 + $400(1 + 0.005) + $400(1 + 0.005)$^2$

**Month 4:** $400 + $400(1 + 0.005) + $400(1 + 0.005)$^2$ +$400(1 + 0.005)$^3$

Each month has a new deposit of $400 that hasn't started to earn interest yet. The $400 that was deposited the month before is worth 1.005 times as much as it was when deposited. After two months, a deposit of $400 is worth $(1.005)^2$, or 1.010025, times it was when deposited (and so on through the years). As you might imagine, the longer a deposit earns interest, the more it's worth. I explain everything you need to know about annuities in the following sections.

Some annuities are *ordinary annuities* and others are referred to as *annuities due*. The main difference between the two types is when the payment is made. In the case of an ordinary annuity, the payments are made at the end of each period. With an annuity due, the payment is made at the beginning of each period.

## Preparing your financial future with a sinking fund

A *sinking fund* is a fund that's set up to receive periodic payments with a particular goal in mind. For instance, you may set up a sinking fund to pay off a note, or even better, to finance the purchase of some large piece of equipment 10 years from now. Why 10 years from now? Well, you know that the current equipment will need to be replaced, so you figure out how much you'll have to spend to replace it, and then you start saving toward that purchase.

With a sinking fund, you're making regular payments, and the payments you're making are earning interest; so in reality you're setting up an annuity. Remember that you have an ordinary annuity if you're making payments at the end of each payment period and an annuity due if you're making payments at the beginning.

Sinking funds were used in 14th-century Italy and in 18th-century Great Britain in order to reduce the national debt. Perhaps the *sinking* expression came in because the countries decided to retire the debt or *go under*?

Here's an example you can work that deals with setting up a sinking fund: Imagine that you have a piece of equipment that's expected to last for another 5 years. You set up a sinking fund in anticipation of having to purchase a new piece of equipment to replace the current one that will be outdated. At the end of each of the next 5 years, you deposit $20,000 in the sinking fund that earns 4.5% compounded quarterly. After those 5 years, how much will you have in that fund to help you purchase the replacement equipment?

What's described here is an ordinary annuity in which the payments and the compounding don't coincide. So now you know you have to set up a sinking fund and use the following formula to determine the future value:

$$A = R \frac{(1+i)^{nt} - 1}{(1+i)^{n/p} - 1}$$

All you have to do is plug the numbers you know into the equation. The regular deposits, $R$, are $20,000. The interest rate per compounding period is 4.5% divided by 4, which is 1.125%, or 0.01125. The number of times the interest is compounded each year is $n = 4$, and that compounding goes on for $t = 5$ years. The number of payments per year is $p = 1$. So with everything plugged in, the future value is

$$A = 20,000 \frac{(1+0.01125)^{4(5)} - 1}{(1+0.01125)^{4/1} - 1}$$

$$= 20,000 \frac{(1.01125)^{20} - 1}{(1.01125)^{4} - 1}$$

$$= 20,000 \frac{0.250751}{0.045765}$$

$$= 20,000(5.479100) = 109,582$$

In 5 years, you accumulate about $109,600, which includes about $9,600 in interest.

# Determining the payment amount

With a sinking fund (which is described in the previous section), you contribute a regular amount of money each time period and accumulate a sum to be used some time in the future. But what if you need a particular amount of money? What if you have a specific goal in mind for a particular project? In that situation, you want to determine how much has to be contributed *now,* in regular payments, in order to have a specific amount of money accumulated *then*.

The amount of the regular payment needed to accumulate $A$ dollars in an ordinary annuity is found with this equation:

$$R = \frac{Ai}{(1+i)^n - 1}$$

where $R$ is the payment amount, $i$ is the interest per pay period, and $n$ is the number of periods over which the amount will accumulate.

Wrap your brain around this equation by trying out this example: Suppose you want to accumulate $450,000 over the next 6 years in order to buy a new building for your business. You plan to make monthly payments into an ordinary annuity that earns 4.5% compounded monthly. What do your payments need to be?

The interest per pay period is the annual rate of 4.5% divided by 12, which gives you 0.375%, or 0.00375, per period. Monthly payments for 6 years add up to 72 payments ($12 \times 6 = 72$). By substituting these numbers into the formula, you get this math:

$$R = \frac{450,000\,(0.00375)}{(1+0.00375)^{72} - 1} = \frac{1,687.5}{0.309303} \approx 5,455.81$$

So this means that you need to make monthly payments of about $5,456. Multiplying the payment amount by the 72 months, you get a total contribution of about $392,832. Add to this the almost $60,000 in interest, and you have what you need for the new building.

# Finding the present value of an annuity

The *present value* of an annuity is the amount of money that you would have to deposit today (in one lump-sum payment) in order to accumulate the same amount of money produced by contributing regularly to the annuity over some particular period of time. These types of annuities are referred to as single-payment annuities. In practice, some people prefer them. After all, if you could make one big payment today, you'd be all done with the payments. But coming up with a huge lump sum of money isn't always possible.

In this section, I compare two different ways of accumulating a particular amount of money: one way is with a lump sum payment (that you have to determine), and the other is with regular payments made to an annuity. You get to decide whether to make the lump sum payment today and let it earn interest over a certain number of years or to make regular payments into an annuity for that same number of years.

The present value of an ordinary annuity is found with this equation:

$$P = R\left[\frac{1 - (1 + i)^{-n}}{i}\right]$$

where $P$ is the principal (lump sum) or present value needed to be invested at this time, $R$ is the amount that would have to be contributed regularly with the annuity, $i$ is the periodic interest amount, and $n$ is the number of periods for which regular payments will be made.

Go over this example for some practice: What's the present value of an ordinary annuity earning 4% compounded quarterly if payments of $500 are made every 3 months (quarterly) for 10 years?

First you have to find out how many payments you'll pay in 10 years. To do so, multiply the number of years by the number of payments per year: $10 \times 4 = 40$ payments. If you divide the interest rate by 4, the rate per quarter is 1%. Using the formula, you get

$$P = 500\left[\frac{1 - (1 + 0.01)^{-40}}{0.01}\right] = 500\left[\frac{0.328347}{0.01}\right]$$
$$= 500[32.8347] = 16,417.35$$

If you deposited a lump sum of $16,417.34 in an account that grows at 4% compounded quarterly, you'd have the same total amount of money in 10 years that you would have if you decided to pay $500 every three months into the account. The following two formulas show you the figures.

Compound interest:

$$A = P\left[1 + \frac{r}{n}\right]^{nt} = 16,417.35\left[1 + \frac{0.04}{4}\right]^{4\,(10)}$$
$$= 16,417.35[1.01]^{40} = 24,443.18$$

## The easy way to sum up a series

Formulas used to find the total amount in an account after a number of years are based on the sum of a series of numbers. A *series* in mathematics is the sum of a list of numbers that are different from one another by some constant number or by some ratio. For instance, the sequence of numbers: 1, 4, 7, 10, 13, 16 . . . is an *arithmetic sequence* with the subsequent numbers found by adding 3 to the previous one. The sequence of numbers: 1, 3, 9, 27, 81, 243 . . . is a *geometric sequence* in which you multiply by 3 to get the next number in the sequence. Both types of sequences have formulas that give you the sum of the numbers you want to add. These formulas are handy because who wants to have to compute $1 + 4 + 7 + 10 + \ldots + 3001$ by hand (or even with a calculator!)?

To find the sum of the first $n$ terms of an arithmetic sequence, use this equation:

$$S_n = \frac{n}{2}\left[a_1 + a_n\right]$$

where $a_1$ represents the first term in the sequence, and $a_n$ is the $n$th term.

You can find the sum of the first $n$ terms of a geometric sequence with this equation:

$$S_n = a_1\left[\frac{1 - r^n}{1 - r}\right]$$

where $r$ is the ratio or multiplier getting you from one term to the next.

So if you want to add up the first 20 terms in the sequence 1, 4, 7, 10, 13 . . . 58, you use the formula:

$$S_n = \frac{20}{2}[1 + 58] = 10\,[59] = 590$$

To add up the first 10 terms in the sequence 1, 3, 9, 27 . . . 19,683, you use:

$$S_n = 1\left[\frac{1 - 3^{10}}{1 - 3}\right] = 1\left[\frac{-59048}{-2}\right] = 29{,}524$$

Future value of an ordinary annuity:

$$A = 500\left[\frac{(1 + 0.01)^{40} - 1}{0.01}\right] = 500\left[\frac{0.488864}{0.01}\right]$$
$$= 500\,[48.8864] = 24{,}443.19$$

The amounts are off by a few cents, which is due to rounding, of course. (If you need a refresher on computing compound interest, refer to Chapter 9.)

# Computing the Payout from an Annuity

As I mention earlier in this chapter, an annuity is set up to collect money and allow that money to earn interest over some period of time. Annuities allow people to make donations to organizations; regular payouts are made as donations over many years — or forever. Scholarships and business-starting grants are often funded with annuities.

When you have an annuity, you can start the payout program as soon as the money is deposited, or you can defer the payments for a number of years, allowing the initial deposit to grow even more and the account to increase in size. You can even set up an annuity with payments *in perpetuity,* meaning that they pay out forever; as long as only the interest is paid out, the annuity can keep providing money.

The payouts of an annuity are just the reversal of the payment into an annuity. In the earlier section, "Finding the present value of an annuity," you see how a lump sum can be equal to making regular payments. In fact, depositing a certain amount of money can accumulate as much in an account as paying in over a number of years. The payout of an annuity is like taking that lump sum and parceling it out in regular, periodic payments, until it's gone. The diminishing amount in the account still earns interest, but the amount of growth of the account gets smaller and smaller over time.

## Receiving money from day one

If you make a lump sum contribution into an annuity, and if you want to withdraw regular amounts of money from the annuity, how much can you take out, and how long will the money last? Both of these questions are dependent on each other, and both are dependent on how much money is in the account. But one thing is obvious: The less you take out each time, the longer the money will last. Be frugal, my friend!

### Figuring out the amount of upfront money that's needed

When an endowment (a type of annuity) is made, the benefactor often has a goal in mind — an amount of money that he or she would like to see given each year for a certain number of years. With that goal in mind, the benefactor then determines how much to invest or deposit so that the desired allocations can be made.

Consider this example: Imagine that an entrepreneur wants to give an endowment to a small business catalyst to help local fledgling businesses get a good start. The arrangement says that the entrepreneur will give $5,000 quarterly to young businesses over the next 5 years. The endowment is invested in an account earning 5% compounded quarterly. Upfront, how much did the entrepreneur have to put into the account?

The present value of 20 payments of $5,000 earning interest at 1.25% per quarter (5% ÷ 4 = 1.25%) is found with the formula for the present value of an ordinary annuity (see the earlier section, "Finding the present value of an annuity," for more on this formula):

$$P = R \left[ \frac{1 - (1 + i)^{-n}}{i} \right] = 5,000 \left[ \frac{1 - (1 + 0.0125)^{-20}}{0.0125} \right]$$

$$= 5,000[17.599316] = \$87,996.58$$

It looks like the benefactor gave approximately \$88,000 to help the new businesses.

### Discovering how much time the annuity will last

If you deposit a lump sum into an annuity, regular amounts can be withdrawn from the account over a period of time until the money is gone. So you'll likely want to figure out how long that annuity will last based on your withdrawals.

Say, for example, that Hank gets a large insurance settlement and deposits it in an account earning 6% compounded monthly. He arranges for monthly payments of \$5,000 to be made from that annuity to help offset his business expenses. If the insurance settlement was for \$400,000, how long will he be able to get the monthly payments?

Using the formula for the present value of an ordinary annuity, you can solve for the number of payments, $n$. The interest rate per month is 6% ÷ 12 = 0.5%, which is 0.005. The regular payments, $R$, are the amounts being paid to Hank each month. Here's what your equation should look like:

$$400,000 = 5,000 \left[ \frac{1 - (1 + 0.005)^{-n}}{0.005} \right]$$

Simplify the equation by dividing each side by 5,000 and then by multiplying each side by 0.005, like so:

$$\frac{400,000}{5,000} = \frac{5,000 \left[ \frac{1 - (1 + 0.005)^{-n}}{0.005} \right]}{5,000}$$

$$0.005 \times 80 = \left[ \frac{1 - (1 + 0.005)^{-n}}{0.005} \right] \times 0.005$$

$$0.4 = 1 - (1 + 0.005)^{-n}$$

$$(1 + 0.005)^{-n} = 0.6$$

Now take the natural log of each side of the equation, which allows you to bring the exponent, $-n$, down as a multiplier. Divide each side by $\ln(1.005)$, and use a scientific calculator to do the computation:

$$\ln(1.005)^{-n} = \ln(0.6)$$
$$-n\ln(1.005) = \ln(0.6)$$
$$-n = \frac{\ln(0.6)}{\ln(1.005)} \approx -102.42$$
$$n \approx 102.42$$

Are some of these steps a bit unfamiliar to you? If it has been a while since you've seen an algebraic solution like this, refer to "Doubling your money, doubling your fun," earlier in this chapter, where you can see a similar process used to solve an equation.

The value of $n$, the number of monthly payments, is about 102.42. Divide that amount by 12 months, and you see that Hank will receive about 8.5 years of monthly payments of $5,000 from his insurance settlement. $5,000 × 102.42 = $512,100, which is the total amount of money that Hank will receive during those 8.5 years. The settlement was for $400,000; the difference between $512,100 and $400,000 is $112,100. So about $112,100 of that total is interest.

## Deferring the annuity payment

You can deposit an amount of money into an account with the understanding that regular payments will be made, but only after several years. You may have deposited the money in the account all at once, or you may have accumulated it over a period of years as an annuity. In any case, the amount in the account grows with compound interest before being disbursed. To figure the present value of a deferred annuity, you need to determine the number of payment periods in the payout and the number of payment periods the payout will be deferred before starting the disbursement.

The present value of a deferred annuity is found with this equation:

$$P = R\left[\frac{1 - (1+i)^{-n_2}}{i(1+i)^{n_1}}\right]$$

where $R$ is the regular payout amount, $i$ is the interest per period, $n_1$ is the number of periods the annuity is deferred, and $n_2$ is the number of periods that payments are to be made from the annuity.

The formula for deferred payments combines two other formulas: the present value of an ordinary annuity (to determine the amount of money needed at the beginning of the disbursement period) and the present value of that present value (to determine how much you need to deposit right now to have the required amount some time in the future).

Check out this example: A trust fund is set up to pay for the college education of a 2-year-old boy. With the arrangement, $60,000 will be paid to the student per year for 5 years, starting when he's 18 years old. The fund earns 5.5% compounded quarterly. The payout will also be quarterly. What's the present value of this fund?

The quarterly payout amounts are $60,000 ÷ 4 = $15,000. The interest rate per quarter is 5.5% ÷ 4 = 1.375%, which is 0.01375 in decimal form. The number of periods that the annuity is deferred is 16 × 4 = 64, and the number of periods over which the payment will be made is 5 × 4 = 20. Putting all the numbers in their proper places, you get this equation:

$$P = 15,000 \left[ \frac{1 - (1 + 0.01375)^{-20}}{0.01375(1 + 0.01375)^{64}} \right]$$

$$= 15,000 \left[ \frac{0.2390035}{0.0329517} \right]$$

$$= 15,000[7.253146] = 108,797.19$$

So the present value is $108,797.

# Chapter 11

# Understanding and Managing Investments

*O*ne of the first things you hear on the evening news is how the stock market has behaved (or misbehaved) during the day. The ups and downs of the market seem to act as a barometer of how the country and the world are doing — on all fronts. And it doesn't seem to take much to instigate a change in the market. When change happens, people become either sellers or bargain-seekers.

You may need to understand the basics of investing so you can answer questions from investors, help employees sort through investment choices, know how the competition is doing, and manage your own stocks. You want to understand the basic premises of stocks and bonds and be able to make the computations necessary to further your understanding.

The daily price quotations in the newspaper and scrolling across your television screen are interesting — but only with the right perspective. And the stock averages at day's end serve to summarize the day's trading — but even those numbers need a little explanation.

So, in this chapter, I show you how to compute and understand some earnings ratios and daily quotes and determine the percent change from the day's activities. Bonds are a way of investing in a company, too, so I take a brief look at the earnings involved with this investment tool.

# Interpreting the Daily Stock Market Quotations

Current technology allows for instantaneous information on all sorts of investment topics: the price of various stocks, the number of shares that have traded, and the performance of the different stock indexes. These indexes include the Dow Jones Industrials, NYSE Composite, NASDAQ, S&P 500, and others. Choose the average that makes you happiest — at that particular point in time. Blink and it's changed, anyway.

## Getting to know the stock quotations

The daily price quotations for stocks appearing in the newspapers and on television are much easier to decipher than they used to be. Until the Common Cents Stock Pricing Act was passed in 1997, the prices used to be in dollars and fractions, with halves, quarters, eighths, and sometimes even sixteenths and thirty-seconds. Beginning in August 2000, the stock market began changing the prices to decimals until all of the stocks and markets were converted.

Converting these fractions to pennies when the decimal value of the fraction had more than two places was especially difficult. For example, the fraction $\frac{1}{8}$ is equal to 0.125, which is $12\frac{1}{2}$ cents. I don't know about you, but I can't recall seeing many half-penny coins around in my time. Luckily, the new, enlightened quotation method uses pennies instead of fractions of dollars. Table 11-1 is set up something like the daily stock quotations that you see in the newspaper.

| Table 11-1 | Example Stock Market Report on Selected Stocks | | | | | | |
|------------|------|------|------|------|--------|--------|-----------|
| *Name* | *Ex* | *Div* | *Yld* | *PE* | *Last* | *Chg* | *YTD % Chg* |
| Allstate | NY | 1.52 | 3.0 | 6 | 50.84 | −.36 | −21.9 |
| Archer Daniels | NY | 0.46 | 1.3 | 11 | 36.41 | +.14 | +13.9 |
| Boeing | NY | 1.40 | 1.6 | 17 | 87.86 | −.41 | −1.1 |
| Caterpillar | NY | 1.44 | 2.1 | 13 | 68.28 | +.08 | +11.3 |
| Daimler | NY | 2.00 | 2.0 | . . . | 98.04 | +4.09 | +59.6 |
| Google | NASD | . . . | . . . | 51 | 648.54 | +22.69 | +40.8 |

| Name | Ex | Div | Yld | PE | Last | Chg | YTD % Chg |
|------|----|----|----|----|------|-----|-----------|
| Harley-Davidson | NY | 1.20 | 2.7 | 12 | 45.09 | −2.02 | −36.0 |
| McDonald's | NY | 1.50 | 2.6 | 31 | 58.48 | −.12 | +31.9 |
| Wal-Mart | NY | 0.88 | 1.9 | 15 | 45.50 | +.03 | −1.5 |
| Wiley | NY | 0.44 | 1.1 | 20 | 39.76 | −0.15 | −0.1 |

Table 11-1 shows a sampling of ten companies, most listed on the New York Stock Exchange (NYSE) and one on NASDAQ. Here's what each of the columns refers to:

- ✔ *Ex* indicates the exchange
- ✔ *Div* tells you the current annual dividend per stock
- ✔ *Yld* is the stock yield ratio
- ✔ *PE* is the price earnings ratio
- ✔ *Last* is the column that has the price of the stock at the end of trading on that day
- ✔ *Chg* tells you how much the price of the stock changed during the day's trading
- ✔ *YTD % Chg* gives you the percent change in the price of the stock since the beginning of the year

## Counting on stocks with different indexes

Even though the Dow Jones Industrial Average (DJIA) and Standard & Poor's 500 are the most recognizable stock indexes, several other indexes are used, observed, and studied. The NASDAQ index, for example, includes many smaller companies. Russell 2000 and Wilshire 5000 include smaller companies as well. The 2000 and 5000 in the names tell you the number of companies included in the index. These big numbers are a far cry from the 30 companies measured by the DJIA. Some other indexes are the NYSE Composite, Amex Market Value, S&P MidCap, Dow Transportation, and the Dow Utility Index. Just as the last two names imply, they include transportation companies and utility companies, respectively. Even though it isn't as well-known, the Transportation Index is looked at with great interest, because when it's doing well, you know that goods are being transported. And that, of course, is great for the economy. Conversely, when the Transportation Index is doing poorly, goods likely aren't being moved, which is bad for the economy. The different indexes measure different types and sizes of companies. And the companies included in any of these indexes are in flux — they can change over time.

In general, you compute the previous day's value of a particular stock by either adding to or subtracting from the price of the stock at the end of the day's trading (*Last*). If the change is negative, you add that much to the end value. If the change is positive, you subtract the change from the end value. I know that this seems backward, but that's what you're doing: going backward! To determine the value of the stock at the beginning of the year, divide the current value (*Last*) by (1 + *YTD% Chg*).

**GO FIGURE**

Refer to Table 11-1 and determine the price of the Harley-Davidson stock at the beginning of the day's trading.

The price of the stock finished at $45.09, and the price went down $2.02 during the day of trading. To get the beginning price, simply add $45.09 + $2.02 = $47.11. The stock started at approximately $47.

**GO FIGURE**

Refer to Table 11-1 and the information involving the Google stock. What was the price of the stock at the beginning of the year?

The Google stock finished at $648.54, which reflects a 40.8% increase from the beginning of the year. To solve, let the price of the stock at the beginning of the year be represented by $x$. The increase in the price since the beginning of the year is figured with this equation: $648.54 - x$. Set the fraction equal to 40.8%, which is 0.408 as a decimal. (Changing from percents to decimals is covered in Chapter 2.) To get the percent increase, you divide the difference by the starting price, $x$, like this:

$$\frac{648.54 - x}{x} = 0.408$$

Now solve for the beginning price by multiplying each side by $x$, adding $x$ to each side, simplifying, and dividing. Your math should look like this:

$$\cancel{x} \cdot \frac{648.54 - x}{\cancel{x}} = 0.408 \cdot x$$
$$648.54 - x = 0.408x$$
$$648.54 = x + 0.408x$$
$$648.54 = 1.408x$$
$$\frac{648.54}{1.408} = \frac{\cancel{1.408}x}{\cancel{1.408}}$$
$$460.61 \approx x$$

And, essentially, you've divided the current value by (1 + the percent change). As you can see, the stock started out at a little more than $460 at the beginning of the year.

# *Computing percent change*

As a part of the daily reports on the stock market, you're told not only how much the stocks went up or down but also the percent change. Basically, *percent change* is the change (how much the price went up or down) divided by the starting amount. If the price of the stock goes up, the percent change is a positive number. If the price goes down, the result is negative.

The percent change is different from stock to stock and from index to index, because each item has a different base amount. Dividing various changes by different base amounts gives different decimal values and percentages.

Refer to Table 11-1 and compute the percent changes in the prices of the Harley-Davidson stock and the Caterpillar stock.

To compute the percent change, you divide the change in the stock by the beginning value. The Harley-Davidson stock went down $2.02 during the day of trading, so it started at $45.09 + $2.02 = $47.11. The percent change is the change of –$2.02 divided by $47.11:

$$\frac{-2.02}{47.11} \approx -0.042878$$

The change in this case is negative because the stock went down that day.

The Caterpillar stock went up 8 cents that day, so it started at $68.28 – $0.08 = $68.20. In this case, the percent change is the change of $0.08 divided by $68.20:

$$\frac{0.08}{68.20} \approx 0.0011730$$

The Harley-Davidson stock went down by about 4.3% and the Caterpillar stock went up by about 0.1%. The less the stock costs, the more changes have the potential to affect the percent change. For instance, if the Caterpillar stock had cost $6.80 instead of $68, the percent change would have been closer to 1% than one-tenth of a percent.

Here's another example: Say that you read in the newspaper that Wiley stock went up $12 on Monday, which is a 25% increase, and then it went down $12 on Tuesday, which is a 20% decrease. You get out your trusty pen and paper and start checking the computations (because they just simply can't be correct).

Here's what you determined: The price of Wiley stock started at $48 at the beginning of trading on Monday morning. The price increased by $12, so the percent increase is:

$$\frac{12}{48} = 0.25 = 25\%$$

The price of the stock at the end of trading on Monday is $48 + $12 = $60. On Tuesday, the price of the stock falls by $12. So the percent decrease is:

$$\frac{-12}{60} = -0.20 = -20\%$$

The stock is now back to $48, but the percent increase was based on $48 and the percent decrease on $60. The larger the denominator of the fraction, the smaller the percentage (when the numerators are the same).

## Using the averages to compute prices

The different stock market indexes (or averages) all give interpretations of the status of the price of stocks. For instance, the Dow Jones Industrial Average (DJIA) is based on the prices of 30 representative stocks — but not their exact stated prices, because a *factor* enters into the picture. The factor, or dividing number, is used on the actual prices to try to standardize the prices when stock splits are involved; you get a more accurate picture of the value of the stock when increasing the number of shares is taken into account.

The S&P 500 is the Standard and Poor's average of the 500 biggest companies in the world. These companies are weighted by their size. Other indexes include NASDAQ, Amex Market Value Russell 2000, and Wilshire 5000. Each has its own formulation for determining its value. And the companies used with any of the indexes today are probably not those used several years ago, because the companies often come in and go out of the various indexes. Then the situations of the stocks change, which in turn changes their impact on the average.

In the following sections, I show you how to use the DJIA and the S&P 500 to compute stock prices. I include only these specific indexes because they're the most recognizable, and the techniques used can be adapted to computations in other indexes.

### Employing the Dow Jones Average

The Dow Jones Industrial Average (DJIA) isn't really an *average* of the values of the 30 selected stocks; it's a sum of the worth of those stocks. And the DJIA doesn't consist of just industrial stocks anymore. You have technology and energy stocks and a smattering of the different types of commonly held stocks, such as McDonald's, Coca-Cola, Microsoft, Home Depot, and Wal-Mart.

Over time, stocks being used by the DJIA may split, making the price per share lower. In this case, a factor, or dividing number, is applied to the sum of the stock values to keep the comparison consistent and meaningful. The factor at any time divides the total cost of the 30 selected stocks. So, if the factor is greater than 1, the DJIA comes out to be lower than the actual total cost of the stocks. In late 2007, for example, the index was 0.13532775, which is between one-eighth and one-seventh. Dividing the total cost of the stocks by this rather small number makes the DJIA much larger than the actual cost of the stocks.

A stock *split* is a way that a company makes more shares of its stock available to the trading public. For example, if you had 100 shares of Company Z worth $20 per share, after a split you'd have 200 shares of the Company worth $10 per share. You haven't lost any money; you just have more shares that cost less. This process of splitting stocks was introduced to lower the cost of a share and make it more affordable for more people.

If the DJIA is quoted as being 13,010.14 and the index is 0.13532775, what's the actual total price of the 30 stocks being used in the index?

Divide the actual total price of the stocks by 0.13532775 to get 13,010.14. Let $x$ represent the total price of the 30 stocks. Then solve the equation for $x$ by multiplying each side of the equation by the index value. Your math for this problem should look like this:

$$\frac{x}{0.13532775} = 13{,}010.14$$

$$0.13532775 \cdot \frac{x}{0.13532775} = 13{,}010.14\,(0.13532775)$$

$$x \approx 1{,}760.63$$

The total price of the 30 stocks is about $1,760.63. So the average price of the 30 stocks used is $1,760.63 ÷ 30 ≈ $58.67. On average, each stock is worth about $60. What's the impact on the DJIA if the price of one stock goes up by $2.00?

At first, you want to say that the DJIA increases by 2. But don't forget the factor. You divide the $2.00 by the index, 0.13532775, and get roughly $14.78. So, when you read that the Dow Jones went up by $14.78 in one day, it could have been that one stock increased by just $2.00.

## Weighting share values with the S&P 500

The Standard & Poor's 500 Index (S&P 500) is one of the more widely watched indexes, and is probably the most popular after the Dow Jones Industrial Average. The 500 stocks that are represented in the S&P 500 are mostly those of American corporations. Another qualification of the index is that the S&P 500 uses only shares in a company that are available for public trading.

The numerical value of the index is computed by *weighting* the share values (giving more *weight* or importance to the companies whose stock is worth more). The weighting is based on the corporation's total market valuation — and is extrapolated (using what it was worth some time ago) back in time — to when the first S&P Index was introduced in the 1920s. The *market*

*valuation* of a company is its share price multiplied by the number of outstanding shares. The reason that a corporation's shares are weighted is
because price movements of companies that have a larger market valuation
have more of an influence on the index than price changes in smaller companies. The following example shows you how a weighted average works.

If Company A has 2,000,000 shares selling at $45.90 each, Company B has
1,000,000 shares selling at $60.80 each, and Company C has 500,000 shares
selling at $120.40 each, what's the weighted average of one share of each
stock?

If you want to average just the prices of the stocks, you add the three prices
together and divide by 3. For example, you first add: $45.90 + $60.80 +
$120.40 = $227.10. Now divide that sum by 3, like so: $227.10 ÷ 3 = $75.70.
Unfortunately, this average isn't a good representation, because too much
emphasis is given to the price of the stock of Company C, which has one-
fourth the number of shares that Company A has. Instead, you have to find
the weighted average.

One way to find the weighted average is to multiply each number of shares
by its price and divide by the total number of shares. Another method is to
find the percent shares that each company has, multiply the percentage by
the share price, and add the products together. I show you both ways in the
following sections for contrast and clarification.

The first method, multiplying the number of shares by the prices, introduces
large numbers that can become difficult to manage. The second method of
using the percentages involves dreaded decimals, but it also gives you a
better perspective on the actual weighting and relative size of the company's
stock offering.

### Method 1: Multiplying the number of shares by price

To find the weighted average of the stock prices with this method, multiply
each number of shares by the price of the stock:

**Company A:** 2,000,000 × $45.90 = $91,800,000

**Company B:** 1,000,000 × $60.80 = $60,800,000

**Company C:** 500,000 × $120.40 = $60,200,000

Now add the products together and divide by the total number of stocks.
First off, find the sum: $91,800,000 + $60,800,000 + $60,200,000 = $212,800,000.
Then find the total number of stocks: 2,000,000 + 1,000,000 + 500,000 =
3,500,000. Now divide the two: $212,800,000 ÷ 3,500,000 = $60.80. The
weighted average is quite a bit lower than the $75.70 obtained by just averag-
ing the three prices.

### Method 2: Finding the percentage of the stocks

As you discover in the previous section, the three companies offer a total of 3,500,000 stocks. The percentage held by each of the individual companies is found by dividing the number of stocks a company has by the total number of all three:

**Company A:** $2,000,000 \div 3,500,000 \approx 0.571$

**Company B:** $1,000,000 \div 3,500,000 \approx 0.286$

**Company C:** $500,000 \div 3,500,000 \approx 0.143$

So Company A has about 57%, Company B has about 29%, and Company C has about 14%. Multiply each percentage (its decimal equivalent) by the price of the respective stock and add the results together, like this: $(0.571 \times \$45.90) + (0.286 \times \$60.80) + (0.143 \times \$120.40) = \$26.2089 + \$17.3888 + \$17.2172 = \$60.8149$, or about $60.81. The average with this method is about one cent different from the previous method because of the rounding that's done with the fractions.

# Wrangling with the Ratios

The different stock averages or indexes offer a measure of the activity and worth of stocks as a whole. *Ratios,* such as earnings per share, price earnings, yield, return on investment, return on assets, and profit ratio, offer information on the individual stocks. The numbers associated with each ratio allow you to compare the stocks to one another on another front. In this section, I show you the various computations necessary for the ratios.

A ratio is a fraction in which one quantity divides another. The ratio is represented by the fraction, such as ⅞, or by a corresponding decimal, in this case 0.875. When the fraction's *numerator* (the top number) is larger than the *denominator* (the bottom number), the ratio has a value greater than 1. For instance, the ratio ¾ = 2.25.

## Examining the stock yield ratio

The *stock yield ratio,* abbreviated *Yld* on a stock report, is the ratio of the annual dividend per share and the current value per share of the stock. The stock yield ratio changes when either the dividends change or when the price of the stock changes. By referring to Table 11-1, you can see columns for the dividend (*Div*), the yield (*Yld*), and the current price of the stock (*Last*). If you divide the dividend by the stock price, you get a percentage — which is how the value is reported under the yield.

For instance, from Table 11-1, Allstate stock has a dividend of $1.52 per share and is currently trading at $50.84. To get the stock yield ratio for this stock, divide $1.52 by $50.84:

$$\frac{1.52}{50.84} \approx 0.0298977$$

The ratio rounds to about 0.03, or 3 percent.

Using the information in Table 11-1, determine how much the stock yield ratio changed for Daimler during the day's trading.

According to the table, the Daimler stock has a dividend of $2 and, currently, a stock yield ratio of 2%. The stock yield ratio is determined by dividing the dividend, $2, by the price of the stock, $98.04. During trading that day, the stock went up by $4.09, so you know that the previous price of the stock was $98.04 – $4.09 = $93.95. You can now compute the previous stock yield ratio by dividing the dividend by $93.95:

$$\frac{2.00}{93.95} \approx 0.021288$$

The ratio rounds to 2%. So it appears that the increase in price didn't change the ratio. The apparent lack of change has to do with the rounding. Dividing $2 by $98.04 gives you a decimal value of about 0.0203998; dividing $2 by $93.95 gives you the decimal value of about 0.0212879. The decimals are different by about one thousandth (or one tenth of one percent).

The daily fluctuations in a stock's price usually have little or no impact on the yield ratio. Over the long run, however, the change is more significant and apparent. If your stock value goes up and the dividends stay about the same, the ratio will decrease. You have to decide what you want from your stock: growth or income.

## Earning respect for the PE ratio

The *price earnings ratio,* or PE ratio, is determined by dividing the price per share by the earnings per share. The earnings in this case aren't the dividends paid to the stockholder. The earnings used in the PE ratio are the profits of the particular company. So the PE ratio is determined by dividing the number of shares the company is offering by the profit associated with those shares. Try out the following example to better understand what I mean.

One problem with the PE ratio is that the earnings figures may not be absolutely up-to-date. Even with computerized reporting and instantaneous figuring, the numbers may not truly reflect the exact value if the information input is several days old and the price of the stock is that day's price.

Refer to the Table 11-1 entries for Wal-Mart. The PE ratio is given as 15, and the price per share is $45.50. What are the earnings per share, based on this information?

You can find the PE ratio by dividing the price per share by the earnings. Let *x* represent the earnings in the equation, and then multiply each side by *x* and divide by the ratio. Here's what your math should look like:

$$PE \text{ Ratio} = \frac{\text{Price per share}}{\text{Earnings per share}}$$

$$15 = \frac{45.50}{x}$$

$$15x = 45.50$$

$$x = \frac{45.50}{15} \approx 3.0333$$

So you can see that the earnings are slightly over $3 per share. If you know the total number of shares involved, you can multiply the $3 by the number of shares to get the total earnings of the company.

## Working with earnings per share

You can determine the earnings per share of stock if you have the price of the stock and the PE ratio. (Refer to the previous section, "Earning respect for the PE ratio," for details on how to compute the PE ratio.) You may be wondering whether you can determine the earnings per share if you don't have the PE ratio available.

The answer is yes! Why? Well, by definition, the ratio is obtained by dividing the earnings (after taxes) by the number of available shares. So, if a company's earnings (after taxes) come to $6,000,000 and 300,000 shares are available, the earnings per share ratio is:

$$\frac{6,000,000}{300,000} = 20$$

So the earnings per share is $20 per share.

## Calculating profit ratios

When investing in a business or corporation, the investor is usually interested in the prospect of profits. The revenue, cost, and profit values themselves don't always give you the information that you're seeking when deciding on an investment. But with standard *profit ratios,* you can compare one company's ratio with another. Or you can just keep track of how the ratio is changing within one company.

The return on investment ratio (ROI), the return on assets ratio (ROA), and the profit ratio provide three different — but somewhat connected — bits of information about a company. I explain each of them in the following sections.

### The return on investment ratio

A company's *ROI ratio,* which measures its ability to create profits for its owners or stockholders, is computed by dividing the net income of the company by the equity in the company. The ratio as a percentage represents the dollars in net income earned per dollar of invested capital.

For some types of businesses, an ROI of 15% is satisfactory. After a particular target ROI is established, a business can adjust its prices and operations to best meet that mark.

A company has equity of $1,000,000 and a net income of $40,000. What's its ROI?

To find out, all you have to do is divide the net income by the equity:

$$ROI = \frac{net\ income}{equity}$$

$$ROI = \frac{40,000}{1,000,000} = 0.04 = 4\%$$

The ROI, as you can see, is 4%. Depending on the business, this percentage is either satisfactory or unsatisfactory.

### The return on assets ratio

The *ROA ratio* is quite similar to the ROI (from the previous section), but there are a few differences, which can be better explained by an accountant. The ROA is computed by dividing the earnings (before interest and taxes) by the net operating assets. In general, both ratios are reported as percentages, and, unless unhealthily so, the higher the ratio the better.

A company's ROA is 8%, and its earnings (before interest and taxes) are $45,000,000. What are the company's net operating assets?

Because the ROA is computed by dividing the earnings by the assets, you find the assets by dividing the earnings by the ROA percentage. Dividing $45,000,000 by 0.08 you get $562,500,000. Again, you determine, depending on the business, as to whether this ROA is acceptable for your company or not.

### The profit ratio

The *profit ratio* is a percentage indicating the profit that's generated for each dollar of revenue (after all normal costs are deducted). You can determine the profit ratio by dividing net income by net sales. The *net income* in this case is the income after all expenses; the *net sales* represents the total revenue less any returns.

Some companies, such as appliance or software companies, expect a profit ratio of 30% to 40%. A grocery store, however, expects a profit ratio of 2% to 3%. Why does a grocery store expect less? Well, because it handles a lot of different merchandise that a lot of people need. Plus, the food found in grocery stores has already had additions to the cost of the raw product because of processing, transporting, storing, and stocking on the shelves.

What's the profit ratio of a company whose gross revenue is $50,000,000 with 5% returns expected? The net income, in this case, is 30% of the gross revenue.

First determine the net revenue by reducing the gross amount by the returns. If 5% is returned, then 95% of the gross revenue constitutes the net revenue. So, to reduce the gross amount by the returns, you multiply the gross revenue by the net revenue: $50,000,000 \times 0.95 = $47,500,000. Now find 30% of the gross income by multiplying the gross revenue by the decimal form of 30%: $50,000,000 \times 0.30 = $15,000,000. Now you're all set to find the profit ratio:

$$\text{Profit ratio} = \frac{\text{net income}}{\text{net revenue}}$$

$$= \frac{15,000,000}{47,500,000} \approx 0.3157898$$

The profit ratio is about 31.6%.

# Making Use of Your Broker

A *stockbroker* or *financial consultant* is someone you can turn to if you need financial or investment advice. The brokers and brokerage houses have access to all sorts of information and resources to help you with your investing activity. The Internet offers lots of opportunities for individuals to do their own investing, but that means they have to do all the research and dirty work (which can be difficult).

Those who prefer to go with a professional for advice and resources can take advantage of such options as buying stocks on margin and using puts, calls, straddles, and spreads. When you contract with a broker for a put or call, you pay a commission, because a deal has been *brokered* — someone had to put the two parties together to make the transaction work.

Because brokers charge for their expertise, you also need to be familiar with the commission schedule and what this means to your bottom line. I explain many of the basics you need to know about investing in the brokerage world in the following sections.

## Puts, calls, straddles, and spreads

I'll bet you're thinking that I've switched from math to a football game. No such luck. However, the calls made in investing *are* like the plans made by the quarterback. But football games can't wait for days or weeks to see the results of the call. And, similarly, a spread in bond prices may act like a point spread prediction of a sportscaster; the investor is more leery when the spread is too great. Okay, enough of the analogies.

A *put* is a contract that allows the owner of a particular stock to *sell* that stock if it reaches a particular price in a designated amount of time. A *call* is a contract that gives someone the right to purchase a given stock at a particular price — if it reaches that price in a designated amount of time. An investor uses a *straddle* if he contracts for an equal number of puts and calls with the same price and the same expiration date (playing it both ways). And a *spread* is the difference between the amount in the contract and the current asking price.

Puts are used by investors who are looking for a profit from a rise in the stock prices, and calls are used by investors who expect a profit from a fall in the stock prices. Confusing? Well, just remember that it's more confusing than the rules for football, but less confusing than the rules for soccer.

## *Buying stocks on margin*

The usual procedure for buying some commodity is to make an offer, strike a deal, pay for the item, and then take it home. In the case of stock purchases, you make an offer — usually what the market price is at the time — pay for the item or items, and then receive a certificate showing your ownership of the stock. There are some variations on making the offer that your stockbroker can tell you about: calls, puts, and straddles are a few of these variations.

Your broker can also explain the process of buying stock *on margin*. The margin purchasing option is sort of a cross between buying and borrowing — at the same time. Essentially, when you purchase stocks on margin, you arrange to buy a certain number of shares of stock, but you don't pay the full price for the stock. Instead, you pay a percentage of the price. The percentage you pay is usually 50%, but it can fluctuate between 50% and 100%, depending on the circumstances. So, if you only pay 50% of the price of the stock, who pays for the rest? The brokerage house that you happen to be dealing with. It loans you the money for the rest of the purchase, and it charges interest, of course. The math involved in margin accounts can be a bit tricky, but that's why I provide you with the upcoming example.

The only downside to buying on margin is if the stock goes down instead of up. In that case, you'll be asked by the brokerage firm to settle and pay off the loan or to add additional cash to the account to bring your investment up to the desired percentage.

Suppose you establish a margin account with your broker. You arrange to purchase 100 shares of your favorite stock, which is now selling for $70 per share. The brokerage house sends you a bill for 50% of the cost of the stock, which you pay. You agree to pay 7% interest on the difference until the stock goes up 10% in value (which ends up taking four months). As soon as the stock increases in value by 10%, you sell the stock and pay back the broker. How much money do you make on this transaction?

First of all, you determine how much money is being borrowed at 7%. One hundred shares of stock at $70 per share is $100 \times \$70 = \$7,000$. Because you paid 50% of this bill and borrowed the rest, you paid $3,500 and borrowed the other $3,500. The interest on $3,500 at 7% is about $20.42. (Computing interest on loans is covered in detail in Chapter 12.) You pay this amount of interest monthly for the four months until the stock goes up to the level you want.

If the stock's value increases by 10%, it's then worth the original amount multiplied by the original 100% plus the increase of 10%: $7,000 \times 110\% = \$7,000 \times 1.10 = \$7,700$. (You can find out more about percent increases in Chapter 3.)

You pay your broker $3,500, the amount borrowed. So your net profit is the $7,700 selling price of the stock minus the cost of the stock minus the interest payments. In other words, your net is: $7,700 − $7,000 − 4($20.42) = $700 − $81.68 = $618.32. You usually also have to pay broker fees, but you still end up with a nice profit.

## Paying a commission

You pay bankers, mechanics, and accountants for their expertise. Likewise, you can expect to pay a stockbroker for advice, guidance, and other services provided by the brokerage firm. Until a few years ago, stock was bought in lots of 100 (multiples of 100); those that weren't purchased in lots of 100 were deemed *odd lots*. In this case, a price differential was then applied to the odd lots to make up for the inconvenience of not dealing with nice, round numbers. In other words, the broker received the usual commission, based on the purchase price, and an extra charge was added on for the strange number of stocks purchased.

Nowadays, any number of stocks can be purchased, and the commission is a percentage of the purchase price. Different brokerages have different commission scales, but most charge somewhere between 1% and 3% of the purchase price. The more stock you purchase and the greater the price you pay, the smaller the percentage you're charged.

Table 11-2 is a possible commission schedule. The number of stocks being purchased is multiplied by the cost per share to determine the *purchase price*. The maximum commission is $500 and is paid for any transaction whose price is $50,000 or more.

| Table 11-2 | Percent Commission Based on Purchase Price | | | | |
|---|---|---|---|---|---|
| **Purchase $** | **0–1,000** | **1,001–5,000** | **5,001–20,000** | **20,001–50,000** | **+50,001** |
| % Commission | 3% | 2.5% | 2% | 1.5% | ($500 max.) |

What's the commission paid on the purchase of 250 shares of stock that cost $47.90 per share?

The purchase price is $250 \times \$47.90 = \$11,975$. An $11,000 purchase puts you in the 2% category. So the commission is $\$11,975 \times 2\% = \$11,975 \times 0.02 = \$239.50$.

# Investing in the public: Buying bonds

A *bond* is a way of investing in a corporation, university, or some other public project. Essentially, you're lending money to the institution with the expectation of getting back your principal plus some interest. In most cases, the bond and interest are all paid back at the end of the agreed-upon time period. In some cases, however, the bond is *discounted* — the interest is deducted from the amount loaned and is paid to the bondholder up front.

Most bonds are arranged for with your stockbroker. You can purchase treasury bonds or savings bonds at a bank, but the others are negotiated through a brokerage firm.

Bonds are different from stocks, because a stockholder is actually a part-owner of the enterprise when they hold the stock certificates. A *bondholder* is just a lender. Many issues of bonds are available on the principal exchanges, which means that the stockbroker is aware of and can make recommendations as to their availability and advisability.

Bonds come in all shapes and sizes and arrangements of payments. You have tax-free bonds, zero coupon bonds, premium bonds, and discount bonds. You decide with your broker how you want to arrange your profit — equally over the term of the bond, more upfront, or more toward the end. The variations are many.

Check out this example to see how to calculate bond interest: Suppose you invest $10,000 in bonds at a 5% annual rate of interest. The agreement stipulates that the company make semi-annual interest payments until the bond is repaid. How much interest is paid each year?

The annual interest of 5% on $10,000 earns $0.05 \times \$10,000 = \$500$ each year; half is paid each six months. Not too bad!

# Chapter 12

# Using Loans and Credit to Make Purchases

*L*oans are a necessary part of life. They allow consumers to buy houses, cars, and furniture, and they help to pay for college. Loans allow business owners to expand their facilities, improve their inventories, and even weather tough times. The specifics of the loan depend on many things. Some of the questions you have to ask yourself are: Is the loan secured? (Do I have some asset that the lender will get if I don't pay the money back?) How long will I be borrowing the money? Will I pay the money back all at once, or will I pay it back in installments?

Loans come in one of several forms. For instance, they can be in the form of a *promissory note* or a *discount note.* Discount notes are called such because the interest amount is deducted from the amount that's borrowed. A credit card balance is also a form of loan. This type of loan doesn't have a set end to it — you pay back the balance, but not necessarily for any fixed amount of time. *Installment buying* is another form of loan. And, finally, there are personal property or business loans.

You need to arrange for the type of loan that best fits your needs. And you better know the ins and outs of the numbers involved. After all, you don't want to get involved in a loan that costs more than it should or that puts you into a position that you don't want to be in. This chapter helps with all this and more.

# Taking Note of Promissory and Discount Notes

A *promissory note* is a legal document that requires the borrower to pay back the amount he or she owes at specific time intervals and at a specific fixed rate of interest. With this type of note, you determine the amount of interest based on the *face value* — the amount borrowed — and you determine how all the money will be paid back over time. The terms of the note may include the deduction of interest from the amount borrowed — in which case the note is called a *discount note*.

Promissory notes may be used when a business borrows from other individuals and groups rather than banks or other institutions. The note is the legal, official statement of the terms. The terms of the note may be better than what you can get at the bank — if you're dealing with good friends or relatives — or not necessarily all that good. But you may have no other option. The following sections describe both full face value notes and discount notes.

## Facing up to notes that have full face value

A promissory note in which the interest isn't deducted has full *face value*. In other words, you get the full amount of money, and nothing is deducted from what you're given. When you pay back what you owe, you add the face value (the amount of money borrowed) and the interest and then make arrangements for payment of that sum. The upcoming example shows you the best way to calculate payment amounts when it comes to full face-value loans.

Here's a clue that will help you with the example problem: *Simple interest* is determined with $I = Prt$, where $I$ is the amount of interest, $P$ is the principal or amount borrowed, $r$ is the rate of interest, and $t$ is the time in years. (You find lots of information on simple and compound interest in Chapter 9.)

Say, for example, you borrow \$20,000 for 3 months in order to pay for some holiday inventory. You agree to make 3 monthly payments and pay 9% interest compounded quarterly. How much is each of your payments?

First you have to compute the interest using the simple interest formula that I mention earlier in this section. To do so, you simply have to plug the numbers you know into the formula. For instance, you replace the interest rate, 9%, with its decimal equivalent, 0.09. You also replace the amount of time with 0.25, which is one-fourth of a year (3 months). And, of course, you know the loan amount, \$20,000, which is put in place of the principal.

So, here's what your math looks like: $I$ = \$20,000 $\times$ 0.09 $\times$ 0.25 = \$450. Add the interest to the principal, and you owe a total of \$20,450. When you divide the total by 3 (to get the amounts of the payments), the answer doesn't come out evenly. So, instead, you can make two payments of \$6,816.67 and one payment of \$6,816.66.

# Discounting the value of a promissory note

Suppose you want to borrow \$10,000. You arrange for the transaction and sign a promissory note. The note is then discounted — the amount of the interest is deducted from what you're given. In other words, you don't have the full \$10,000 to work with, but you have to pay the face value, \$10,000, back to the lender. Discounting a note isn't an unusual practice, so make sure you know what you're getting into. The payments for a discount note are fairly straightforward to determine. You may want to rethink the transaction, though, and increase the amount of the note so that you have more money to work with.

One of the more well-known types of discount notes is the U.S. Treasury bill, in which you're the one loaning money to the U.S. Government for a period of time.

### Working with a discounted note

Say that you're borrowing money for a short period of time and agree to the terms of a discount note. Now you need to determine what the payback looks like. To do so, you have to subtract the interest owed from the amount borrowed. Here's an example to get you started.

A discount note is arranged so that \$10,000 is borrowed for 6 months at 12% compounded monthly. The interest is deducted from the amount of the note, and you agree to pay the money back in two installments — one at the end of 3 months, and the other at the end of the 6 months. How much money do you actually get to work with, and how much do you pay back?

First determine the amount of interest using the interest formula, $I = Prt$. The interest rate of 12% is written 0.12 as a decimal; the 6-month period is half a year, or 0.5. So the interest on the note is figured like this: $I$ = \$10,000 $\times$ 0.12 $\times$ 0.5 = \$600. With this loan, you're actually given \$10,000 – \$600 = \$9,400. You pay back the full \$10,000, so your two payments are each \$5,000.

### Increasing the amount of the discount note to cover expenses

If you're borrowing money using a discount promissory note, you may have a target amount in mind. To ensure that you get the amount you need after the discounting, you may have to increase the face value of the note.

For instance, if you need $50,000 to cover expenses for 3 months, and you arrange for a discount note charging 10% interest, you don't get the full $50,000 to work with. Instead, you get $50,000 minus the interest charge. So, instead of borrowing $50,000, you need to increase the amount of the note to an amount that allows you to end up with $50,000 after the interest is subtracted.

To find out how much you need to increase the note by, consider the following: The $50,000 you want is the difference between the face value of the note and the interest paid. Let $f$ represent the face value and $I$ the amount of interest. Three months is one-quarter of a year, or 0.25.

So here's what the equation would look like:

$$\$50,000 = f - I$$
$$= f - (f \times 0.10 \times 0.25)$$

You replace the amount of interest, $I$, with the face value times the interest rate of 10% times the quarter year. Now you're ready to solve for the value of $f$:

$$50,000 = f - (f \times 0.10 \times 0.25)$$
$$50,000 = f - 0.025f$$
$$50,000 = 0.975f$$
$$\frac{50,000}{0.975} = \frac{0.975f}{0.975}$$
$$51,282.05 \approx f$$

As you can see, if you arrange to borrow about $51,282, you'll have your $50,000 to work with during those 3 months.

# Borrowing with a Conventional Loan

A *conventional loan* may come in the form of a mortgage on a building, a car or truck loan, a building equity loan, or a personal loan. Loans are available through banks, credit unions, employers, and many other institutions. Your task is to find the best possible arrangement to fit your needs. If you're comfortable doing the legwork and computations yourself, get out paper, pencil, and a calculator and start searching the newspaper and Internet. If you don't trust yourself quite yet, find an accountant or financial consultant whom you trust and feel comfortable talking to.

You can compute the interest and payments of a loan by using a book of tables and your handy, dandy calculator, or you can even find a loan calculator on the Internet. The variables involved in the computation of loan payments are

- ✔ Amount of money borrowed
- ✔ Interest rate
- ✔ Amount of time needed to pay back the loan

Each of these variables (values that change the result) affects the result differently — some more dramatically than others.

Dealing with a bank or financial institution may seem more impersonal than dealing with an acquaintance or small lender, but the trade-off is more flexibility in payment schedules, availability of more money, and, usually, better rates and terms of the loan.

## Computing the amount of loan payments

You can determine the amount of your loan payments using tables of values or a loan calculator. If you're more of a control person (like me), you probably want to do the computations yourself — just to be sure that the figures are accurate.

The periodic payment of an amortized loan is determined with this formula:

$$R = \frac{Pi}{1 - (1 + i)^{-n}}$$

where $R$ is the regular payment you'll be making, $P$ is the principal or amount borrowed, $i$ is the periodic interest rate (the annual rate divided by the number of times each year you're making payments), and $n$ is the number of periods or payments.

Try your hand at determining a payment with this example: What's the monthly payment on a loan of $160,000 for 10 years if the interest rate is 9.75% compounded monthly?

The interest rate per month is 9.75% ÷ 12 = 0.8125%, or 0.008125 per interest period. The number of payments, $n$, is 10 years × 12 = 120. Using all the values in the formula, you get:

$$R = \frac{160,000\,(0.008125)}{1 - (1 + 0.008125)^{-120}} = \frac{1,300}{0.621319} = 2,092.32$$

So the monthly payments are about $2,100. To find the total payback, simply multiply the monthly payment by the number of payments: $2,092.32 × 120 = $251,078.40. So more than $91,000 is paid in interest. (The interest is the total payback minus the amount borrowed: $251,078.40 − $160,000 = $91,078.40.)

## Considering time and rate

There's no question that the amount of money borrowed affects the conditions of a payback. After all, the more you borrow, the more you're going to owe. But the rate of interest and amount of time needed to repay the money act a bit differently in the overall payback. I explain both of these in the following sections.

### Taking your time to repay a loan

Spreading your loan over a longer period of time makes your regular payments smaller. But what does spreading out those payments do to the total amount of the payback? And how does spreading out the payments affect the total amount of interest? Well, it only makes sense that the longer you take to repay a loan, the more you're going to pay for the privilege of doing so.

Table 12-1 gives you an example of how time affects payback — both in the monthly payments and the total amount repaid. The figures in Table 12-1 are all based on a loan of $40,000 at 9% interest compounded monthly.

| Table 12-1 | Stats on a Loan for $40,000 at 9% Interest in Monthly Payments | | |
|---|---|---|---|
| **Number of Years** | **Monthly Payments** | **Total Payback** | **Total Interest** |
| 10 | $506.70 | $60,804.37 | $20,804.37 |
| 15 | $405.71 | $73,027.19 | $33,027.19 |
| 20 | $359.89 | $86,373.69 | $46,373.69 |
| 25 | $335.68 | $100,703.60 | $60,703.60 |
| 30 | $321.85 | $115,865.70 | $75,865.70 |
| 35 | $313.60 | $131,710.82 | $91,710.82 |
| 40 | $308.54 | $148,101.40 | $108,101.40 |

As you can see from Table 12-1, the payments are much smaller if you pay back in 40 years rather than in 10 years. But the amount of interest paid is enormously greater when spread over the longer period of time.

### Showing an interest in the interest rate

As astounding as increasing the amount of time to repay a loan can be in terms of interest paid, an increase in the interest rate tends to be just as astonishing. For instance, increasing the rate by only 1% makes a significant difference when you're paying over a multiyear period.

Table 12-2 shows you what happens to a $40,000 loan as the interest rate increases. In the table, you see the total payback and interest paid if the loan is for 10 years and if the loan is for 30 years.

| Table 12-2 | Stats on a Loan for $40,000 for 10 Years and 30 Years | |
|---|---|---|
| *Interest Rate* | *Total at 10 Years (Interest)* | *Total at 30 Years (Interest)* |
| 6% | $53,289.84 ($13,289.84) | $86,335.28 ($46,335.28) |
| 7% | $55,732.07 ($15,732.07) | $95,803.56 ($55,803.56) |
| 8% | $58,237.25 ($18,237.25) | $105,662.10 ($65,662.10) |
| 9% | $60,804.37 ($20,804.37) | $115,865.70 ($75,865.70) |
| 10% | $63,432.35 ($23,432.35) | $126,370.30 ($86,370.30) |
| 11% | $66,120.01 ($26,120.01) | $137,134.60 ($97,134.60) |
| 12% | $68,866.06 ($28,866.06) | $148,120.20 ($108,120.20) |

Table 12-2 confirms it: The higher the interest rate, the more you pay in interest. At 6% and 30 years, you pay a little more than twice what you borrowed after adding interest. At 12% and 30 years, you pay over $100,000 more than you borrowed just in interest.

## Determining the remaining balance

If you borrow money using a 20- or 30-year loan and then, all of a sudden, you have a windfall and want to clear all your debts, you need to know what's left to pay on your loan. You may not expect to win the lottery, inherit a fortune, or receive some other type of windfall when you take out a loan, but you really should avoid any contracts that penalize you for making an early payment. Get rid of that type of clause, if you can. It won't hurt anything if you do get that windfall, and you'll save money if you're able to settle up before the end of the loan term. By paying early, you avoid paying some of the interest. However, most amortized loans build in high interest payments at the beginning of the payback. So you want to get that windfall early in the game.

You can find the approximate balance remaining on a loan by using this formula:

$$B = R \left[ \frac{1 - (1 + i)^{-(n-x)}}{i} \right]$$

where $B$ is the remaining balance on the loan, $R$ is the regular payment amount, $i$ is the periodic interest rate, $n$ is the number of payments, and $x$ is the number of payments that have already been made.

Say that you've been paying off a 20-year loan and are 10 years into the payments. You think you have close to half the loan paid, right? Wrong! Check out the following example to see why you aren't so close to having your loan paid off.

Ten years ago, you took out a 20-year loan for $60,000 at 7.25%. You make monthly payments of $475. But now you've inherited some money and want to pay off the balance of the loan. How close is the balance to being halfway paid off?

To find out, you use the formula to estimate the remaining balance. First figure out the values of all the known variables. The value of $i$ is 7.25% ÷ 12 ≈ 0.604167%, or 0.00604167; $n = 20 \times 12 = 240$; and $x = 10 \times 12 = 120$. Now you're ready to plug all these numbers into the formula and solve. Your math should look like this:

$$B = 475 \left[ \frac{1 - (1 + 0.00604167)^{-(240-120)}}{0.00604167} \right]$$

$$= 475 \left[ \frac{1 - (1.00604167)^{-120}}{0.00604167} \right]$$

$$= 475 \left[ \frac{0.514618}{0.00604167} \right] = 475 [85.178105]$$

$$\approx 40,459.60$$

Halfway through the loan period of 20 years, only one-third of this loan has been paid off. Yikes!

## Paying more than required each month

Say that you've taken out a loan and arranged for monthly payments over a 20-year period. Then you realize that you're able to pay a little more than the monthly payment. Is it a good idea to do so? Absolutely! Making payments larger than necessary is a splendid idea. You'll finish paying off the loan much sooner, and you'll save lots of money in interest.

Making larger-than-necessary payments reduces the amount of the principal still owed, which in turn reduces the interest that needs to be paid. Why? Because the monthly interest is figured based on the current principal, and the interest is paid first. Anything left goes toward reducing the principal. At the beginning of an amortized loan schedule, most of the payment goes toward interest.

Table 12-3 shows you how much of a payment on a 20-year, $100,000 loan at 6% goes toward interest and how much goes toward the principal. The monthly payments are about $716.44.

| Table 12-3 | Principal and Interest on a $100,000 Loan | |
| --- | --- | --- |
| *Principal* | *Interest That Month* | *Payment – Interest = Applied to Principal* |
| $100,000 | $500 | $716.44 – 500 = $216.44 |
| $99,783.56 | $498.92 | $716.44 – 498.92 = $217.52 |
| $99,566.04 | $497.83 | $716.44 – 497.83 = $218.61 |
| ↓ | ↓ | ↓ |
| $2,128 | $10.64 | $716.44 – 10.64 = $705.80 |
| $1,422.20 | $7.11 | $716.44 – 7.11 = $709.33 |
| $712.87 | $3.56 | 0 |

You see that most of the payment goes toward interest at the beginning of the 20 years. Toward the end, most of the payment goes toward the principal. The last payment is only $712.87. By increasing the payment to $800 each month, the additional $83.56 goes toward reducing the principal. This way, the loan is paid off in about 16.5 years instead of 20. And the total amount of interest paid is reduced.

To figure out how much interest is saved, multiply the payment amount by the total number of payments you'll make in 20 years ($20 \times 12 = 240$ payments): $716.44 \times 240 = $171,945.60. The loan is for $100,000, so $71,945.60 is paid in interest. The number of payments in 16.5 years is $16.5 \times 12 = 198$. By increasing the payments to $800, you end up paying $800 \times 198$ payments = $158,400. The interest paid this time is $58,400. The savings is over $13,500.

How did I get the number of payments needed at $800? The easiest way to figure this out is with a spreadsheet and formulas that you type in for the interest and payments. I explain how to set up spreadsheets in Chapter 5.

# Working with Installment Loans

An *installment loan* is a loan in which the principal and interest are paid back in installments. What makes the installment loan different from a conventional loan or promissory note is that the amount paid in each installment and the number of installments are what drive the interest rate. The interest may or may not be clearly indicated. The *annual percentage rate,* or APR, is the actual rate of interest charged for the privilege of having the loan. The stated rate and the actual rate are usually completely different things. For example, the stated rate may be 4%, but if you take into account the effect of monthly compounding, that rate is really 4.07415%. The difference may not seem like much, but it adds up when you're dealing with lots of money. A formula allows you to determine the annual percentage rate. I introduce this formula in the following section.

Installment loans are commonly used when making purchases from retailers who are anxious to move their merchandise and earn a profit not only from the markup but also from the interest on the loan. You may be willing to purchase an item using an installment loan if your only option is to let that merchant be your banker. If you aren't in the position to pay cash or borrow more money from a bank, the installment plan allows you the use of the merchandise immediately.

## Calculating the annual percentage rate

The APR isn't something that's widely broadcast on many installment loan contracts; you'll see the stated interest amount, but the APR takes a little more figuring. It takes into account the effects of compounding and any fees and extra charges. If you've ever tried to find a formula for determining APR, you've probably discovered that it's difficult to find one. Most Web sites just want to do the calculations for you. It must be assumed that you really don't want to tackle a formula or that it's beyond your capabilities. Well, I know better than that! Of course you want to do the math!

The best formula I've found is a variation on Steve Slavin's formula in *Business Math* (Wiley). Here it is:

$$APR = \frac{2m\,(\text{finance charge})}{P\,(n+1)} = \frac{2m\,(Prt)}{P\,(n+1)} = \frac{2mrt}{n+1}$$

where *m* is the number of payments made each year, *P* is the principal or amount borrowed, *r* is the interest rate, *t* is the number of years involved, and *n* is the total number of payments to be made. After you determine your variables, you're all set to plug in and solve. Try out an example.

What's the APR of a loan for $6,000 for 2 years at 10.125% interest with monthly payments?

The number of payments per year, $m$, is 12; the interest rate, $r$, is 10.125, or 0.10125; the amount of time, $t$, is 2; and the number of payments, $n$, is $2 \times 12 = 24$. Plugging these numbers into the formula for APR and solving gives you an answer:

$$\begin{aligned} APR &= \frac{2mrt}{n+1} \\ &= \frac{2\,(12)(0.10125)(2)}{24+1} \\ &= \frac{4.86}{25} = 0.1944 \end{aligned}$$

The APR of 19.44% is much higher than the stated interest rate of 10.125%. You don't have any choice on the rate — the APR is what's used in the computations. You just need to know what you're getting into when you agree to a particular stated rate.

## Making purchases using an installment plan

Making purchases with an installment plan allows you to make use of the item or items you're purchasing while you're paying for them. Making installment purchases is often a necessary part of doing business. For instance, you may need machinery or supplies to provide services or produce products. The profit from those services or products allows you to make the installment payments. The trick is to be sure that you aren't paying more than you intended for the machinery or supplies. The stated costs or payments may be deceptive. You may think that you're getting a good deal — until you do the math and determine that the total payback far exceeds the value of the item.

For example, an advertisement claiming that you can purchase a riding tractor for just $2,500 down and $200 each month for the next 5 years may seem like a great deal. But how good a deal is it? Read through the following example to find out.

So, say that you're deciding whether to purchase a $10,000 riding tractor using an installment payment plan in which you put down $2,500 and pay $200 each month for the next 5 years. What's the APR on this plan?

You first need to determine the finance charge, because no interest rate is given. The total cost to you is the down payment of $2,500 plus 5 years of monthly payments of $200. So your cost is $2,500 + 5 × 12 × $200 = $2,500 + $12,000 = $14,500. You'll end up paying $14,500 for a riding tractor that costs $10,000 in cash. So the extra charge is $14,500 – $10,000 = $4,500.

Using the formula for APR, the number of payments per year, $m$, is 12. The principal, $P$, is $12,000 (the 5 years of payments), and the total number of payments, $n$, is 60. By plugging these numbers into the formula, you get

$$APR = \frac{2m(\text{finance charge})}{P(n+1)}$$

$$= \frac{2(12)(4{,}500)}{12{,}000(60+1)}$$

$$= \frac{108{,}000}{732{,}000} \approx 0.147541$$

The interest rate comes out to be more than 14.75%.

# Part IV
# Putting Math to Use in Banking and Payroll

The 5<sup>th</sup> Wave     By Rich Tennant

"How about that folks. A perfectly balanced checkbook – good boy!"

## In this part . . .

*W*hen employees count on their employer to not only pay their salary, but also to help cover their insurance, taxes, social security, and so on, the importance of performing accurate and fair computations becomes abundantly clear. Similarly, a sound budget and reasonable account process make for a successful business venture and satisfied employees. This part shows you how to do all of this and more.

# Chapter 13

# Managing Simple Bank Accounts

. . . . . . . . . . . . . . . . . . . . . . . . . . . . . . . . . . . . . . . . . . . . .

## In This Chapter

▶ Understanding banking basics

▶ Making daily computations of balance and interest

▶ Performing reconciliation on your bank accounts

. . . . . . . . . . . . . . . . . . . . . . . . . . . . . . . . . . . . . . . . . . . . .

*Y*ou can find as many types of bank accounts as you do banks. And keeping the names of the types of accounts straight is as much of a challenge as keeping up with the current name of your bank (you'd think they were required to change every week!). What *is* consistent in banking, though, is the basic mathematics involved when doing the computations. No matter what type of account you have, the math for average balance, daily interest, and additional fees is always the same.

The average balance versus daily balance affects the amount of money earned in interest over a period of time. Your bank's computers do the computations swiftly and automatically. Luckily, though, you can reproduce them with a simple calculator — although maybe not quite as quickly.

Reconciling your checkbook or savings account register is a necessary task. The reconciliation can go smoothly, if you keep good, accurate records. On the other hand, if you throw in a simple reversal of numbers or forget to enter a transaction, you'll be faced with the challenge of making right a wrong. In this chapter, I help you gather the skills and arithmetic techniques needed to keep your banking house in order.

## Doing Business with Banks

Whether you like it or not, banks and other financial institutions are in business to make money. So you end up paying for their services in one way or another — even with free accounts. For instance, you may pay fees for the following reasons:

✔ Excessive check writing

✔ Falling below a predetermined balance

> ✔ Having insufficient funds
>
> ✔ Making telephone transfers
>
> ✔ Obtaining account printouts
>
> ✔ Replacing ATM cards
>
> ✔ Stopping payment on a check
>
> ✔ Transferring funds
>
> ✔ Withdrawing money from an ATM that isn't on your system

As you can imagine, it just makes sense to be familiar with all of your bank's different services — and their corresponding fees. That way you can determine which account best accommodates your needs.

The following sections give you some basic (but important) info on bank accounts, including the types available and how to maintain them.

## Exploring the types of business bank accounts available

Most banks offer many different options for large and small businesses. It's up to the individual business to decide which type of account or accounts to use based on the type of balance the business can maintain and the types of services it uses most frequently. A good business manager will choose wisely when it comes to the types of accounts, and he or she will see that all the minimums and other requirements are met.

The following sections show examples of the many accounts that are available and the services and benefits associated with the accounts. Every bank will have its own set of accounts, and the wise business manager will weigh all the benefits against the possible costs.

### The small business checking account

When a business is relatively small or just starting up, the *small business checking account* makes sense, because it requires a minimum daily balance of only $50 or $100. The small business checking account is usually noninterest bearing, and you assume that you'll make few actual transactions. The bank may set a limit of 100 or 200 transactions, which are free of charge, and then it will charge for any transactions in excess of the limit.

### The basic business checking account

A business that's well established or involved in a moderate level of activity will use a checking account that has a higher minimum balance accompanied by corresponding increased services. These *basic business checking accounts* may have an established monthly maintenance fee that's waived if the minimum balance of $500 or $1,000 is kept in the account. The number of free transactions allowed with such an account may be as many as 400 or 500. After exceeding that number, a fee is charged. Other services that are often provided with such an account include night deposit service, coin processing, and free online bill payments or tax payments.

### The commercial checking account

Accounts for businesses that have large volumes of deposit and check writing activity fall into a *commercial* category of checking accounts. The minimum daily balance may be as low as $100, but the potential for earning interest on the funds in the account occurs if the daily balance stays above some higher amount, such as $1,500 or $2,000. Commercial accounts usually offer unlimited credit and debit activity.

### The business interest checking account

Businesses that can depend on having large daily balances — and that need to keep funds fluid — can arrange for a *business interest checking account.* With this type of account, businesses can get increasing interest rates depending on the amount of money kept in the account. For example, a rate of 0.37% may be paid if the balance is less than $25,000, and the increments may increase to 0.55% if the balance exceeds $250,000. Fees may or may not be charged on these accounts.

### The community checking account

Most banks offer special accounts, called *community checking accounts,* for nonprofit, religious, and other community organizations that have a moderate level of deposit and check-writing activity. Such accounts usually waive any fees as long as the business keeps a minimum balance of $50 or $100 in the account. However, businesses often have the potential for earning interest if the amount in the account exceeds a predetermined amount such as $1,500 or $2,000.

### The business savings account

When a business wants to earn decent interest on deposited money but still wants to keep it relatively fluid (not invested in a bond or CD), the *business savings account* is the way to go. The rate of interest depends on the amount in the account, and the business can avoid service charges if a minimum balance is maintained.

## Understanding the importance of account management

Managing your bank account is important for a couple of reasons: It enables you to avoid unnecessary fees, and more important, it allows you to make smart decisions about your money. You choose the type of bank account that works best for the type of business that you have. You need to consider how many deposit and payment transactions you expect to make each month. And you need to determine how much money you can afford to have out of your reach. In other words, you need to know how much money you can have invested and earning interest or maintaining a minimum balance. After all, you want your money to be working for you, not against you. The saying that *money begets money* is true in the world of banking — as long as you know how to invest it properly.

Depending on the type of account in which you have your money, you may be subject to some fees for the privilege of doing business with the financial institution. Some banks provide accounts that waive a monthly fee if you keep a certain minimum daily balance in the account. However, if you go below that minimum, even for a day, you end up paying the fee that month. Another charge that you may incur is a per-check fee after you've written a certain number of checks for the month.

Keeping your money in an interest-bearing account can help offset some of the fees, but you need to determine whether having the minimum amount of money in the account is worth it. You may be earning interest in that interest-bearing account, but the rate of interest may be too small to consider having a large amount of money in that account. In other words, you may be better off investing the minimum elsewhere and simply paying the fees.

When managing your accounts, it's important to keep an eye on all the factors involved, and not just on your deposits and withdrawals. For instance, to manage your account successfully, you need to

- ✔ **Know your bank's fee policies.** Determine which services are free and which have a charge (and under what conditions the charge is applicable). Consider the different types of accounts and the fees associated with each type.

- ✔ **Determine which options are less costly.** Compare the cost of fees for services with the loss of interest income from having substantial cash in a noninterest-bearing account. Consider the cost of convenience in terms of travel and business.

- ✔ **Keep track of your account balances.** Arrange for a daily review using Internet access. Chart normal balances during different times of the month.

Suppose you have a checking account in which you only pay the $7.50 monthly fee if your daily balance falls below $500 during any day of the month. Is this a good deal? Or are you better off paying the $7.50 monthly fee and keeping the $500 in a savings account that pays 1% interest?

You're probably better off keeping the $500 in the checking account. Why? Well, first of all, the total amount of interest earned on $500 at 1% interest, compounded quarterly, is slightly more than $5. And that's for the whole year. Even if you tried to keep most of your money in a savings account and periodically transferred some of it over to the checking account as needed, you'd be hit by other fees for doing all the transferring. Just plan well, and don't let the account balance fall below the $500 amount.

Here's an example to consider: Imagine that you have a checking account exactly like the one shown in Table 13-1. For this account, you pay a fee of $9 per month if the balance falls below $1,000 during any day of the month. You also pay a per-check fee of $0.25 per check for each check in excess of 35 for the month. Your account is interest-bearing, paying 1.51%, which is credited to your account monthly. Suppose you wrote 40 checks (which all cleared) during April. What's your balance on May 1, after all these transactions?

| Table 13-1 | Account Activity in April | |
|---|---|---|
| *Date* | *Transaction* | *Balance* |
| April 1 | No transaction | $4,321 |
| April 2 | Deposited $345 | $4,666 |
| April 5 | Withdrew $2,456 | $2,210 |
| April 10 | Deposited $1,234 | $3,444 |
| April 17 | Withdrew $3,000 | $444 |
| April 20 | Deposited $2,345 | $2,789 |
| April 22 | Withdrew $1,000 | $1,789 |
| April 24 | Deposited $4,567 | $6,356 |
| April 29 | Deposited $1,234 | $7,590 |
| April 30 | Withdrew $2,000 | $5,590 |

To figure out the balance, here are the things you need to take into consideration:

- ✔ From Table 13-1, you see that the balance was below $1,000 for three days, so the monthly $9 fee will be applied.

- ✔ In Table 13-2 (which you can find later in this chapter), you can see the interest earned at 1.51% is computed and comes out to be $4.51 for the month of April.

- ✔ Because you wrote 40 checks, you have to pay $0.25 for each of the last 5 checks, which totals $1.25.

To find the net result, you first have to add all the fees and interest, like this:

$$-\$9.00 + \$4.51 - \$1.25 = -\$5.74$$

Now subtract that total from the balance of $5,590 on April 30, which gives you $5,584.26 at the start of May.

# Balancing Act: How You and the Bank Use Your Account Balance

Your bank or financial institution keeps track of the balance in your accounts on a daily basis. In fact, with today's availability of online banking, the bank computers can tell you minute by minute what your account balance is. They do this in response to additions and subtractions to your account.

The balance in your account is an important number to both you and the bank. Here's why:

- ✔ The bank needs to know your balance in order to know how much interest to assign your account (if it's interest-bearing). It also needs the balance to determine whether you need to be charged for letting your balance fall beneath a predetermined level.

- ✔ You need to know your balance in order to decide whether you have enough money to pay those pesky bills. Plus, you want to know whether you're earning the maximum amount of available interest.

You and the bank also both need your balance to compute the average daily balance, which is used in determining whether any fees need to be charged for services rendered. And once your average daily balance is computed, you can use it to figure out the interest you're earning on your account. Read on for details.

## Computing your average daily balance

An *average daily balance* is used by banks and credit card companies because the amount of money in an account (or the amount of money charged on a credit card) can fluctuate greatly from day to day or week to week. The bank would prefer that you have approximately the same amount of money in your account throughout the month so that it can accommodate business transactions and lending more easily. But the reality is that businesses need cash to pay bills and employees at certain times of the month; so those payment times may not coincide with the revenue deposits.

When your bank uses an *average daily balance* for any of its reports or computations dealing with your account, it simply finds the balance in the account for each day and then averages them. In other words, the daily amounts are added up, and the sum is then divided by the number of days. The bank does this because it's the fairest and best way of determining interest and fees. Instead of using the largest or smallest amount or the most frequent amount, the average amount is the figure used in the computations.

You can figure the average daily balance one of two ways:

✔ You can do a simple average by adding all the daily balances together for a given month and then dividing by the number of days.

✔ You can use the method that employs a weighted average.

A *weighted average* is useful because, with many accounts, the balance remains unchanged for several days in a row — or even for a week or so. Plus you have fewer numbers to deal with if you do a weighted average. Instead of having about 30 different numbers in the computation (depending on the number of days in the month), you have only as many numbers as there are different amounts in the account. Because of these advantages, I focus on the weighted average for purposes of this chapter.

To find the weighted average, multiply each different balance by the number of days the account was at that amount, and then divide by the number of days.

To practice with an example, check out Table 13-1 (you can find it earlier in the chapter), which shows an account and the activity in that account during the month of April. Using the info in that table, find the weighted average. *Note:* Because the numbers are all in whole dollars, they really aren't all that realistic. But illustrating the average balance doesn't require nitpicking with pennies, so I made the computations easier.

April, as you know, has 30 days. To find the weighted average you have to account for the balance on each of the 30 days, add up the numbers, and then divide by 30. To determine the number of days at each value or balance, you subtract the date numbers. For example, the balance was $444 for three days (April 20 – April 17 = 3 days).

The weighting part of the weighted average comes from multiplying each balance by the number of days that the account had that balance. For example, here's what the math for the example should look like:

$$\$4,321 \times 1 + \$4,666 \times 3 + \$2,210 \times 5 + \$3,444 \times 7 + \$444 \times 3 + \$2,789 \times 2 +$$
$$\$1,789 \times 2 + \$6,356 \times 5 + \$7,590 \times 1 + \$5,590 \times 1 = \$108,925.$$

Now divide this weighted total by 30 to get the average daily balance:
$$\$108,925 \div 30 \approx \$3,630.83.$$

## Go Phish?

E-banking has increased in popularity over the past ten or so years because of certain services that e-banks provide. Consider these popular services:

✔ Online bank statements and daily transactions

✔ Instantaneous transfers from account to account

✔ Online bill payment, which helps you avoid checks, envelopes, or stamps

✔ Online deposits

Accompanying the ease and convenience of e-banking are the opportunities for attacks by the unscrupulous. *Phishing* is done when scam artists send e-mails that are created to look like they're from your financial institution. You're told that you need to verify personal information to protect your account. However, the very e-mail

is the attack upon you and your finances. Even clicking on the link could subject you to spyware, key-logging software, or Trojans being installed on your computer. You pay dearly for your curiosity.

In your e-banking research, you may have come across the terms *float* or *floating*. What does *floating* have to do with online banking (besides adding to the rather tenuous theme)? Again, with the convenience of e-banking comes the blessing and curse of rapid banking processes. With instantaneous processing, you have less time to get money into a checking account (assuming it's not already there).

The key to using online banking effectively is to keep careful track of your account balances. Also safeguard your personal information at every turn, and don't be tempted to answer unsolicited or unfamiliar e-mails.

If you look at Table 13-1 again, you'll notice that for 11 days in April, the actual balance was higher than the average daily balance, and the other 19 days had an average daily balance that was lower than this average. Why is this important? Well, different accounts have different qualifications, and the average daily balance and exact daily balance each play a part in the fee figures. Because computerized computations are available, financial institutions can figure fees and interest on the balance day-by-day rather than using the average. When you're figuring your average daily balance for your own purposes, though, you don't want to deal with that many numbers; the weighted averages are more efficient and quicker for you to compute.

## Determining interest using your daily balance

The interest earned on money in your account usually accrues for a certain amount of time — for a month or three months, for example — and then that interest is credited to your account as a deposit. To determine the amount of interest earned daily (or over several days, when your balance stays the same for a while), you have to divvy up the stated interest rate into a daily amount.

The decimal values get pretty small when you divide them into daily amounts, so you have to carry some extra decimal places in your computation. And you have to be sure not to round off the numbers too soon. For example, when dividing 4.5% by 360 days, you get $0.045 \div 360 = 0.000125$. If you rounded to the nearer ten-thousandth, you'd have 0.0001, which would make the results too low when used in multiplications.

To get some practice, use the numbers from Table 13-1 (shown earlier in the chapter) to find the interest earned by the money in the account for each balance shown. Assume that the bank in this case is paying 1.51% interest.

The first step is to get a daily interest rate. To do so, divide the overall interest rate, in decimal form, by 365. (Refer to Chapter 2 if you need a reminder on how to change 1.51% to its decimal equivalent.) Here's what your daily interest rate should be:

$$\frac{0.0151}{365} \approx 0.00004136986301$$

The result of the division doesn't come out evenly, which is why you see quite a few digits in the decimal answer. Keep all of the decimal places for now, and be sure to use them in your computations.

Now, to compute the interest earned for each different balance, you multiply the amount of money by the number of days the account showed that balance by the interest rate per day.

An electronic spreadsheet will make short work of the computations, even when using all those decimal digits. Refer to Chapter 5 for more on using spreadsheets for your calculations.

Table 13-2 gives the account balances from Table 13-1, and it shows you the interest earned for each different balance and each different length of time. To save room in the table, I let *I* represent 0.00004136986301 in each case. The results are rounded to nine decimal places.

| Table 13-2 | | Computing the Interest Earned in April | |
|---|---|---|---|
| *Date* | *Balance* | *Number of Days* | *Balance × Days × Interest Rate* |
| April 1 | $4,321 | 1 | $4,321 × 1 × *I* = 0.178759178 |
| April 2 | $4,666 | 3 | $4,666 × 3 × *I* = 0.579095342 |
| April 5 | $2,210 | 5 | $2,210 × 5 × *I* = 0.457136986 |
| April 10 | $3,444 | 7 | $3,444 × 7 × *I* = 0.997344657 |
| April 17 | $444 | 3 | $444 × 3 × *I* = 0.055104658 |

*(continued)*

**Table 13-2 (continued)**

| Date | Balance | Number of Days | Balance × Days × Interest Rate |
|------|---------|----------------|-------------------------------|
| April 20 | $2,789 | 2 | $2,789 \times 2 \times I = 0.230761096$ |
| April 22 | $1,789 | 2 | $1,789 \times 2 \times I = 0.14802137$ |
| April 24 | $6,356 | 5 | $6,356 \times 5 \times I = 1.314734246$ |
| April 29 | $7,590 | 1 | $7,590 \times 1 \times I = 0.31399726$ |
| April 30 | $5,590 | 1 | $5,590 \times 1 \times I = 0.231257534$ |

Now, you just have to add up all the interest amounts for the month, which gives you $4.506212328. And that means the interest earned is about $4.51 for the month of April. Great work!

# Reconciling Your Account

All bank accounts that you put money into (deposits and interest, for example) and that you withdraw from, need to be reconciled. *Reconciling* your checking or savings account means that you're performing an audit or check of the figures over a particular time period. The word *reconcile* literally means to bring into agreement or harmony — to make compatible or consistent. So, in other words, you want the money in the account to agree with what you think should be there.

Reconciling isn't a critical factor when it comes to stable accounts such as CDs, which are primarily accruing interest. However, checking accounts (and some savings accounts), which see a lot of regular action need to be reconciled on a daily or weekly basis. The following sections show you how to go about reconciling your account and correcting errors when you find them.

Banks issue monthly reports to both aid your reconciliation process and to prompt you to do this periodic audit. However, the monthly printed reports that you receive in the mail are often out of date in terms of recent activity. These printed statements don't reflect the past few days' worth of deposits and withdrawals. Luckily, many banks provide online access in case you need records that reflect accountings to that day.

## Making reconciliation simple

To reconcile your banking records, you need the statement from the bank and you need your own accounting records, such as a checkbook register or other bookkeeping device. It also helps to have a good head for numbers (or a calculator if you aren't number savvy). Your bank statement will likely list your current balance, transactions during the previous time period (usually a month), and other tidbits of information, such as average balance, a numerical listing of cleared checks, and postings of fees and interest.

Table 13-3 shows a shortened version of a checking account register. It has a column for checkmarks indicating that a transaction has shown up on the bank statement.

| Table 13-3 | | A Portion of a Checking Account Register | | | | |
|---|---|---|---|---|---|---|
| Check Number | Date | Transaction | Debit | ✓ | Credit | Balance |
| 1309 | 4/1 | Allstate Insurance | $617.15 | ✓ | | $2,234.56 |
| 1310 | 4/1 | Bank of America | $411.11 | ✓ | $1,823.45 | |
| | 4/3 | Paycheck | | ✓ | $3,245 | $5,068.45 |
| | 4/5 | Utility Company (trans) | $313.07 | | | $4,755.38 |
| | 4/7 | Insurance settlement | | | $5,200 | $9,955.38 |
| 1311 | 4/10 | Eat-A-Lot Foods | $245.16 | ✓ | | $9,710.22 |
| | 4/10 | Transfer to savings | $4,000.00 | ✓ | | $5,710.22 |
| 1312 | 4/15 | US Treasury | $3,303.07 | | | $2,407.15 |
| | 4/17 | Checks | $22.20 | ✓ | | $2,384.95 |

Notice the checkmarks in the check register in Table 13-3. You insert these marks into your register to indicate the transactions that appear on the bank statement. For example, as shown in Table 13-3, at the time of the reconciliation, the automatic payment to the utility company and the check for taxes haven't yet cleared the bank. The insurance settlement check isn't showing on the statement yet, either.

Most bank statements include forms on the back that you can use for your reconciliation computation. To actually compute the reconciliation, you need to add up all the *outstanding* checks or other debits (those that haven't yet been cleared by the bank), and then you add up all deposits not shown on the statement. The general format of the reconciliation looks like this:

+ Bank balance shown on statement

+ Deposits not yet appearing on the statement

– Checks and other debits outstanding

= Balance showing in your register

Try working a reconciliation as practice. Reconcile the account shown in Table 13-3, assuming that the bank balance on the statement is $801.09 (you don't see the bank statement; just use this number for convenience).

Here are the steps to follow:

1. **Add up all the deposits in the register that aren't shown on the bank statement.**

 You see only one such entry, the insurance settlement of $5,200.

2. **Add up the debits not appearing on the statement.**

 The two such debits are: $313.07 + $3303.07 = $3,616.14.

3. **Do the appropriate adding and subtracting.**

 Your calculations should look like this:

| | |
|---|---:|
| +Bank balance shown on statement | $ 801.09 |
| +Deposits not yet appearing on the statement | +$5,200.00 |
| | $6,001.09 |
| –Checks and other debits outstanding | –$3,616.14 |
| =Balance showing in your register | $2,384.95 |

The numbers agree, so your accounting in the register appears to be correct. Good work!

## Finding the errors

Unless you're a really perfect person, not all reconciliations of a checking or savings account come out correctly the first time, every time. In fact, even the most nimble of number-smiths make occasional errors when adding or subtracting — especially when the person doing the computation is in a hurry or distracted by something. Or you may input one wrong digit in an amount, which subsequently throws you off. No matter what the problem is, if the amount that's supposed to be in the account doesn't coincide with the amount that you have in your register (or spreadsheet), you have to backtrack, do some checking, and find out where the error is. The following sections show you the different errors you might make and how to find them.

### Adding incorrectly

One of the first places I check when my numbers don't jibe is the listing of outstanding checks and debits. Because I tend to add two consecutive checks in my head and write the total (supposedly to save time), errors can and do occur. If, after checking your addition, you get the same sum for the outstanding checks, go back and be sure that you didn't miss one or write down a check that's already cleared.

### Reversing the digits

Another common error in check registers involves reversing digits. Many people (including yours truly) have a tendency to reverse numbers when writing them down or entering them in a calculator or computer. I find that mumbling the numbers to myself helps keep them in the right order. (I do get funny looks at times, so it's best to only mumble to yourself when you're alone.)

One indication that your error might be from reversing digits is when the sum of the digits from the number you're supposed to have in your reconciliation computation and the sum of digits from the balance in your register are the same, even though the numbers are different. (Refer to the nearby sidebar, "Casting out nines," for an explanation as to why this can be.)

For example, if your check register shows a balance of $94.99 and the computations for the reconciliation come out to $95.08, you add up the digits in $94.99 to get 9 + 4 + 9 + 9 = 31. Then add up the digits in 31 to get 3 + 1 = 4. Now do the same thing with $95.08: 9 + 5 + 0 + 8 = 22, and 2 + 2 = 4. Having the same sum-of-digits doesn't for sure mean that you reversed digits, but it gives you a place to start when looking for the error.

### Homing in on the difference

Another way to find an error in your check register — if that's where the discrepancy comes from — is to find the difference between the balance in your register and the balance that you think you're supposed to have. For instance, if the difference between the two is $48, you can then look through your register and the bank statement for an item that's $48 or for one that's half that amount, $24. If you've missed checking the amount off or entered it in the wrong place, you'll find your error more quickly if you know what number you want.

# Casting out nines

A quick, neat trick for checking whether the addition in a long list of numbers is correct is called *casting out nines*. The procedure in casting out nines works like this:

1. **Cross out all the 9's and groups of digits that add up to 9 from the numbers being added.**

2. **Add up the digits that are left.**

If the sum has more than one digit in it, add these digits together — and keep repeating until you have just one digit left.

3. **Perform Step 2 for the sum.**

If the digit you got from the numbers being added and the digit in the sum aren't the same, you know that the sum is incorrect.

For example, look at the addition problem and the method of casting out nines side by side:

|  |  |  |
|---|---|---|
| 3419 | 341~~9~~ | Cross out the 9. |
| 6223 | ~~6~~22~~3~~ | Cross out 6 and 3. |
| 5401 | ~~5~~40~~1~~ | Cross out 5 and 4. |
| +8322 | +~~8~~322 | Cross out 8 here and the 1 above. |
| 23365 | 23~~3~~6~~5~~ | Cross out 3 and 6. |

The more obvious 9's and the numbers adding up to 9 are crossed out on the right. (Other combinations and groupings can be crossed out, too, but enough digits have been eliminated to illustrate the method.) Look at the digits that haven't been crossed out in the numbers being added. Their sum is: $3 + 4 + 1 + 2 + 2 + 0 + 3 + 2 + 2 = 19$. Cross out the 9 in 19, and you're left with 1. Or you can add $1 + 9 = 10$ and then $1 + 0 = 1$. You've been able to reduce all the digits above the addition line to a 1. Keep that number in mind; it's your target or goal for the digits in the sum. Now look at the sum (the number below the addition line). Add up the digits in the sum that haven't been crossed out: $2 + 3 + 5 = 10$, and $1 + 0 = 1$. The digits remaining above and below the addition line, after casting out nines, are the same. So you know that your addition is correct.

Now, to be honest with you, the casting out nines method isn't a fool-proof one. If the two numbers that you're comparing (the digits you got from the numbers and the digit you got from the sum) are different, you can be sure that your sum is wrong. However, if the two digits are the same, there's still a slight chance that your sum is wrong. For instance, you'll get a false positive if your incorrect answer is off by 9. However, this method is so quick and easy that it's worth the slight chance of an error in your error.

 And last, but not least (are you getting the impression that I have been there and done that with my check register?), when the numbers don't agree, check to see if you've remembered to subtract automatic payments. It's convenient to have bills taken out automatically, but you have to be careful that you subtract them from your register.

# Chapter 14

# Protecting Against Risk with Insurance

. . . . . . . . . . . . . . . . . . . . . . . . . . . . . . . . . . . . . . . . . . . . . . . .

. . . . . . . . . . . . . . . . . . . . . . . . . . . . . . . . . . . . . . . . . . . . . . . .

*I*nsurance is a *risk pool.* What's that, you ask? It's something that's formed to reduce losses in the case of some huge, catastrophic occurrence. For example, insurance companies in Florida, where hurricanes are fairly common, may form a risk pool with Midwestern insurance companies, where you find seasonal tornadoes. The companies share the risk and the expense when claims are filed after a bad storm in one place or the other. Being a risk pool, an insurance company *accepts* risks. Accepting risks, however, isn't the same as *taking* risks. The risks involved in the insurance business are closely tied to the chance that a loss will occur. The loss is based on a monetary amount for property, health, life, or some other entity that has value. And basic to this chance of loss is determining the probability that a loss will occur.

When you start talking about probability of loss, that's when *actuaries* come in. These folks are mathematician-accountants who study conditions and histories to determine the *probability* (percentage chance) that a particular occurrence will transpire; this numerical probability helps determine the amount of the insurance premium and even whether the insurance company wants to do business with the client at all.

Insurance policies are many, and they're all varied. The pages of small print on most policies spell out all the details — what's covered and what's not. In this chapter, you see how to do some of the computations you need in order to be best served with insurance policies by the insurance companies. The math involved in this chapter is no more than percentages and ratios and some multiplication and division. However, it's the application of those numbers and operations that sometimes gets a bit sticky. You won't be on par with an actuary after reading this chapter, but you'll definitely be able to keep up with the information from and structures of insurance coverage.

# Surveying the Types of Insurance Available

In order to cover all of the important things in life, insurance comes in many different flavors. Here are the main types of insurance that you'll run across in your quest for coverage:

- **Liability insurance:** Liability insurance covers claims made by others (not the insured) for losses caused by the insured. This type of insurance is probably most well known in terms of medical professionals.

- **Health insurance:** This insurance is used to make financial payments to hospitals, clinics, physicians, and other health professionals on behalf of the insured.

- **Property insurance:** This type of insurance involves buildings, various structures, contents of buildings, and vehicles. The risks included in property insurance are fire, storms, wind, earthquakes, water, and collision. Any or all of these risks may be covered by insurance — although sometimes with a high premium. I cover property insurance in more detail in the later section, "Protecting Yourself from Loss by Insuring Your Property."

- **Life insurance:** Life insurance, which seems as if it's misnamed, is payable at the *death* of an individual. Life insurance is designed to replace the loss of wages (due to an untimely death) of an income-producer. The funds or payout provided by a life insurance policy are intended to provide the income needed by the recipient or recipients of the policy until such funds are no longer needed. In order to establish how much life insurance needs to be purchased, you need to determine how much yearly income is needed and for how long. (See the upcoming section, "Living It Up with Life Insurance," for further details.)

Insurance comes in many different varieties, but the common theme of all of them is the quest to reduce loss if something catastrophic happens. You really hope that you'll never need the insurance; but you have the insurance just in case you have to cover the expense of replacing what's lost. Also common to the different types of insurance are the various computations necessary. Many of the mathematical procedures work much the same for the different types of insurance.

# Living It Up with Life Insurance

Many types of life insurance are available — and each type has its own contingencies and variations that are designed to fit the particular needs of an individual or a company. In this section, you find explanations of the types of life insurance that are designed to help individual employees. I also explain the type that benefits the company. So, whether you're deciding on a program for your employees or some protection for the company, here you find out how to deal with the options and mathematics of the decisions. Here are the types of insurance you're likely to see:

- ✔ **Ordinary life insurance** is insurance that doesn't expire before the death of the insured. This type has an initial premium that doesn't increase, and it has a cash value. An advantage to this insurance is that you can budget a set amount for the insurance and not have to worry about the premium increasing as a person gets older.

- ✔ **Participating whole life insurance** provides premiums to the policyholder. If the insured expects to have income from the paid premiums, this insurance is an option.

- ✔ **Endowment insurance** has a term of a specific length (either in years or up until a certain age) and pays the premium if the person dies. This insurance also returns the premiums to the insured if they survive the term of the insurance.

Each type of insurance has its pros and cons, can be purchased individually or with a group, and is suitable for different situations. One of the main considerations for any of these types of insurance is the cost. Ordinary life, for instance, may have cheaper premiums, but you don't get the income. So you have to decide whether the income is worth the additional cost. You need to do the math and determine whether, in your specific situation, a type of insurance makes sense.

## Insuring with a group

*Group insurance* is arranged by a company or other organization (virtually any group of people) with the goal of decreasing the ratio of the overall risk — and consequently, the total amount of the premiums. Some companies offer more than one type of life insurance to their employees, but most find one type that fits their situation (and budget) the best.

For example, a company can offer life insurance to its employees as a fringe benefit. The total amount of insurance provided to an individual is usually tied to earning level or longevity, and companies often offer options for employees to increase the amount of coverage if they want to. The insurance rate or cost is determined by the ages or conditions of the people being covered, and, because many people are involved, the risk is spread out (averaged) to make the cost less than if the policies were purchased individually. The policies may also have a provision that decreases the amount of life insurance to be paid out after an employee reaches the age of 65 or 70.

When determining the cost of providing life insurance for employees, part of the computation involves the age of the individuals involved. You find the number of people in each age group and compute the cost at each level.

Another factor to consider is the length of service of different employees. You can reward long-serving staff by providing more coverage. Again, the computations involve how many are at each level and the projections of cost into the future.

Start with a basic example. Say that Ajax Company offers life insurance for its employees as a fringe benefit. Each employee has $10,000 in life insurance for the first $25,000 in yearly earnings. The amount of insurance increases by $5,000 for each additional $10,000 earned by the employee. How much life insurance do employees Tom, Dick, and Mary get if their yearly earnings are as follows:

Tom: $27,543

Dick: $47,343

Mary: $63,255

Tom only qualifies for a $10,000 life insurance policy. He needs to earn at least $35,000 to get an additional $5,000 added to the basic policy. Dick, on the other hand, qualifies for a $20,000 policy. He earns $47,343, which is $22,343 more than the $25,000 yearly earnings base. The difference of $22,343 is actually two $10,000 increments — each increases the policy by $5,000 for a total increase of $10,000. Mary qualifies for a life insurance policy of $25,000. If you subtract $63,255 – $25,000, you get $38,255. That's three increments of $10,000 (plus the remainder), so she gets $15,000 added to the base policy of $10,000.

If you have a large number of employees in your company, you'd be better off making a chart of the earnings ranges and insurance amounts. That way you don't have to compute each person's insurance coverage individually. Here's an abbreviated example of what I mean:

| *Earnings* | *Amount of life insurance* |
|---|---|
| $25,000 through $34,999 | $10,000 insurance |
| $35,000 through $44,999 | $15,000 insurance |
| $45,000 through $54,999 | $20,000 insurance |
| $55,000 through $64,999 | $25,000 insurance |

Try out this example, which has a bit of a twist: Beta Company offers its basic life insurance package to all employees and also offers supplemental coverage that's paid by both the employee and the company. The cost of the supplemental coverage is based on the employee's age. Say that there's also a provision that the maximum amount of supplemental coverage may not exceed three times your salary. Having said all that, say that Moe, Larry, and Curly apply to increase the amount of their life insurance policies. Determine how much each pays for his supplemental coverage. Use Table 14-1 and the following information to determine your answer:

Moe, who's 46 years old, earns $64,000 each year and requests $100,000 of additional life insurance.

Larry, who's 29 years old, earns $83,000 each year and wants $200,000 of additional insurance.

Curly, who's 60 years old, earns $57,000 and asks for $250,000 of additional insurance.

| Table 14-1 | Rates for Supplemental Life Insurance (Cost for Each $1,000) | |
| --- | --- | --- |
| *Age* | *Company Contribution* | *Employee Contribution* |
| Less than 25 | 0.002 | 0.068 |
| 25–33 | 0.002 | 0.078 |
| 34–42 | 0.003 | 0.097 |
| 43–51 | 0.005 | 0.175 |
| 52–60 | 0.012 | 0.468 |
| 61–69 | 0.073 | 0.687 |

The rates given are per thousand dollars, so you multiply each rate by the number of thousands in the supplementary policy. Here's how the example would break down:

✔ Moe pays 0.175 per thousand dollars. Because he wants an additional $100,000, he pays $0.175 \times 100 = \$17.50$ per month for the additional coverage. The company pays $0.005 \times 100 = \$0.50$ per month for his supplementary insurance.

✔ Larry pays $0.078 \times 200 = \$15.60$ per month for an additional $200,000 life insurance. The company pays $0.002 \times 200 = \$0.40$.

✔ Curly is denied his request. Why? Well, according to the table, he earns $57,000. Because $3 \times \$57,000 = \$171,000$, the $250,000 life insurance exceeds his allowed amount (because he can't have any amount over three times his salary). If he goes for the $171,000, he'll pay $0.468 \times 171 = \$80.03$ per month for the extra coverage. The company will pay $0.012 \times 171 = 2.052$, or $2.05 per month.

**GO FIGURE**

Ready for a more advanced example? Consider this: Happa Kappa Company offers life insurance coverage for its employees to the tune of three times the salary of that employee for that year. However, the life insurance coverage starts to decrease, on a graduated scale, at age 65. Stella is 69 years old and earns $250,000 this year. If she gets a 6¼% raise next year, what will her death benefit be? Table 14-2 shows the payout schedule for Happa Kappa Company.

| Table 14-2 | Life Insurance Rates |
|---|---|
| *Age* | *Death Benefit (Percentage of Three Times Annual Salary)* |
| Less than 65 | 100% |
| 65–69 | 80% |
| 70–74 | 67% |
| 75–79 | 45% |
| 80 and older | 30% |

First determine Stella's earnings for next year. A 6¼% raise means that she'll earn $250,000 × 1.0625 = $265,625. Because she'll be 70 next year, her death benefit would be 67% of three times that salary: $265,625 × 3 × 0.67 = $533,906.25.

How does this payout compare to the benefit if she dies this year at her current salary, and the policy pays at 80%? To find out, simply multiply like so: $250,000 × 3 × 0.80 = $600,000.

Tripling her raise doesn't make up for the decrease in percentage. (If you need a refresher on percentages and percent increases, go to Chapters 2 and 3.)

## Protecting your business with endowment insurance

Life insurance is designed to replace the income or wages of a person who has dependents, partners, or a business. The insurance doesn't replace the person and what he or she has to offer in physical production or guidance, but the monetary relief helps the family or business continue financially.

*Endowment insurance* is a type of life insurance that can be used to protect a business partner in the event of the other's death. With endowment insurance, you choose a particular term, such as 20, 25, 30, or more years. Or you pick a particular age of the person being insured to be the end date of the insurance, such as age 50 or 65. Then if the person dies, a set amount is paid to the partner. If the person survives the term or age, the premiums are repayable to the insured.

When determining whether you want term insurance or endowment insurance, you consider several things:

✔ How much would it cost to replace the person and keep the business running as usual?

✔ How much does each type of insurance cost, and can you afford either type?

✔ How likely is it that you'll need the insurance? (Of course, you don't know any more than the insurance companies actuarially, but you do have a better feel for the health and well-being of the people involved.)

Try this example on for size: Stan and Ollie just opened a new pet shop and have invested most of their life savings in the endeavor. They have two other salespersons, but Stan and Ollie will only take a minimal salary until the business picks up. They decide to insure each other's life for 10 years in order to cover the cost of hiring additional help should one of them die. Their choices are between term insurance and endowment insurance. At their age, term insurance costs $11.60 per thousand per year, and endowment insurance costs $44.60 per thousand per year. Which is the better choice?

First, determine the cost of each over the 10-year period. To do so, assume that the policy is for $200,000. The term insurance costs $11.60 × 200 × 10 = $23,200. The endowment insurance costs $44.60 × 200 × 10 = $89,200. If one of them dies and the other collects the $200,000, then it seems that the term insurance is the much better deal. After all, it's more than $60,000 cheaper. But with the endowment insurance, the premiums are returned if they both survive. They'd each have $89,200 to spend at the end of 10 years. So, in this case, the endowment insurance is probably the way to go.

As you can see, the difference in cost between term insurance and endowment insurance can be significant. So what if you decide to become your own "broker" and offset the end-of-term settlement that's available with endowment insurance by banking the difference in cost? Of course, you need to have enough discipline to keep your hands off the money you're putting in an account. One way to help you with that self-discipline is to contract for an annuity — you'll be more apt to make the regular payments with this type of account. In Chapter 10, I discuss all the details of annuities. For now, I just offer the basic equation:

$$A = R \left[ \frac{(1+i)^n - 1}{i} \right]$$

where $A$ is the total amount of money that's accumulated, $R$ is the regular periodic money payment, $i$ is the interest rate for each payment period (written as a decimal), and $n$ is the total number of payment periods.

Using the scenario from the previous example, consider what would happen if Stan and Ollie put the difference between the two policies in an annuity during the 10-year period (in other words, they pay for the term insurance and save the difference). How would the money in the annuity compare to the lump sum payment?

The difference between the two policies is $33 per thousand dollars or $33 × 200 = $6,600 per year. If you put this in an annuity at 4% interest, the amount at the end of 10 years is

$$\text{Amount} = \$6{,}600 \left[ \frac{(1 + 0.04)^{10} - 1}{0.04} \right] \approx \$79{,}240$$

The \$79,240 doesn't quite match the repayment on the endowment policy, but if both partners survive for all or most of that 10 years and they bought the term insurance, they'd have the money in the annuity (they don't get the refund if the insurance is paid out). It all depends on who you want to take the risk. In other words, are you willing to pay the higher amount to the insurance company and lose the chance of the lump sum if something happens to the insured? Or would you rather pay the insurance company a smaller amount and faithfully bank the difference so that you're sure to have a lump sum at the end of the time period? That's what insurance is all about: risk and chance and who takes the risk and chance.

# Protecting Yourself from Loss by Insuring Your Property

Owning property brings with it a lot of possibilities: pleasure, extra income, security, and, of course, risk. Many folks tend to focus on the risk because all pieces of property are at the mercy of the actions (or inactions) of nature and people. But luckily, insuring your property passes the risk of loss to another entity: the insurance companies. The insurance companies determine the risk of loss in your particular situation. Then they factor in the cost of operating expenses, add in some profit, and set an amount that you pay them for taking the risk.

Different types of buildings have different rates of insurance. These rates are all based on what the buildings are constructed of, where they're located, and what type of work takes place in them (storage or manufacturing, for example). Some buildings are constructed of brick and stone and others mainly of wood. Some are on the sea coast and others are in tornado-prone areas. Some buildings house families, others house retail enterprises, and yet others contain manufacturing activities. Actuaries, which I explain earlier in this chapter, weigh all of these factors and determine the risk for specific buildings.

After the rates are set, you may consider coinsurance, multiyear contracts, multiple building insurance, and maybe some sort of deferred payment plan. I explain each of these in the following sections.

## Considering coinsurance

When you insure your property for just part of its value, you're considered a *coinsurer.* You're taking on part of the risk, and the insurance company is taking on the rest. A common coinsurance contract is the *80% coinsurance clause.* This clause stipulates that if the insurance covers at least 80% of the value of the property, the insurer will pay all of the losses up to the amount on the policy. An 80% coinsurance policy only applies if the property is insured for 80% of its value or less. If the property is insured for less than 80% of its value, the insurer will pay an amount that's found with the following equation:

$$\text{Payment} = \frac{\text{Amount of insurance} \left( \text{value of policy} \right)}{80\% \text{ of value of property}} \times \text{loss}$$

The following two examples show you the difference between insuring your property for the full 80% of its value as opposed to insuring it for less than 80% of its value. You use the same equation for each, but you see that insuring for the full amount results in the first fraction just becoming the number 1.

Here's an example to try: Nancy has a summer home at Lake of the Ozarks worth $100,000, and she has it insured with an 80% coinsurance clause. Say a fire caused $40,000 in damage. How much can she expect in payment if she has insured the home at $80,000?

After plugging the numbers into the formula, you get the following:

$$\text{Payment} = \frac{\$80,000}{\$80,000} \times \$40,000 = \$40,000$$

You can see that Nancy will collect the full $40,000, because she gets full coverage as long as the loss is less than the $80,000 face value of the policy.

Now consider this example: Say that Chuck's neighbor decides to economize and reduce the cost of insurance on his summer home. His 80% coinsurance clause has his $100,000 summer home insured at $50,000. If a tornado does $60,000 in damage, what will the insurance payment be?

If you plug the appropriate numbers into the formula, your math looks like this:

$$\text{Payment} = \frac{\$50,000}{\$80,000} \times \$60,000 = \$37,500$$

As you can see, Chuck's neighbor probably regrets being so frugal.

Coinsurance is also used when you insure a property for less than it's worth. You use coinsurance when a property appraises for less than what was paid for it. In general, coinsurance is used to save on the cost of the insurance policy. However, the insured is taking on some of the risk and responsibility for a portion of the loss if it occurs. A person who wants to insure property for the full value pays a higher rate; this higher rate is called the *flat rate.*

Say, for example, that Helen bought a building for her quilting business and it cost her $100,000. The building was then appraised at $72,000, so she insured it for $72,000 (the full value of the building). One month after moving in, a fire caused $50,000 damage to the building. What did she collect under 80% coinsurance?

---

## The effects of the Great Chicago Fire

The city of Chicago, Illinois, was growing rapidly in the mid-1800s. In fact, the building industry was struggling to keep up with expansion. Much of the city at that time was constructed of wood. The buildings weren't the only parts made of wood, either; most of the paved streets and sidewalks consisted of pine or cedar planks as well.

On October 8, 1871, in the middle of a particularly dry Indian summer, the wind was blowing strongly from the southwest. A fire started in the O'Leary barn (because of a well-placed cow hoof, as legend has it) and rapidly spread to the northeast, leaping across the Chicago River and

burning for more than a day. By the time the fire died out (it started to rain on Tuesday), it had burned a swath four miles long and three-quarters of a mile wide. Almost 20,000 buildings were destroyed, and 100,000 people were left homeless. Nearly 300 people died; it was a miracle that the number wasn't greater.

Only about half the properties that were burned were insured. And only about half of those insured buildings were ever paid for. The insurance companies were not only overwhelmed with claims, but many were also burned down themselves. The fire destroyed roughly one-third of Chicago's property value.

---

Using the formula, you would work the problem like this:

$$\text{Payment} = \frac{\$50,000}{\$80,000} \times \$50,000 = \$45,000$$

$45,000 is nine-tenths of her loss.

## *Examining multiyear contracts*

*Multiyear contracts* are contracts for more than one year. Multiyear contracts are for two, three, or more years in which the rates are a specified amount. Either the same rate is charged each year, or a graduated rate is agreed on in the contract. Buildings and their contents are usually eligible for these types of contracts at reduced rates. Why? Well, insurance companies want to increase the number of years a contract is binding because it provides stability for their business. Businesses like multiyear contracts because they help them with their financial planning, and because the reduced rates save them money.

Consider the following multiyear rates for premiums of a particular insurance coverage (you can assume that the premiums are paid at the start of the insurance term):

| Length of coverage | Rate |
|---|---|
| 1 year | 1.00 |
| 2 years | 1.85 |
| 3 years | 2.70 |
| 4 years | 3.55 |
| 5 years | 4.40 |

Using these rates, compute the following: If the cost for 1 year's insurance is $2,000, what's the cost for 3 years of insurance? How much is saved?

To find your answer, multiply the 1-year cost by the 3-year rate: $2,000 × 2.70 = $5,400. So, if 3-year-long policies were purchased, the cost would be: $2,000 × 3 = $6,000.

As you can see, the savings is $600 because $6,000 – $5,400 = $600. Of course, that's assuming the insurance can be purchased for the same amount 3 years in a row. It's definitely nice to lock in the cost!

## Taking advantage of multiple building insurance coverage

When a person or company owns several buildings, it's usually advantageous to insure all the buildings and the contents of the buildings under a single contract. This way, you spread the risk over all the different properties.

Formulas for determining the overall cost of such a contract incorporate these items:

- ✔ Multiple building volume credit
- ✔ Multiple building dispersal of risk
- ✔ Loss ratio

The credits and ratios are just small parts of the total calculation needed to determine the premiums. Computers are fed all the vital information and claim history, and the amount of the premium is spit out.

But for you, knowing the parts that some of the listed items play in the overall picture helps you determine which factor is having the greatest impact on the cost of insurance. Knowing the impact helps with future plans and economizing efforts. I explain each of the parts in the following sections.

### Determining the multiple building volume credit

A *multiple building volume credit* is an amount of money credited against the total premium. (Remind you of income tax credits?) In other words, if the annual premium exceeds a certain amount, the percentage of the premium in excess of the maximum figure is deducted.

Consider this example: Suppose that the amount of a multiple building premium subject to the volume credit is the amount of the premium exceeding $3,000. The credit is 5% of the premium excess. If the annual premium is $14,000, what's the credit?

First find the amount of the excess by subtracting $3,000 from the premium amount: $14,000 − $3,000 = $11,000. Multiply the excess by 5% to get the volume credit: $11,000 × 0.05 = $550. So the premium of $14,000 is reduced by $550 due to the multiple building volume credit.

### Exploring the multiple building dispersal risk

The point of taking advantage of a multiple building rate is that the risk is spread over several properties. But if one property is worth considerably more than the others, the risk is centered too much in one place. For example, if you have ten properties worth $10,000 each and one property worth $1,000,000, most of the worth is found in just one of the properties. The risk (value) isn't spread out very evenly. This is an extreme example, but I think it proves my point.

Tables that are used to determine how much of a credit is given for multiple buildings incorporate the number of buildings involved and the maximum percentage value that any one building has. It's best if no one building represents more than 20% of the total value of all the buildings.

Consider again the extreme example of owning ten properties worth $10,000 each and one property worth $1,000,000. Of course, the value of the million-dollar property has the greatest percentage of the worth. But what exactly is that percentage?

To find the percentage value of the million-dollar building, first find the sum of all the values of all the properties: 10 × $10,000 + $1,000,000 = $1,100,000.

Now divide $1,000,000 by that sum in order to get 0.909, or 90.9%. The million-dollar building is worth too much — it's almost 91% of the value of all the properties added together. You won't get any percentage credit for that package, because the one building is worth so much more than all the others together.

Now consider a more reasonable grouping of properties. Figure the percentage of the property worth the most. The eight buildings involved are worth the following amounts:

A: $500,000

B: $600,000

C: $700,000

D: $800,000

E: $1,200,000

F: $2,300,000

G: $2,400,000

H: $6,000,000

The total value of all eight buildings is $14,500,000. Property H has the greatest value, so divide its worth, $6,000,000, by the total value: $6,000,000 ÷ $14,500,000 ≈ 0.414, or 41.4%

You may qualify for a small discount. However, that building is still worth much more than the others.

### Computing the loss ratio

The *loss ratio* is a percentage determined by dividing the total losses by the total premiums during a particular time period. Here's the formula:

$$\text{Loss Ratio} = \frac{\text{Total Losses During the Time Period}}{\text{Premium Amount} \times \text{Number of Years}}$$

The loss ratio is used when determining the *loss modification factor,* which is an adjustment to the premium amount based on the loss history from the past few years. The idea is for the loss modification factor to result in a reduction in the premium.

Check out this basic example: Say that a company is renegotiating its multiple building insurance contract. The agent needs the loss ratio as part of the loss ratio modification factor. In the past 5 years, the company has had $60,000 in losses and has paid premiums of $20,000 per year. Using these figures, determine the loss ratio.

To compute the loss ratio, you divide the losses by the total premiums for the 5 years like this:

$$\text{Loss Ratio} = \frac{\$60,000}{\$20,000 \times 5} = \frac{\$60,000}{\$100,000} = \frac{6}{10} = 0.60 = 60\%$$

So, as you can see, the loss ratio is 60%. Depending on what the projected or expected losses were — and how much of a variance from those losses is considered above or below normal — an adjustment may be considered.

# Deferring premium payments

As you probably know, insurance is usually paid in advance. You pay for one year of coverage at the beginning of that year (or three years of coverage at the beginning of the three years). But, even though you sign up for insurance for three years, you may be able to defer payments for the second two years until the beginning of each of those years. Of course, the insurance company prefers that you pay for all three years at the beginning, so the company makes it more attractive for you to pay upfront with one payment. However, rest assured that if you aren't in the position to pay the total amount all at once, you can opt for three installments instead.

Before committing to one payment plan or another, you want to determine how much more it will cost you to pay in several installments than to pay all at once, at the beginning. The following example shows you what I mean.

Suppose that a three-year contract for property insurance has a total premium of $20,000. Two different deferred payment plans are possible: one for three payments and the other for six payments. The three-payment plan calls for 35% of the premium at the beginning of each of the three years. And the six-payment plan has you paying 20% of the premium biannually at the beginning of each six-month period. How much more do you pay for the insurance if you opt for one of the deferred payment plans (as opposed to paying it all upfront)?

If you pay three times, you pay 35% of $20,000 three times. You can find the amount of each payment by multiplying like this: $20,000 × 0.35 = $7,000. Three payments of $7,000 comes to a total of $21,000. You pay $1,000 more for the convenience.

If you pay six times, you pay 20% of $20,000 six times. Multiply to find the amount of each payment: $20,000 × 0.20 = $4,000. Six payments of $4,000 is $24,000. That's $4,000 more than you'd pay if you made one payment at the beginning of the three years.

# Chapter 15

# Planning for Success with Budgets

*A* *budget* is really just a plan. It incorporates what's anticipated in terms of revenues and expenses and provides a basis for comparison to what actually transpires. The figures in a budget are based on history and reasonable projections into the future. A budget helps a company make plans and motivate its employees, and it serves as a standard to measure performance.

Budgets usually are created for an entire year, but some companies prepare new budgets more frequently in order to make necessary revisions. A variance analysis (which I discuss later in this chapter) is used to measure just how different the budgeted items are from the actual. With the figures, the variance analysis also helps determine future actions.

In this chapter, you see different types of budgets at work, and I help you decide how to best use the one or ones that are right for you and your business. You also see different ways to compute, massage, manipulate, and analyze budgets and budget items.

## Choosing the Right Type of Budget

A budget can take on many different forms, so you have to pick the one or ones that work best for you. You can choose from several different types of budgets, including operating budgets, sales budgets, production budgets, cash budgets, and capital expenditure budgets. A budget can be as simple or complex as necessary to keep track of your company's financials and to provide the needed information in a timely fashion.

The following are some types of budgets that you may find used by companies or organizations. The different budgets can then be controlled by a master budget. The processes and mathematics found in each type of budget are similar, so you find examples from any that can be applied to the others. Here are some of the common types of budgets:

- **Operating budget:** An *operating budget* contains information on the resources that are needed to perform the operations and to provide reimbursement for services provided. A school system uses an operating budget to outline the salaries paid, cost of transportation and utilities, and so on.

- **Sales budget:** A *sales budget* outlines the projected sales levels and the amount of revenue expected to be collected. The sales budget helps production divisions determine how much product to provide.

- **Production budget:** A *production budget* includes beginning inventory from an earlier production period and projects the amount of production necessary to meet the new needs. The production budget is closely tied to the sales budget.

- **Cash budget:** A *cash budget* is responsive to the rest of the budgets in a company, because it's used to estimate the cash flow needs of the different divisions of the company.

- **Capital expenditure budget:** A *capital expenditure budget* contains not only amounts of money to be used for purchases during the current time period, but it also outlines the accrual of money for purchases in the future.

- **Master budget:** A *master budget* can be used to keep track of the different divisions of a company through their individual budgets. It can be used to do periodic measurements of how the different departments are doing with respect to their projections.

## *Cashing In on Cash Budgets*

A *cash budget* is used to plan for the anticipated sources of cash and the consequent uses of it. In other words, this type of budget consists of numbers reflecting the cash that comes in and the cash that goes out. A cash budget, which is used to ensure that enough cash is on hand, shows the following items:

- Entries of cash on hand
- Expenditures
- Receipts of cash

A cash budget allows a financial manager to determine whether a loan is necessary to meet the payment needs of the company — to compensate employees and cover bills, for example.

## Looking at an example cash budget

Table 15-1 shows an example cash budget for the first three months of the year.

| Table 15-1 | Cash Budget for LMNOP, Inc. | | |
|---|---|---|---|
| *Item* | *January* | *February* | *March* |
| Beginning cash balance | $50,000 | $120,000 | $240,000 |
| Budgeted cash receipts | $350,000 | $400,000 | $350,000 |
| **Total cash available** | **$400,000** | **$520,000** | **$590,000** |
| Disbursements – operations | $280,000 | $280,000 | $285,000 |
| Disbursements – capital expenditures | 0 | 0 | $400,000 |
| **Total cash disbursements** | **$280,000** | **$280,000** | **$685,000** |
| Minimum cash balance desired | $50,000 | $50,000 | $50,000 |
| **Total cash needed** | **$330,000** | **$330,000** | **$735,000** |
| Excess or deficiency | A: _____ | B: _____ | C: _____ |
| Financing – bank loan | 0 | 0 | D: _____ |
| **Ending Cash Balance** | **$120,000** | **$240,000** | **$50,000** |

The first three lines of the cash budget in Table 15-1 show the cash balance (at the beginning of the month), the anticipated receipts, and the sum of the beginning balance and receipts. The next three lines show *disbursements,* or expenditures. The operational disbursements are roughly the same throughout the three months — at about $280,000. A big capital expenditure (new equipment) accounts for the much larger total disbursement in March. The total cash needed is the sum of the disbursements plus the desired minimum cash desired. The excess or deficiency entries are missing from the table on purpose. The following example shows you how to determine these figures when you have the rest of the budget worked out.

Refer to Table 15-1 and determine the excess or deficiency for each month. Then determine the amount of money that needs to be borrowed in March to keep the cash balance at the desired level.

The amount that belongs on line A is the difference between the total cash available in January and the total cash needed in January. So subtract it out like this: $400,000 − $330,000 = $70,000.

Similarly, the amount that belongs on line B is the difference between the total cash available in February and the total cash needed. Subtract the figures: $520,000 − $330,000 = $190,000.

The ending balance for both January and February is the sum of the excess or deficiency and the desired $50,000 cash balance. Notice that the ending balance for each month is carried up to the beginning cash balance of the next month.

The amount on line C is a negative number, or deficiency. When you subtract the total cash needed from the total cash available, you get: $590,000 − $735,000 = − $145,000.

Negative numbers are written in parentheses in a balance sheet, so you write "($145,000)" on line C. By borrowing $145,000 and entering that number on line D, the deficit is taken care of, and the ending balance is exactly what you want it to be: 0 + $50,000 = $50,000. Now you can carry the $50,000 to the top entry for April.

## Comparing budgeted and actual cash receipts

Preparing a budget is one thing, but if you never look at it to see how the actual business is behaving compared to the projections, you won't end up making the necessary adjustments or corrections for the next year or business period. You want to know why the actual amounts don't match those that were budgeted. Is the difference minor or significant? Is someone not doing his or her job? Comparing the budgeted amounts to the actual amounts can help you gain a better idea of what's going on with the company or organization.

The quickest way to compare budgeted items with the actual amounts is to prepare your budget with adjoining columns that show each entry. You can also make separate columns for each month in a year's budget, with the *year-to-date amount* (the accumulated amount) in its own column. Table 15-2 shows an example expense budget with the first three months' actual numbers and the accumulated total for each item.

| Table 15-2 | Selling Expense Budget | | | |
|---|---|---|---|---|
| **Account** | **Budgeted for the Year** | **January** | **February** | **March** | **Year-to-Date** |
| Salaries | $39,600 | 3,300 | 3,300 | 3,300 | $9,900 |
| Commissions | $124,000 | 10,000 | 10,200 | 10,800 | $31,000 |
| Employee Benefits | $16,000 | 1,350 | 1,375 | 1,375 | $4,100 |
| Telephone | $5,000 | 400 | 500 | 550 | $1,450 |
| Travel | $80,000 | 6,000 | 6,000 | 6,000 | $18,000 |
| Supplies | $20,000 | 1,500 | 1,600 | 1,500 | $4,600 |
| **Total** | **$284,600** | | | | **$69,050** |

As careful as you may try to be, errors can occur in budget entries and totals. (You are human after all!) Amounts can (and should) be checked for accuracy. In Table 15-2, for instance, you can check that the year-to-date total is correct by first adding up the columns for January, February, and March and then by finding the sum of those three sums.

You also can do a reality check to see if the expenses are on track with what has been budgeted. Compare the first quarter actual expenses with one-quarter of the budgeted amount for the full year. However, remember that this type of comparison only works when expense items stay roughly the same each month. I further explain how to keep an eye on the accuracy of your budget in the following sections.

## Checking the addition

A good method to use when checking the addition in several columns is to compare the horizontal sums with the vertical ones. In a year-to-date accounting, you already have the horizontal sums all listed in the right-most column. After that, all you have to do is add the numbers in each vertical column, find the sum of all those results, and then compare the sum of all the vertical sums with the total from the right-most horizontal computation.

Here's an example to help you better understand what I mean: Check the year-to-date total in Table 15-2 by first adding up the expenses for the three months given, and then find the grand total so far.

The expenses in January add up to $22,550. The totals for February and March are $22,975 and $23,525, respectively. The sum of the expenses for those three months is: $22,550 + $22,975 + $23,525 = $69,050. This jibes with the sum in the last column.

### Determining whether you're on track for the year

If expenses or income are fairly level throughout a particular time period, you can budget accordingly and do periodic checks to see whether you're on track. You can check to see whether the percentage of the number of months that have passed corresponds to the same percentage of the budgeted amounts. (This system doesn't work as well when amounts fluctuate significantly from month to month.) The following example will help you understand how to perform this reality check.

Use Table 15-2 to determine how close the actual figures are to the budgeted figures for the year. The year-to-date totals in the right-hand column are for the first three months (which is the first quarter of the year). Compare each total with one-fourth of the entire budgeted amount for the year.

All you have to do to perform this check is divide each budgeted item by 4:

$$\frac{39,600}{4} = 9,900 \qquad \frac{124,000}{4} = 31,000 \qquad \frac{16,000}{4} = 4,000$$

$$\frac{5,000}{4} = 1,250 \qquad \frac{80,000}{4} = 20,000 \qquad \frac{20,000}{4} = 5,000$$

The salaries and commissions are right on target — exactly what was budgeted. Employee benefits and telephone expenses are slightly under the budgeted amount. Travel and supplies are also under the budgeted amount. It's always nice to have expenses be less than budgeted, but the manager may be asking why. Refer to the later section, "Measuring Differences with Variance Analysis," to see just what the differences may mean.

For those businesses that don't have a relatively steady revenue or expense stream, this check involves the different budgeted figures and their individual percentages instead of what part of the year has passed.

# Varying with a Flexible Budget

The term *flexible budget* doesn't exactly represent the type of budget that it really is. Budgets are pretty rigid. You have set numbers that are used for goal-setting, planning, and measuring performance. A *flexible* budget shows the expected behavior of costs at various levels of volume. A flexible budget allows you to put a *then* after each *if*. For instance, you say: "*If* the volume is 100,000, *then* the income is $4,000,000." Or you might say: "*If* the volume is at 60%, *then* the expenses are $40,000." A flexible budget allows you some flexibility in a rigid world.

Table 15-3 shows an example of a flexible budget in which various levels of production are represented by percentages and overhead costs are represented in dollars.

| Table 15-3 | | March Overhead Budget | | | | | |
|---|---|---|---|---|---|---|---|
| *Expense* | *80%* | *90%* | *100%* | *110%* | *120%* | *130%* |
| Office salaries | $2,500 | $3,000 | $3,000 | $3,000 | $3,500 | $3,500 |
| Office supplies | $1,400 | $1,600 | $1,800 | $2,000 | $2,200 | $2,400 |
| Utilities | $2,200 | $2,300 | $2,500 | $2,800 | $3,200 | $3,700 |
| Labor | $5,832 | $6,480 | $7,200 | $7,920 | $8,710 | $9,585 |
| Depreciation | $4,000 | $4,000 | $4,000 | $4,000 | $4,000 | $4,000 |

The increases in each category aren't *linear*. In other words, they don't go up by an equal amount for each percentage increase. Instead, the numbers are usually determined by history and various formulas based on the volume.

For percent volumes not shown on the table, you can interpolate or extrapolate to estimate the amount in the category. *Interpolation* and *extrapolation* are computations designed to determine a value between two given values or to project a value outside a list of given values — based on the given pattern. I explain each of these computations in the following sections.

### Interpolating for an in-between value

As I mention earlier, Table 15-3 shows various expenses based on percentages of volume. The percentages increase by 10% as the columns move from left to right, so if you need to estimate an expense when the percentage isn't a multiple of ten, you use interpolation. To understand how interpolation works, check out the following example.

Use the information in Table 15-3 to estimate the cost of labor if the production volume is 115%.

The production volume percentage, 115%, is halfway between 110% and 120%. So you need to find the expense amount that's halfway between the respective amounts on the table. The labor expense at 110% is $7,920; the expense at 120% is $8,710. To find the halfway point, you simply average the two numbers. In other words, add the amounts together and then divide by 2. So, first you add the amounts: $7,920 + $8,710 = $16,630. Then you divide that sum by 2, which gives you $8,315. Now you have your labor expense at 115%. Great work!

Not all production levels fall exactly halfway between two given values. You may want to determine an expense that's one-third or one-fourth of the way between two given values.

### Going off the chart with extrapolation

Interpolation is usually a bit more reliable than extrapolation. Especially when you make an estimate of a value that's significantly greater or smaller than any number you already have on a table, you run the chance of assuming something that just can't hold. However, sometimes you absolutely have to find an estimate of some value outside the table. To do so, you simply set up a proportion comparing the differences of the entries. Check out the following example to see what I mean.

Using Table 15-3, extrapolate the total overhead expenses if production is at 154%. *Tip:* The total of all expenses isn't given in the table, but adding any column gives you the total for that percentage.

To solve this problem, first find the total overhead expense at a production level of 130%. Then compare the difference in percentages between 130% and 154% with the differences between expenses at 130% and 154%. The difference between 154% and 130% is 24%. Let the unknown expense be represented by $x$.

So the total overhead at 130% is: $3,500 + $2,400 + $3,700 + $9,585 + $4,000 = $23,185. The decimal equivalent for 130% is 1.30, and the decimal equivalent for 24% is 0.24. Now you're ready for the proportion:

$$\frac{0.24}{1.30} = \frac{x - 23,185}{23,185}$$
$$0.24\,(23,185) = 1.30\,(x - 23,185)$$
$$5,564.40 = 1.30x - 30,140.50$$
$$35,704.9 = 1.30x$$
$$x \approx 27,465.31$$

Your estimate for the total overhead expense is $27,465.31.

# Budgeting Across the Months

It's great when you can budget income or expenses week by week or month by month without having to worry about one number intruding on the next. But realistically, this ideal situation doesn't necessarily happen in business. After all, the sales that you make during one month may not show revenue until the next month or even later. And the inventory that you amass during one time period may be predetermined by looking two or three months into the future — in other words, by looking to projected sales during that future time.

Of course, you don't create a budget using a crystal ball (but wouldn't it be great if you could?). Instead, you simply have to make educated projections. And these projections are important because a good budget helps you have a successful business. And creating a budget that incorporates occurrences overlapping during the year is useful in the planning. The following sections explain.

## Using revenue budgets to deal with staggered income

A company may budget for a certain volume in terms of sales. In this case, the inventory needs to be available, and the product must be paid for. However, not all sales are paid for immediately. Some sales result in cash payments, but others are credit sales that come with the promise to pay within a particular time period. This is where revenue budgets come into play.

A *revenue budget* indicates the revenue collected from sales over a period of time. It shows the total revenue for a particular month and the ways in which the revenue is expected to come in over time. Companies use these budgets to calculate the amount they need to assign as revenue from sales. I show you exactly what I mean in the following example.

Suppose a company prepares a revenue for the first four months of the year. Historically, half of the customers pay cash (they pay upon delivery). Of the other half, 60 percent of the credit customers pay during the month in which they make their purchase. Thirty-eight percent of the credit customers pay the next month, and half of those who haven't paid in that second month finally pay two months after the purchase. The remaining sales are considered bad debts and get written off.

If the projected sales for January, February, March, and April are $88,000, $90,000, $96,000, and $104,000, respectively, what should the company budget as revenue from those sales during the first four months of the year? What amount should be written off as a bad debt? In other words, how much does the company expect to never be paid for?

To determine the final figures, prepare a table that looks like Table 15-4, which shows the following items:

- ✔ The projected sales for each month

- ✔ The amount collected in that month (50% of the cash sales plus 60% of credit sales = 50% + 60%(50%) = 80%)

- ✔ The amount collected in the next month (38% of the credit sales from the previous month = 38%(50%) = 19%)

- ✔ The final amount collected two months later (1% of the credit sales = 1%(50%) = 0.5%)

- ✔ The amount written off — the bad debt (1% of the credit sales = 1%(50%) = 0.5%).

| Table 15-4 | | Revenue Collected from Sales | | | | |
|---|---|---|---|---|---|---|
| Month | Total Sales for Month | Collected in January | Collected in February | Collected in March | Collected in April | Bad Debt |
| January | $88,000 | $70,400 | $16,720 | $440 | | $440 |
| February | $90,000 | | $72,000 | $17,100 | $450 | $450 |
| March | $96,000 | | | $76,800 | $18,240 | |
| April | $104,000 | | | | $83,200 | |
| Total | | $70,400 | $88,720 | $94,340 | $101,890 | |

The amount collected in January represents 50% of the sales in cash and 60% of the credit sales. The total comes from this math:

$$50\%(\$88,000) + (60\% \times 50\%)(\$88,000) =$$
$$0.50(\$88,000) + (0.60 \times 0.50)(\$88,000) =$$
$$\$44,000 + \$26,400 = \$70,400$$

The amount collected in February comes from three sources: 38% of the credit sales from the previous month, 50% of the cash sales from February, and 60% of the credit sales from February. So the total is determined like this:

$$(38\% \times 50\%)(\$88,000) + (50\% \times \$90,000) + (60\% \times 50\%)\$90,000 =$$
$$(0.38 \times 0.50)(\$88,000) + (0.50 \times \$90,000) + (0.60 \times 0.50)\$90,000 =$$
$$\$16,720 + \$45,000 + \$27,000 = \$88,720$$

A computer spreadsheet and formulas are invaluable when setting these budget amounts. (Refer to Chapter 5 for more on the use of computer spreadsheets.)

## Budgeting for ample inventory

When you set a sales budget, you have high hopes for your inventory. You probably imagine that it will fly off the shelves and result in great revenue and, consequently, wonderful profit. But to have this type of success, you have to have enough in stock to meet your customers' needs. If you don't have the item on hand, your customer will go elsewhere.

It's difficult to plan exactly what your inventory needs will be, but a good budget based on past sales and reasonable projections goes a long way toward keeping your inventory at its proper levels. To get some practice with this type of planning, take a look at the following example.

Garry's Gifts prefers to carry an ending inventory amounting to the expected sales of the next two months. Expected sales for April, May, June, July, August, and September are the following (respectively): $40,000, $44,000, $56,000, $60,000, $72,000, and $66,000. If the inventory at the end of March was $84,000, what purchases need to be made in April, May, June, and July?

Start by preparing a table showing the sales and inventory needed (use the information given in the problem description to fill in the blanks). Table 15-5 shows you exactly what I mean. All the information given so far is included in this table. You enter the end needs by finding the sum of the projected sales for the next two months.

| Table 15-5 | Projected Sales and Needed Inventory | | | | | |
|---|---|---|---|---|---|---|
| Item | April | May | June | July | August | September |
| Beginning inventory | $84,000 | | | | | |
| Sales | $40,000 | $44,000 | $56,000 | $60,000 | $72,000 | $66,000 |
| End needs | $100,000 | $116,000 | $132,000 | $138,000 | | |

After you've created your initial table, find the purchases needed to make the beginning inventory equal to the budgeted sales for the next two months. Fill in the beginning inventory blanks by adding the sales for the two months after the blank (which is the same as the end needs amount from the previous month). For example, to find the beginning inventory for May, you add $44,000 to $56,000, which gives you $100,000. (***Note:*** Because you don't have the information for October, you won't be able to find the beginning inventory for September.) Table 15-6 shows each of the beginning inventory amounts you should have come up with.

Now add a row in the table to show the inventory at the end of the month. (You can label this row "Net.") The net inventory is obtained by subtracting the sales from the beginning inventory. Then add another row for the purchases that are needed to produce the needed inventory at the beginning of the month. (Label this row "Purchases.") Table 15-6 shows how your table should look with all this new information. Note that you can't fill in the last entries for August and September without the sales information from the months that follow.

| Table 15-6 | Projected Sales and Needed Inventory | | | | | |
|---|---|---|---|---|---|---|
| Item | April | May | June | July | August | September |
| Beginning Inventory | $84,000 | $100,000 | $116,000 | $132,000 | $138,000 | |
| Sales | $40,000 | $44,000 | $56,000 | $60,000 | $72,000 | $66,000 |
| Net | $44,000 | $56,000 | $60,000 | $72,000 | $66,000 | |
| End Needs | $100,000 | $116,000 | $132,000 | $138,000 | | |
| Purchases | $56,000 | $60,000 | $72,000 | $66,000 | | |

# Measuring Differences with Variance Analysis

Computations are made to determine the difference between budgeted values and the actual values that occur. The results of the computations are in the actual numbers or percentages. The numbers tell you how much of a difference or variance there is. It's up to you to determine the *why* or *who* (if a department head isn't performing as he/she should) of the variance and just how significant the difference is in the big picture.

You'll come across all sorts of variances when creating budgets. *Revenue variances* are the differences between budgeted and actual revenues. *Sales volume variance* is the difference between the actual and budgeted sales volume. *Cost variances* measure differences in cost.

In the following sections, I show you exactly how to compute variances and create a variance range that's acceptable for your business.

## Computing variance and percent variance

The variance between the budgeted estimate and the actual value is the number or dollar amount difference between what was anticipated and what actually happened. Numerically, variances can be positive or negative, meaning that they can have either a positive or negative impact on your bottom line. For instance, a negative variance is a good thing if you're talking about expenses (you spent less than expected), but a negative variance isn't so good when you're talking about income.

A better way to measure variances than with actual numbers is by using percentages. Why? Well, consider this: A difference of $10,000 is huge if the budgeted amount is $20,000, but a $10,000 variance doesn't have all that much impact if the budgeted amount is several million dollars.

Whether you're dealing with actual numbers or percentages, you always subtract the actual amount from the budgeted amount. You get a negative result if the actual number is larger than the budgeted number and a positive result if the actual number is smaller than the budgeted number.

Here's an example you can work for practice: Say that Suds and Stuff Enterprises budgeted $300,000 for production expenses, $200,000 for overhead, 150,000 units of production, and 450 visitors for April. The actual figures were $350,000 for production expenses, $250,000 for overhead, 200,000 units of production, and 200 visitors. What's the variance and percent variance for each item?

To work this problem, make a table showing the items (budgeted and actual), the variance, and the percent variance. To get the variance, you simply subtract the actual from the budgeted. To get the percent variance, divide the variance (which you got by subtracting) by the budgeted amount. (Percent differences — percent increases and decreases — are covered in Chapter 3 if you need a refresher course.) Table 15-7 shows you what your table should look like with all the information filled in.

| Table 15-7 | | | | April Variances |
|---|---|---|---|---|
| *Item* | *Budgeted* | *Actual* | *Variance* | *Percent Variance* |
| Production Expenses | $300,000 | $350,000 | +$50,000 | $\frac{50,000}{300,000} = \frac{1}{6} \approx 0.167 = 16.7\%$ |
| Overhead | $200,000 | $250,000 | +$50,000 | $\frac{50,000}{200,000} = \frac{1}{4} = 0.25 = 25\%$ |
| Production Units | 150,000 | 200,000 | +50,000 | $\frac{50,000}{150,000} = \frac{1}{3} \approx 0.333 = 33.3\%$ |
| Visitors | 450 | 200 | −250 | $\frac{-250}{450} = \frac{-5}{9} \approx -0.556 = -55.6\%$ |

As you can see in Table 15-7, even though the variances for production expenses, overhead, and production units are the same number, the percentages are actually quite different.

# Finding a range for variance

Making a budget is a matter of best guesses, good computations, and a bit of luck. Some differences are tolerable, but others may be too far off to be acceptable and run a good business. In order to monitor your variances without going nuts with anxiety, you can simply set a particular maximum difference or variance. That way you only have to be concerned when the difference is more than that maximum difference.

A budget is just a plan, and you're no crystal ball-gazer, so you don't really expect to be right on the penny with actual figures. You should expect to be a little above or below the budgeted amount when the final figures come in. How much of a difference is reasonable or acceptable? Businesses set a tolerance or range of values that they expect the actual values to fall into when the final numbers are tallied. For example, you may say that the revenues for April will be $40,000, plus or minus $2,000. So your range is from $38,000 to $42,000. As long as the actual numbers fall in that range, you don't really need to take another look at the revenue-producing units in the business.

Try out this example: Say that Best Builders has set a budget for the third quarter of the year and has determined that a 5% variance (up or down) on budgeted items is acceptable. Anything greater than 5% needs to be examined. The budgeted items for the third quarter are

- ✔ 1,000 units of raw material purchased
- ✔ 800 units of raw material put into process
- ✔ 500 hours of labor in production
- ✔ 400 hours of indirect labor
- ✔ $4,000 indirect costs
- ✔ $6.00 standard price of one unit of raw material

Determine the acceptable amounts for each unit if 5% variance is acceptable.

To accomplish this, find 5% of each item. Then subtract and add the 5% amount from the budgeted amount to determine the acceptable range. Table 15-8 shows the computations and results.

| Table 15-8 | Range Allowable with 5% Variance | | | |
|---|---|---|---|---|
| **Item** | **Budgeted** | **5%** | **–5%** | **+5%** |
| Raw materials | 1,000 units | 50 | 950 | 1050 |
| Raw materials processed | 800 | 40 | 760 | 840 |
| Hours of labor | 500 | 25 | 475 | 525 |
| Hours of indirect labor | 400 | 20 | 380 | 420 |
| Indirect costs | $4,000 | $200 | $3,800 | $4,200 |
| Price of material per one unit | $6.00 | $0.30 | $5.70 | $6.30 |

With the range of acceptable variances in place, you can quickly compare the actual values to see if they fall between the numbers desired.

# Chapter 16

# Dealing with Payroll

. . . . . . . . . . . . . . . . . . . . . . . . . . . . . . . . . . . . . . . . . . . . . . . . . . . . . . . . . . . . .

## In This Chapter

▶ Figuring employee earnings

▶ Deducting taxes, insurance, and other withheld expenses

. . . . . . . . . . . . . . . . . . . . . . . . . . . . . . . . . . . . . . . . . . . . . . . . . . . . . . . . . . . . .

*P*ayday can come monthly, bimonthly, weekly, or even daily, depending on your business's chosen pay schedule. On the other hand, some royalty payments for books or movies come only twice a year. And with some contract jobs, an employee gets paid only after finishing an entire project. In any case, monetary payment involves more than just the gross pay amount that a job is worth. After subtracting taxes, Social Security contributions, insurance, and whatever else appears in the deductions list, the net pay usually doesn't much resemble the lofty income figure.

In this chapter, you find multiple ways of figuring earnings. The main payment types you have to choose from are salaries, hourly paychecks, and commissioned payments. Salaried employees usually get a set fraction of the total amount each pay period. Hourly workers are paid according to the time put in during that pay period. Commissioned employees, on the other hand, usually are all paid differently. The variations on commission payments are many, and some can be rather creative.

Also important in this chapter are those pesky deductions. Figuring the amount to subtract from the base pay varies from deduction to deduction. Some deductions are percentages of the gross pay, and others are flat amounts. Some federally mandated deductions have caps or limits to the amount that can be deducted; others don't. You see a lot of multiplying and subtracting (unfortunately, *lots* of subtracting) in this chapter on how to determine net or take-home pay.

# Pay Up: Calculating Employee Earnings

Earnings from being gainfully employed are accumulated in many ways. Here's a rundown of the types of employees and how they earn their pay:

- ✔ **Salaried employees:** These folks earn a set amount of money per year, and the amount is divided evenly between the number of pay periods. Salaried employees sometimes earn extra pay for extracurricular activities, or they may receive end-of-year bonuses.

- ✔ **Seasonal or part-year salaried employees:** These employees may opt to take their salaries over the entire 12 months, or they may choose to be paid just during the 3 months or 9 months that they work.

- ✔ **Hourly employees:** The hourly worker is paid a set rate per hour, but there's often the opportunity to work overtime at a higher rate.

- ✔ **Commissioned employees:** A commissioned employee is paid based on some performance. They aren't paid according to the number of hours worked, however. Instead, they're paid according to income activity or production. Salespersons are frequently paid on a commission basis; a base pay may be included and added to the commissions earned. One variation on a flat percentage rate of commission is having built-in increases for meeting certain levels of production.

Determining the earnings for all these folks may sound daunting, but don't worry. The following sections show you exactly what you need to know to calculate all four types of earnings.

## Dividing to determine a timely paycheck

To find out how much a person working for a set salary should earn each pay period, you simply have to use your division skills. You also have to determine how often you plan to pay your employees. Most employers pay monthly, bimonthly, or weekly. You can find pros and cons for each of these payment arrangements. Check out this list to get an idea of each arrangement:

- ✔ **Monthly:** Paying monthly means issuing paychecks 12 times each year. Most bills come monthly as well, so monthly payments make budgeting and keeping track of bills much easier. The downside to monthly payments is that the months don't have the same number of days, so payday comes on different days of the week. Also, for some employees, a month between paychecks may seem like a really long time.

- ✔ **Bimonthly:** Bimonthly paychecks provide fresh cash to employees more frequently, but it also means more bookkeeping for the employer.

> ✔ **Weekly:** Paying weekly allows for more frequent paychecks on the same day of the week, which is good for the employee. But, like bimonthly checks, this arrangement can be a bookkeeping nightmare for the employer. A bonus for employees on this payment plan: Four months each year have five pay periods.

Don't forget that even though salaried employees receive a set amount each pay period, they also may be eligible for extra pay or bonuses that are added to just one or a few paychecks.

Try your hand at calculating the amount of a monthly paycheck with this example: Sandy is a manager of Save-A-Bunch, and she has been a wonderful employee for the store this year. Sandy filled in for a sick evening manager for six weeks (in addition to her regular job), and she was paid extra for the added responsibility. She's also eligible for a bonus at the end of the year.

Using the following information, determine the amount that Sandy is paid each month: Sandy's yearly salary is $47,700, and her bonus, which she receives in December, is $6,000. She was paid $3,600 for taking on the extra work for six weeks. The extra work covered all of November and half of December.

To determine Sandy's monthly paycheck amount, first find her gross base pay by dividing $47,700 by 12, which gives you $3,975 per month. In November, Sandy gets extra earnings for the additional evening work. So divide the extra amount she was paid by the number of weeks she worked to get the amount per week: $3,600 ÷ 6 = $600.

Now, multiply that $600 by 4 (the number of weeks in November) to determine her extra November payment: $600 × 4 = $2,400. To calculate her total November paycheck, simply add her gross base pay with the extra cash: $3,975 + $2,400 = $6,375.

Don't forget that in December Sandy gets two more $600 payments totaling $1,200 plus the bonus. So her December paycheck is $3,975 + $1,200 + $6,000 = $11,175. Merry Christmas, Sandy!

Now try the following example, which has a bit of a twist on interim work.

Mike is a professor at a local university. He has an opportunity to teach a summer interim course for extra pay. The arrangement with the university is that extra compensation is paid at the rate of 7% of a person's base salary for each three-credit course. Mike currently makes $63,000 per year, and the course is a four-credit course. How much will he be paid to teach during the interim?

Sounds like a toughie doesn't it? Don't worry; it really isn't that bad. First, determine the percentage rate for a four-credit course. Use a proportion that has the number of credits in the left-hand fraction and the percentages in the right-hand fraction. (Refer to Chapter 4 if you need a review of setting up proportions.) Let $x$ represent the unknown percentage. Cross-multiply and solve for $x$. Here's what your math should look like:

$$\frac{3 \text{ credits}}{4 \text{ credits}} = \frac{7\%}{x\%}$$

$$\frac{3}{4} = \frac{0.07}{x}$$

$$3x = 0.07\,(4)$$

$$x = \frac{0.28}{3} \approx 0.0933 = 9.33\%$$

So Mike will be paid 9.33% of his salary. To figure out how much he'll be paid to teach the interim course, simply multiply his current salary by the percentage (in decimal form), like this: $63,000 \times 0.0933 \approx \$5,878$.

## Part-timers: Computing the salary of seasonal and temporary workers

Seasonal or temporary workers may be employed for a few weeks, a few months, or a major portion of the year. And depending on what type of workers they are, they're paid differently. Seasonal workers, for instance, are usually paid a salary during the months that they work. Others, such as temporary workers, who typically work for nine or ten months each year, may opt to be paid equal amounts each month, every month of the year.

Suppose Clyde retired from a management position with a large heavy-equipment manufacturing company. Now he does consulting with the company and works for eight months each year. He can choose to be paid monthly during those eight months, or he can even out his payment and be paid each of the 12 months for that 8 months' work. Clyde is being paid $4,500 per month for his consulting. How much would he get per month if he chooses the 12-month plan?

First find the total amount Clyde receives for his work by multiplying the payment per month by the number of months he works: $4,500 \times 8 = \$36,000$. Now, you just have to divide that total by 12 to find out how much he'll be paid per month: $36,000 \div 12 = \$3,000$.

## Determining an hourly wage

Being paid an hourly rate means that for every hour that an employee works, she's paid a set amount — and if she doesn't work, she isn't paid. Sounds pretty simple, but be careful; you can find contingencies and arrangements that make one hourly wage different from another. I explain some arrangements in the following sections.

### Working with an adjustable hourly rate

Businesses often choose an adjustable hourly rate for their employees. With this arrangement, the rates eventually change with the number of months or years of employment. In other words, employees are rewarded for their continued loyalty with wage increases.

Hourly rates are adjusted according to some prescribed formula. The formula may add a set number of dollars after a particular amount of time, or the formula may incorporate a percentage increase.

Say that Pete's Produce has four hourly workers on staff; each of these employees works 35 hours per week. Pete's pays an hourly rate of $8.00 for starting workers, and it increases the rate by $0.50 cents per hour after the worker has worked at the market for 6 months. Workers who have been with Pete's for a year or more get an increase of $0.10 cents for every 6 months they have been employed after the first 6 months. How much do Amy and Ben each make at Pete's if Amy has been employed for two years and Ben for six years?

First, determine Amy's hourly rate. At the end of her first year, she was earning: $8.50 + $0.10 = $8.60 per hour. Add two more increases, and Amy's rate is $8.60 + $0.10 + $0.10 = $8.80 per hour.

Ben has been employed at Pete's for six years, so his rate is: $8.50 + $0.10 + $0.10 + $0.10 + $0.10 + $0.10 + $0.10 + $0.10 + $0.10 + $0.10 + $0.10 = $9.60 per hour.

What if Ben had been there for 10 years or 15 years? Don't you think there's a better way to figure out his rate than adding up all those $0.10-cent additions? Well, sure! You could use the following formula:

$$
\text{Hourly rate} = \begin{cases} \$8.00 \text{ if:} & n < 6 \text{ months} \\ \$8.50 \text{ if:} & 6 \text{ months} < n < 12 \text{ months} \\ \$8.50 + \$0.10 \left[\left[ \dfrac{n-12}{6} \right]\right] \text{ if:} & n > 12 \text{ months} \end{cases}
$$

where $n$ is the number of months and the fraction in the [[ ]] is rounded down to the greatest whole number.

The *greatest integer function* takes a number and rounds it down to the largest integer that's smaller than the number. For instance, performing the greatest integer function on 3.14, you get: [[3.14]] = 3. An example with a fraction is

$$
\left[\left[ 7\frac{8}{9} \right]\right] = 7
$$

So, in effect, the greatest integer function just lops off any fraction or decimal part of a positive mixed number.

### Choosing an hourly rate with an overtime option

Another arrangement you may choose for your business is an hourly rate with the option of overtime. With this option, after a certain number of hours each week, overtime rates kick in. These rates increase the amount per hour an employee earns for the hours worked in overtime. Being paid overtime is a nice perk for hourly workers. However, it's both a blessing and a bane; you get paid at a higher rate, but you *do* have to work more hours.

The overtime rate is usually either a set amount that's greater than the employee's usual rate, or it's a percentage increase. You have to determine which rate works best — especially considering how large a percentage increase will be for employees earning at the top of the rate scale.

Check out this example: A local fast-food restaurant pays its employees $7.50 per hour for the first 40 hours, time-and-a-half for the next 10 hours, and double-time for hours exceeding 50 hours during any one week. How much do Ronald and Donald make if Ronald works 48 hours and Donald works 63 hours during a one-week period?

At $7.50 per hour, time-and-a-half is $7.50 × 1.5 = $11.25, and double time is $7.50 × 2 = $15. Given this information, you can see that Ronald earns $7.50 for the first 40 hours and $11.25 for the next 8 hours (48 – 40 = 8 hours). So Ronald earns ($7.50 × 40) + ($11.25 × 8) = $300 + $90 = $390.

Donald, on the other hand, earns $7.50 for the first 40 hours, $11.25 for the next 10 hours (50 – 40 = 10 hours), and $15 for the last 13 hours (63 – 50 = 13). So Donald makes ($7.50 × 40) + ($11.25 × 10) + ($15.00 × 13) = $300 + $112.50 + $195 = $607.50.

## Taking care of commission payments

Working on a commission means that an employee earns a percentage of some sales or activity. This type of arrangement is usually set up to reward hard work and good salesmanship. Imagine the car salesperson, for example; the more cars he sells, the more he earns.

Commission structures vary. The commission rates may be flat, single rates, or graduated as the total sales increase. A base pay may also be included. And managers sometimes earn commissions based on their salespersons' efforts. I explain each of these structures in the upcoming sections.

### Keeping an even keel with a flat commission rate

A *flat commission rate* is one that stays the same no matter how much or how little a person produces in sales. The *commission* — or payment to the salesperson — is a percentage of the total amount of revenue brought in by the salesperson.

To compute the amount of commission earned by a salesperson, you multiply the total amount of the sale or sales by the decimal number corresponding to the commission percentage. If the commission rate is 10%, for instance, sales of $10,000 earn the salesperson $10,000 × 0.10 = $1,000. A commission acts as an incentive to sell more. With a flat percentage of sales, a salesperson gets the same commission for selling one $100,000 piece of jewelry as she does for selling five $20,000 pieces.

Ready for an example? Take a look at this one: Carole works in a local shoe store and earns a commission of 6% on all the products that she sells. During the five days that she worked last week, she sold 211 pairs of shoes for a total of $8,239.37 in gross sales. What's her commission?

To find Carole's commission, simply multiply her gross sales by her commission percentage (in decimal form): $8,239.37 × 0.06 ≈ $494.36.

### Graduating to a graduated commission schedule

A *graduated commission schedule* rewards the salesperson for the magnitude of their sales. For instance, as the total in sales increases, the commission rate increases as well. Depending on the arrangement, the increase either applies to the sales over a certain number or to all the sales that have accumulated over a time period.

For example, a person's commission rate may be 5% on the first $10,000 in sales, 10% on the next $40,000 in sales, and 15% on all sales over $50,000. Or, the arrangement may be that the salesperson earns 5% on all sales up to $10,000, 10% on all sales if they sell up to $50,000, and 15% on all sales if they sell more than $50,000. In each case, you multiply the sales that qualify for a particular rate by the decimal corresponding to the percentage.

Try your hand at this example: Imagine that Carole has worked in a local shoe store for several years and is negotiating a better commission rate for herself — as a condition of her staying with that store. Carole suggests to her boss that a suitable weekly commission rate would be one of the following:

> 6% commission on the first $5,000 in sales and 10% on all sales exceeding $5,000
>
> 6% commission on the first $5,000 in sales; if her total sales exceed $5,000, she receives an 8% commission on *all* of her sales

If Carole typically sells about $8,500 in shoes each week, which commission rate is better for her?

Using the first rate, which is 6% on the first $5,000 and 10% on the next $3,500 ($8,500 − $5,000 = $3,500), Carole earns ($5,000 × 0.06) + ($3,500 × 0.10) = $300 + $350 = $650.

Using the second rate, where she gets 8% commission on the total when she exceeds $5,000, Carole earns $8,500 × 0.08 = $680. So, she does better with the second rate.

Refer to the previous problem where Carole is negotiating a new commission rate. In that problem, Carole makes $8,500 and is better off taking 8% for her entire commission (rather than 6% on the first $5,000 and 10% on the sales exceeding $5,000). How much does she need to sell each week for the first commission schedule to be better?

To find out, construct an equation in which the commission paid at the first rate (6% on the first $5,000 and 10% on any sales more than $5,000) is equal to the second rate (8% on all sales when she exceeds $5,000). Let the amount in sales be represented by $x$. Your equation and math should look like this:

$$6\% (5,000) + 10\% (x - 5,000) = 8\%x$$
$$0.06 (5,000) + 0.10 (x - 5,000) = 0.08x$$
$$300 + 0.10x - 500 = 0.08x$$
$$0.10x - 200 = 0.08x$$
$$-200 = -0.02x$$
$$\frac{-200}{-0.02} = x$$
$$x = 10,000$$

As you can see, if Carole starts selling more than $10,000 each week, she's better off with the first rate. For instance, if Carole sells $11,000 during the week, the first commission rate earns her $5,000 × 0.06 + $6,000 × 0.10 = $300 + $600 = $900.

Using the $11,000 sales figure and the second rate, where she gets 8% on the total amount, she earns $11,000 × 0.08 = $880.

### Combining a base salary and a commission

Many businesses pay their salespersons a small base salary to be in the store, on the lot, or out in the field, and then they add on a commission to reward for volume of sales. This arrangement of base pay plus commission works best when it's necessary that someone is available to customers whether or not sales are made. Of course, the business doesn't want to pay someone who just stands around and doesn't try to sell anything. But the base pay serves to compensate the salesperson for some of his time.

To compute the total payment for this type of arrangement, you determine how much the person earns in commission fees by multiplying the sales total by the percentage. Then you add the commission to the base salary for that time period.

A cutlery company sells its products through home visits and catalogue sales only. The compensation arrangement is that salespersons receive $15 per sales presentation and 10% of the gross sales. Salespersons also get $5 for each customer referral they obtain during a sales presentation. What does a salesperson earn during a week when he makes 20 sales presentations, obtains 35 referrals, and sells $3,500 in products?

To calculate the salesperson's compensation, multiply the 20 presentations by $15, multiply the 35 referrals by $5, multiply the $3,500 by 10%, and then add the three results together: $(20 \times \$15) + (35 \times \$5) + (\$3,500 \times 0.10)$ = $300 + $175 + $350 = $825.

### Splitting a commission

Splitting a commission occurs when two or more people share in the commission earned for a particular sale or sales. Salespersons work under this arrangement when it's to the mutual benefit of all those involved to share in the payment. Splitting the money with someone who helped accomplish the transaction is better than getting nothing. After all, the sale couldn't have happened otherwise. You see the commission-splitting arrangement in real estate and other high-priced items such as yachts.

When it comes to real estate, different areas of the country have different arrangements for commissions, but the basic premise is that the listing agent (the person who contracts with a seller to sell her property), the selling agent (the person who finds a buyer and sells the property), and any managers and support staff all have a claim to part of the total commission.

The share in a split commission is really just a percent of a percent. You multiply the amount of the sale by the commission rate (changed to a decimal value) by the share of the commission (changed to a decimal value).

Try this real estate problem to get an idea of how you can calculate payments when splits are involved: Say, for example, that in Springfield, the standard commission on a home sale is 7%. Of that percentage, half goes to the selling agent's company and half to the listing agent's company. Now say that in a particular transaction, where a home sells for $230,000, both the selling and listing agents work for the same company. The arrangement with that company is that 20% of all commissions go to the office manager and 15% go to office expenses; the rest go to the agents. What do the selling and listing agents earn for this transaction?

First, find the total commission, which is 7% of the selling price of the home: $0.07 \times \$230,000 = \$16,100$. Now determine the manager's share of the commission and the office expenses share of the commission. The manager's share is 20% of the commission: $0.20 \times \$16,000 = \$3,220$. The office expenses amount, which is 15% of the commission, is: $0.15 \times \$16,000 = \$2,415$. The next step is to subtract the manager and office expense amounts from the total commission: $\$16,100 - (\$3,220 + \$2,415) = \$10,465$. Finally, you can split what's left for the agents. Splitting this difference two ways, each agent gets $5,232.50 for the transaction.

# Subtracting Payroll Deductions

Reality hits home each time an employee receives a paycheck. Why? Because before the check is cut, all sorts of pesky deductions are subtracted from the earnings. Deductions are made *before* the paycheck is issued perhaps to make the process less painful, or maybe to be sure that the payments are made to the insurance company, the federal government, and the other taxing bodies.

Depending on the deduction, the subtraction is either a percent of the total income, a percent of the taxable income, a set monthly amount, or an occasional debit. Each deduction has its effect to the *income* before there's any *outgo*.

## Computing federal income tax

The federal income tax was authorized in 1913 with the passage of the 16th Amendment. And ever since that time, April 15 — tax day — has come to hold special meaning for most Americans.

The rate for federal income tax is a graduated amount going as high as 35 percent of a person's taxable income. The expression *taxable income* has been interpreted and reinterpreted — and likely will be interpreted yet again. What this term refers to is the amount of your income that Uncle Sam can tax. However, you also get to disallow some of your income to be subject to the tax. You just have to keep up with the changes and rules for these adjustments and plan accordingly.

### Figuring income tax on taxable income

An employee's taxable income is arrived at after doing all the computations and manipulations necessary on their income tax form. They subtract deductions and exemptions and come up with as small a number as possible to multiply by the percentage rate. (Well, you do this, don't you? Am I the only one who scrapes the barrel for deductions?) The rates used in the multiplication depend on the employee's status (single, head of household, married) and how much taxable income he has. Worksheets for employers and employees help determine how much should be deducted from their paycheck to pay their federal, state, and/or local income tax.

Be sure to use a tax computation worksheet carefully and accurately. You don't want to take out too much of a deduction, because no interest is earned on the money (even though it's refunded later). And you don't want to take out too little money, because penalties are charged for significant underpayment.

### Calculating corporate rates

Most corporations pay income tax at rates varying from 15% to 39% of their taxable income. After the tax rate is determined (by looking at the total income for the corporation), the income tax rate then applies to the entire income amount. This is unlike the personal income tax rate in which a lower rate is applied to the first few thousand dollars, a higher rate applied to the next few thousand, and so on. Corporations pay the same high rate for the entire amount of their income. For instance, look at these rates:

| *If the taxable income is . . .* | *The tax rate is . . .* |
| --- | --- |
| $0–$50,000 | 15% |
| $50,001–$75,000 | 25% |
| $100,001–$335,000 | 39% |
| $335,001–$10,000,000 | 34% |

To calculate the tax amount, multiply the total income by the decimal value corresponding to the percentage rate.

Using the previous table of rates, determine the difference in income tax paid by a corporation whose taxable income is $330,000 and one whose taxable income is $340,000.

To do so, multiply the taxable incomes by their respective tax rates:

$330,000 × 39% = $128,700

$340,000 × 34% = $115,600

The corporation that makes $10,000 more in taxable income pays about $13,000 less in taxes.

# Determining Social Security contributions

The FICA tax, which is named after the Federal Insurance Contributions Act, consists of both Social Security and Medicare taxes. Social Security taxes and Medicare taxes are paid partly by the employee and partly by the employer — each pays half of the total contribution.

The Social Security part of the tax has an upper limit — the employee pays tax on his or her income only up to a certain level. In 2007, for example, the top amount being taxed was $97,500; so the cap for taxes was 6.2% of $97,500, or $6,045. The Medicare part of the FICA tax has no cap. All earnings are subject to the 1.45% tax rate for Medicare.

So, in 2007, an employee may have had Social Security contributions of up to $12,090 (half contributed by the employee and half contributed by the employer) and no limit on the Medicare contribution, which would equal 2.9% of the income (half paid by the employee and half paid by the employer).

Try your hand at an example for a bit of practice. Say that an employee earns $4,400 in taxable income each month. Using the 2007 rates that I mention earlier, determine how much is deducted for Social Security and Medicare during the month of November.

To solve this problem, first determine whether the employee has reached or passed the upper level of making Social Security contributions. If you multiply $4,400 by the first ten months (which gives you $44,000), you see that the total is nowhere near the $97,500 cap. So the full contribution to Social Security will be made, which means you have to multiply the employee's taxable income by the percentage cap to find the amount that will be deducted: $4,400 \times 6.2\% = \$272.80$.

This amount is deducted from the employee's paycheck, and the business contributes another $272.80. The Medicare contribution in this scenario is 1.45% of $4,400, which comes out to be $63.80. The company matches that amount, too.

# Part V

# Successfully Handling the Math Used in the World of Goods and Services

The 5th Wave          By Rich Tennant

©RICHTENNANT

"Our profit statement shows a 13 percent increase in the good, a 4 percent decrease in the bad, but a whole lot of ugly left in inventory."

## In this part . . .

A business that deals with handling goods or services has to purchase or produce the items, set a price that earns a profit, sell the items, and then report the profit for possible taxation. In addition to these processes, the businesses have to figure discounts and markups and overhead and depreciation. Plus, it pays to keep good track of inventory. The bigger the business, the better the records and processes need to be. Find the mathematical help you need in the chapters of this part.

# Chapter 17

# Pricing with Markups and Discounts

*E*very day radio, television, and newspaper ads bombard customers with wonderful opportunities to spend money. Customers may sometimes feel that it's difficult to separate the good from the bad with these retail opportunities. And for the merchant, pricing can be both a science and art. It's necessary to mark up items to earn a reasonable profit and cover costs, but prices that are too high lead to lower sales. On the flip side, discounting items can bring in more customers. However, too deep a slash can hurt the company's bottom line.

Keeping track of all these price changes can be tricky. So you need to be familiar and comfortable with percents, decimals, and fractions. Being familiar with these will help you catch obvious errors — and save money.

In this chapter, you discover how to determine markups, discounts, and the general game plans for establishing the new prices of items. New prices are most frequently determined by using fractions and percents in such situations as having a half-off sale or giving customers 60% off of every orange-dot item. (If you're feeling percentage anxious, check out Chapter 3, where you find the basics of percent increase and percent decrease.)

## Examining Markups and Retail Prices

The amount of money that a store pays for merchandise is always less than the price that customers pay. It's a known fact that items are marked up from a wholesale price to a retail price. Likewise, a company that provides services may hire a contractor at one amount and mark up the cost of the contractor's work to the end customer. All of this markup is done in the name of profit. In the following sections, I explain everything you need to know in order to deal with markups and retail prices.

This section shows you how to work back and forth from the cost of an item to the price that's charged. To help ease your way, I use *cost* when talking about what the merchant or service provider pays and *price* for what the customer pays. Because *markups* are based on the *price* and profits involve both the *price* and *cost,* I use the two different words to keep things straight.

## Making sense of markups

Unless you've been living in a bubble and have never ventured out to the store, you've had experience with markups. All companies mark up their prices in order to make a profit. The *markup* of an item for sale is the amount of money the retailer adds to the cost of the item to bring it up to the *selling price*. This price is also often referred to as the *retail price* (though for purposes of this book, I only refer to it as selling price). Here's an equation to help you understand this process:

Cost + Markup in dollars = Selling price

Try putting this equation to work with an example.

If the cost of an item is $50 and the selling (retail) price is $90, what's the markup?

To solve this problem, all you have to do is plug the numbers into the equation, which gives you $50 + markup = $90. Using basic algebra, you can see that the markup is $40.

If the markup of an item is $27.50 and the selling price is $90, what's the cost of the item?

To solve, plug your numbers into the equation, like so: cost + $27.50 = $90. After the algebra is hashed out, the cost comes to $62.50.

## Understanding how markups are a percentage of retail prices

Sometimes you may want to figure the markup percentage of an item. For instance, you may be doing some comparison shopping for equipment for your small business. You find the same item at two different stores for two different prices. You assume that the cost to the merchants was the same, so you figure out what the markup is at each store. Knowing the markup that a merchant is using helps you decide where to shop. When a markup percentage is computed, it's based on the retail price, not the wholesale cost. After you find the markup percentage, you can also then find the markup in dollars. I explain how to compute both in the following sections.

### Figuring the markup as a percent

If you know the markup and the retail price of an item, you can figure the markup percentage. (You find information on cost, markup, and retail price in the earlier section, "Making sense of markups.") If you know the retail price and markup percentage, you can figure the markup and the cost of the item. In other words, the cost, markup, markup percentage, and retail price are all intertwined.

Here's the formula showing how to compute the markup percentage:

$$\text{markup percentage} = \frac{\text{dollar amount of markup}}{\text{retail price in dollars}}$$

Here's an example to help you get accustomed to the markup percentage equation: Say that a refrigerator sells for $750. The markup is $250. What's the markup percentage for the refrigerator?

To solve this example, divide the markup amount by the retail price. Here's what the math would look like:

$$\frac{250}{750} = \frac{1}{3} = 0.333\ldots$$

So you have about a 33⅓% markup.

### Calculating the markup in dollars

The markup in dollars is an important calculation, because after you know it, you can determine the cost of the item to the retailer. Here's the formula:

Markup in dollars = Markup percentage × Retail price in dollars

Suppose a car sells for $23,000 and the markup is 5.5%. What's the dollar amount of the markup? After you figure the markup in dollars, determine how much the dealer paid for the car.

To solve, simply plug the numbers into the formula, like so:

Markup in dollars = 5.5% × $23,000 = 0.055 × $23,000

= $1,265

Now you can figure out how much the dealer paid for the car. To do so, use this formula:

Retail price – Markup in dollars = Amount dealer paid

So, simply plug in your numbers to get the amount paid: $23,000 – $1,265 = $21,735.

# Working with the retail price

The process of determining the retail price of an item is somewhat backward from what you may expect. As I note earlier in this chapter, the markup is a percentage of the retail price, so how in the world can you determine the retail price from the markup percentage or the retail price from the cost of the item if you need a percentage of that retail price? It does sound like a conundrum, but the situation really isn't as bad as it sounds. I provide you with a formula to use when determining the retail price.

### Finding the retail price

If you know the cost and markup of an item, you can determine the retail price of the item using the following formula:

$$\text{retail price} = \frac{\text{dollar cost of item}}{1 - \text{markup percentage}} = \frac{\text{cost of item}}{100\% - \text{markup percentage}}$$

Recall that 100% is the same as the number 1. (Chapter 2 shows you how to change percentages to decimal numbers.)

Here's an example for you to try out: Say that it costs you $45 in materials and labor to produce a pair of beaded moccasins. You plan on a 40% markup. What will you charge for the moccasins?

To solve, simply plug the numbers into the formula. Your equation should look like this:

$$\text{retail price} = \frac{\text{cost of item}}{100\% - \text{markup percentage}}$$

$$= \frac{\$45}{100\% - 40\%}$$

$$= \frac{\$45}{60\%} = \frac{\$45}{0.60} = \$75$$

As you can see, you need to sell the moccasins for $75 if you want a 40% markup.

You may be asking: "Isn't that the same thing I'd get by finding 40% of $45 and then adding that percentage to the $45?" Nope, and this is the perfect opportunity to show you why not! After all, take a look at this: If you find 40% of $45, you get 0.40 × $45 = $18. Add $18 to $45, and the sum is $63. This isn't anywhere near the $75 price that was determined properly.

Take a look at another example: What's the retail price of an appliance if the markup dollar amount is $50 and the markup percentage is 25%?

For this problem, use the following formula:

Markup amount in dollars = Markup percentage × Retail price

Substituting in what you know, you get $50 = 25% × retail price. Now divide each side by 25% to solve for the retail price. Your equation should look like this:

$$\frac{\$50}{25\%} = \frac{\$50}{0.25} = \$200$$

As you can see, the appliance is selling for $200. Its markup is $50, which is 25% of the $200 retail price.

## Figuring out the cost of the item

The *cost* of an item is what the retailer pays a supplier or what a manufacturer charges for all the materials and labor. If certain parts of your billing information aren't available, you can still determine the cost of an item based on the other information that you do have.

### Determining cost from retail price and markup

Suppose that you have retail price and markup information. You can definitely determine cost with this information. To do so, play around with the following formula:

Cost + Markup in dollars = Retail price

Check out the following example to see what I mean.

Say that a desk lamp (which would be perfect in your office) has a price tag of $930. The markup on lamps is 60%. What did the lamp cost the supplier?

To solve, you need to determine the markup amount in dollars before substituting into the retail price in dollars formula, which looks like this:

Retail price in dollars = Dollar cost of item + Markup amount in dollars

To find the markup amount in dollars, you use this formula:

Markup amount in dollars = Markup percentage × Retail amount

When you fill your numbers into the markup amount in dollars formula, your equation will look like this: 60% × $930 = $558. Next you go ahead and plug into the retail price in dollars formula to get this: $930 = cost of item + $558. Finally, subtract $558 from each side of the equation to get the cost of the lamp, which is $372.

### Determining cost after an increase

Imagine that you're trying to determine this year's cost of an item based on last year's information, because you want to do an increase in volume this year. This calculation is a bit more complex than the one in the previous section, but it's fairly simple in small steps.

To calculate cost in this situation, all you have to do is determine the actual money amount of the increase based on the percent increase. If you're starting with the retail percent increase, you compute the cost using the appropriate formula. If you're starting with the actual cost, you're home free — you just have to compute the percent increase. (Chapter 3 provides information on percent increases and decreases if you need a refresher.)

Wrap your brain around this concept with an example: Last year, you did $20,000 volume at retail for one particular item in your inventory. This year, you plan to increase your retail volume of this item by 20%. If your markup percentage is 50%, how much will your cost be for this item? After you find the new retail volume, determine how much this will cost and how much more your cost will be this year than it was last year.

First determine what your retail volume will be, factoring in the 20% increase. Twenty percent of $20,000 is $4,000, so you plan on a volume of $24,000. Now you can use the formula for determining retail price from cost, and then solve for the cost. Here's what the equation should look like:

$$\text{retail price} = \frac{\text{cost of item}}{100\% - \text{markup percentage}}$$

$$\$24,000 = \frac{\text{cost}}{100\% - 50\%} = \frac{\text{cost}}{50\%} = \frac{\text{cost}}{0.50}$$

$$0.50 \times \$24,000 = \frac{\text{cost}}{0.50} \times 0.50$$

$$\$12,000 = \text{cost}$$

After multiplying each side of the equation by 0.50, you get that the cost will be $12,000. How much more is your cost of this item than last year? Well, if you did $20,000 and your markup is 50%, the markup amount was 0.50 × $20,000 = $10,000. So you're planning spending an additional $2,000 on the item this year.

# Exploring Discount Pricing

Consumers are attracted to opportunities to save money, and merchants need to keep older inventory moving out the door to make room for shiny new goodies. So you may want to take advantage of this situation and put a sign in the window showing that all items are discounted by 40 percent. You may have a rush on your hands — or so you hope. It all depends on how the discount is computed.

For example, sometimes discounts are computed at the cash register, letting the technology do the work. But customers are more often attracted to items when they see the actual marked-down price on the tag. However, you may run into some problems when showing the actual marked-down price. One problem is that you have to change all the tags back to the original price when the sale is over. Also, you may have a customer who questions the math, and then you need to be able to defend the number — gently and diplomatically, of course. In the following sections, I explain the many ways to discount products.

## Discounting once

You can probably guess what the formula for discount price is, but here it is just in case:

Discount price = Original retail price – Amount of the discount

When the discount is a certain dollar amount, you simply subtract that amount to get the new price. More commonly, though, the discount price is based on a percentage of the original price of the item.

### Subtracting a dollar amount discount from the price

Discounting the price of an item means that you subtract the discounted amount from the full retail price. Unless you sell just one type of item or service, the amount of the discount usually varies from item to item. After all, it wouldn't make much sense to advertise $100 off all items if the merchandise currently varies in price from $1,000 to $200. You may run across some instance where a blanket discount makes sense — the $200 item at $100 off could be your cost leader, for example — but, in general, different items have different discounts.

Suppose a $4,500 big-screen television is going on sale at a 20% discount. What's the new price of the television?

To work this example, you have to find 20% of $4,500. You can do so with the following equation (refer to Chapter 2 to see how to change percentages to decimals): $0.20 \times \$4,500 = \$900$. The discount is $900, so now you subtract that amount from the original retail price for the discounted price of the television: $\$4,500 - \$900 = \$3,600$.

The math is fine and worked well for this example, but what if you have to figure the discount prices of more than 100 items in your store? This method of multiplying each price by 20% and then subtracting the discount from the original retail price is a bit cumbersome. If you have a bunch of items that need discounted, you may be better off using a one-step process. The method shown in the next section makes much more sense when you have many discount prices to figure.

### Multiplying by the net percentage

If you have to find the discount prices of many items, the previous option of doing each computation individually could take days. But don't worry. You have another option. With this option, you multiply by the net percentage, which is done in one step instead of with the two-step process of figuring the discount and then subtracting.

The *net percentage* is the difference between 100% (the full retail price percentage) and the discount percentage. Look at this computation of the discount price in terms of percentages of the discount, the original price, and the new price:

New, discounted price = Original retail price (100%) – Discount (% of original)

Let $x$ represent the original retail price, and you get

New, discounted price = $x$(100%) – $x$(Discount % of original)

To make the equation even simpler, you pull the $x$ out of each term on the right and multiply it by the difference. You get the discounted price by multiplying the original retail price by the difference between 100% and the discount percent of the original retail price, like this:

$x$(100% – Discount % of original)

The value in the parentheses is the net percentage. So, if the discount is 40%, the equation for the new, discounted price is:

New, discounted price = $x$(100% – 40%)

= $x$(60%)

Try this example: Merchandise currently priced at $430, $670, $550, and $375 is to be discounted by 30%. Find the new prices for each.

First, find the percentage multiplier by subtracting 100% – 30% = 70%. Seventy percent has a decimal equivalence of 0.70. Now just plug the numbers into the equation, like so:

$430(0.70) = $301

$670(0.70) = $469

$550(0.70) = $385

$375(0.70) = $262.50

Note that the prices found by multiplying by 70% are the same that you'd get by multiplying the original price by 30% and subtracting the product from that original. For instance, 30% of $430 is worked like this: $430(0.30) = $129.

Then for the subtraction: $430 – $129 = $301. As you can see, you get the same answer with less hassle.

Computations like the ones in the previous example are even easier to do with a spreadsheet. With a spreadsheet you can enter the original prices and a formula to do the multiplication. Chapter 5 has more on using spreadsheets for repeated calculations.

## Successive or multiple discounts

Sometimes a discounted price is discounted again (and again). You can compute a final discounted price that's the result of more than one discount by determining the price from the first discount and then by discounting the discount price. This process can be repeated as many times as you want. Or you can also apply a quick formula to the original price and do the discounting in one fell swoop.

If you're running a sale campaign and will be discounting your discount over the period of days or weeks, you'll want all the gory details — all the discounted prices in steps. See the upcoming section, "Performing successive discounting," for details regarding this situation. However, if you're after the bottom line and want the best and final discount, check out the later section, "Using a formula for multiple discounts." You'll save time and energy with this method.

### Performing successive discounting

If a store is going out of business or a charity sale needs to reduce its inventory by the end of the weekend, successive discounts are made over a period of time. The multiple or successive discounts are each made on the already-discounted price. For example, here are the formulas:

> Original price – Amount of first discount = First intermediate price
>
> First intermediate price – Amount of second discount = Second intermediate price
>
> Second intermediate price – Amount of third discount = Third intermediate price

This method can go on and on as many times as you need.

Ready for an example? Check this one out: Julep's Closet is a large fundraiser held every year at which gently used donated items are sold over a four-day period. On the first day, all items are sold at full price. On the second day, all items are discounted by 10%. On the third day, the remaining items are discounted by another 30%. And, on the final day, everything that's left is discounted by 60%. What's the final price of a coat that was unsold on the last day, if the original price was $60?

Did you notice that the discounts all add up to 100%? Does that mean that the final price is $0? No, of course not. Here's how the math goes:

**Original price – Amount of first discount = First intermediate price**

$60 – 10%($60) = $60 – 0.10($60) = $60 – $6 = $54

**First intermediate price – Amount of second discount = Second intermediate price**

$54 – 30%($54) = $54 – 0.30($54) = $54 – $16.20 = $37.80

**Second intermediate price – Amount of third discount = Third intermediate price**

$37.80 – 60%($37.80) = $37.80 – 0.60($37.80) = $37.80 – $22.68 = $15.12

The final price of the coat on the final day of the sale is just over $15.

Now look at the computations using the same percentages on the same coat, but instead of subtracting the amount of the discount each time, use the net percentage (check out the earlier section, "Multiplying by the net percentage," for more details). The net percentage is found by subtracting the discount percentage from 100%. So, using the information from the example, you get the following net percentages:

100% – 10% = 90%

100% – 30% = 70%

100% – 60% = 40%.

Then you simply fill in the formulas:

**Original price × Net percentage of first discount = First intermediate price**

$60 × 0.90 = $54

**First intermediate price × Net percentage of second discount = Second intermediate price**

$54 × 0.70 = $37.80

**Second intermediate price × Net percentage of third discount = Third intermediate price**

$37.80 × 0.40 = $15.12

### Using a formula for multiple discounts

If you're going to figure a discount of a discount price, and you're in a huge hurry, you can skip the method of successive discounts as shown in the previous section and simply apply a single formula to get the end result — quick, fast, and in a hurry.

Here's how the formula works: Consider an item costing $100 that's first discounted by 10% and then discounted by 20%. The net percentages are 90% and 80%, respectively. Your corresponding decimals are 0.90 and 0.80.

Now you simply multiply the two net percentages together and then multiply the resulting product by the original retail price. For example, your math would look something like this:

$$0.90 \times 0.80 = 0.72$$

$$\$100 \times 0.72 = \$72$$

Try your hand at this example: In the earlier section, "Performing successive discounting," a $60 coat was discounted by 10%, 30%, and finally 60%. Using the easier formula that I discuss in this section, determine the final price.

Multiply the net percentages (90%, 70%, and 40%, respectively) together, like so: $0.90 \times 0.70 \times 0.40 = 0.252$. Now multiply the original $60 by 0.252 and you get $15.12.

## Going for a volume discount

Businesses are often granted a nice reduction in the price of items purchased if they place larger orders for the item or items. In the world of advertising, newspapers and magazines usually discount the price when more lines of advertising are purchased or a larger number of advertisements are contracted for over the period of a year or more.

The discount for buying a larger number of items may apply to all the items purchased, or it may apply just to the items over a certain number. Some discounts could even be retroactive; for instance, you may pay full price and then, when a certain level of purchasing is reached during some time period, the discount is applied to later purchases and you get a rebate on earlier purchases. You may decide that you need to determine discounts like these, so read on for details.

### Discounting the entire order

To grasp the concept of discounting on an entire order, say that a retail supplier provides widgets to several stores in the area. Widgets cost the retailer $4.00 each. If a single order is for more than 300 but less than 500 widgets, the cost per widget for the entire order is $3.80. If the order is for 500 or more widgets, the cost is $3.50 per widget. Look at this pricing using a *piecewise* function, where $n$ is the number of items ordered:

$$\text{Total Cost} = \begin{cases} \$4.00 \times n & \text{if } n \leq 300 \\ \$3.80 \times n & \text{if } 300 < n < 500 \\ \$3.50 \times n & \text{if } n \geq 500 \end{cases}$$

A *piecewise* function is a rule that changes with input volume or amounts. You use piecewise functions to accommodate volume purchases; you build in the discount pricing. Utilities, for example, use piecewise functions where the input is the number of units of electricity or water. The Internal Revenue Service uses piecewise functions where the income is the input and the tax percentage increases with the income.

What's the cost of an order of 470 widgets? How much is saved by ordering all 470 widgets at once, instead of ordering 200 one month and 270 the next month?

An order of 470 widgets allows the retailer to use the middle pricing scheme in the formula (piecewise function) for determining the total price. You simply have to plug in the numbers to get the price of the order of 470 widgets: $3.80 × 470 = $1,786.

If the retailer had ordered 470 widgets in two smaller orders, necessitating using the more expensive pricing, the math would be different: $4.00 × 200 + $4.00 × 270 = $1,880. You can calculate the savings like this: $1,880 − $1,786 = $94.

## Paying in graduated amounts

To better understand the logistics of paying in graduated amounts, check out this example: Suppose a supplier of canned chicken feet (a delicacy in Taiwan) has a pricing scale based on the number of cans bought by a merchant. For instance, if the merchant purchases more than 1,000 cans, the cost per can for all those *after* the first 1,000 is $0.85. The first 1,000 cans cost $0.97 each. If the merchant purchases more than 2,000 cans, the cost per can for all those *after* 2,000 cans is $0.60.

Now suppose that a specialty food shop orders 5,000 cans of chicken feet from the supplier whose pricing formula is shown here:

$$\text{Total Cost} = \begin{cases} \$0.97 \times n & \text{if } n \leq 1,000 \\ \$970 + \$0.85 \times (n - 1,000) & \text{if } 1,000 < n < 2,000 \\ \$1820 + \$0.60 \times (n - 2,000) & \text{if } n \geq 2,000 \end{cases}$$

In the formula, $n$ represents the number of cans ordered. The second line of the formula — $970 + $0.85 × (n − 1,000) if $1,000 < n \leq 2,000$ — incorporates the cost of the first 1,000 cans at $0.97 by multiplying $1,000 × $0.97$. Then it subtracts those 1,000 cans from $n$ and multiplies the difference by the lower cost per can. The same process is used in the third line of the formula: The $1,820 comes from adding $970 for the first 1,000 cans plus $850 for the second 1,000 cans.

Knowing all that information, can you determine what an order of 5,000 cans of chicken feet will cost?

Because the order is for 5,000 cans, you use the third line of the formula. Your math should look like this:

$$\text{Cost} = \$1,820 + \$0.60(5,000 - 2,000)$$
$$= \$1,820 + \$0.60(3,000)$$
$$= \$1,820 + \$1,800$$
$$= \$3,620$$

By dividing $3,620 by 5,000, you get an average of about $0.72 cents per can.

### Getting a retroactive discount

The best way to get a handle on retroactive discounting is to work through an example. So check out the following scenario.

The supplier of a popular bubble gum charges $15 per box for 24 packs of gum. After a retailer has ordered 1,000 boxes of gum in one calendar year, the rest of the boxes of gum cost $14.70 — and the price is retroactive for all purchases that year. Say, for example, that one retailer places orders for 300, 400, and 500 boxes of gum. That retailer paid full price for the first two orders. What's the charge for the third order?

The third order puts the retailer over the 1,000 target, because 300 + 400 + 500 = 1,200. The retailer owes $14.70 per box for the last order of 500 boxes, which comes to $14.70 × 500 = $7,350. The retailer now gets a retroactive discount on the first 700 boxes. The retailer paid $15 per case, but it now gets a discount of $0.30 per case retroactively. Here's what the equation would look like: 700 × $0.30 = $210. Now subtract the $210 from the retailer's current bill of $7,350 to find out what he owes: $7,350 − $210 = $7,140.

### Taking advantage of discounts for lines and pages of advertising

Newspapers and other periodicals would like for you to be a regular advertiser, so discounts are offered depending on how much advertising space you use each year. You may sign a contract so that the discount is in effect from the beginning of the year. Other times, though, you may just pay as you go because you don't know how much of a regular customer you'll be. The discount, of course, is the way to go.

Say, for example, that a newspaper charges $1.00 per line for advertising in the classified section. If you contract for 5,000 lines per year, you get a 10% discount; 10,000 lines per year earns a 12% discount; and 20,000 lines per year gets a 15% discount. What's the total cost of advertising if you contract for 18,000 lines per year?

Contracting for 18,000 lines puts you at the 12% discount level. The net percentage cost per line is 100% − 12% = 88%, or $0.88. To get the total cost of advertising, your math would look like this: 18,000 × 0.88 = $15,840.

---

## Discount Cards

Many local supermarkets, pharmacies, and other retail stores want you to sign up for their free discount cards. The cards are used to identify and reward loyal and frequent customers by giving discounts on select items. But they really do beg the question: Do you really save money? Or do the stores simply increase the prices in general so that the discount really just brings the price down to what it would have been without a card?

The savvy consumer checks out the prices. Consider a curious shopper who wants to determine whether the discount card really is worth the hassle. This shopper purchases about 20 particular items (just normal, weekly purchases)

at the store with the discount card. Then the shopper buys the same 20 items at a store that offers a discount card (but the shopper doesn't have that card). Finally, the shopper buys the same 20 items at a store that doesn't offer a discount card at all.

So, the shopper now has three weeks' worth of purchases, but he's set for almost a month. What do you suppose the shopper determines? Without naming any names, places, or stores, I do declare that the shopper pays the least at the store that doesn't offer the discount card at all. You don't believe me? Try it. Do I have a discount card? Sure, but that isn't what draws me to that particular store for certain purchases.

---

Notice how close you are to the next level of discount. If you increase your contract for just 2,000 more lines, you'd be at the 15% discount level. Will it cost you more or less total advertising money to contract for 20,000 lines?

A 15% discount means the net percentage is 85%. Now multiply to get the total cost of advertising: $20,000 \times 0.85 = \$17,000$. If you subtract the two sets of advertising costs ($17,000 - \$15,840 = \$1,160$), you can see that the cost of the extra 2,000 lines is an increase of $1,160 over what you originally planned to spend. The advertising cost goes up by about 7.3% (divide $1,160 by $15,840) when you add the additional 2,000 lines, but the amount of advertising space increases by about 11% (divide 2,000 by 18,000). It may be worth the added cost to change the contract. (If you need help with percent increases, refer to Chapter 3.)

# Chapter 18

# Calculating Profit, Revenue, and Cost

. . . . . . . . . . . . . . . . . . . . . . . . . . . . . . . . . . . . . . . . . . . .

## In This Chapter

▶ Balancing revenue and cost for a profit

▶ Discovering costs and how to predict them

▶ Breaking even in business

. . . . . . . . . . . . . . . . . . . . . . . . . . . . . . . . . . . . . . . . . . . .

The basic idea of making a profit on a venture comes early to young people. Maybe you had a lemonade stand out in front of your house, selling cups of that sweet stuff to the passing neighbors? Or maybe you were involved in organizations or camps that had you fundraising by selling items such as cookies, popcorn, or candy. And if you weren't lucky enough to be involved in any of those activities, there were always the school supply sales run by the student council or the football game concession stands run by the glee club. The point is that most people have had an opportunity to see profit, revenue, and cost at work in a firsthand situation before they even set foot into the business world.

The general concept of profit is fairly straightforward and relatively easy to understand. A *profit* is simply the difference between the amount of money you receive for goods and the amount it costs you to produce them. What makes the process of determining profit all the more interesting is when you begin to break down the costs into the different types and then investigate the different ways that the costs are handled. This investigation is a far cry from the lemonade stand where you added up how much the lemons, sugar, and cups cost.

Coming out ahead is a definite (and pretty obvious) goal for the business-person. And, before you get ahead, you have to *break even*. In other words, you want to earn at least what you're spending. You want to reach the break-even level of sales and exceed it by as much as possible. Determining the break-even point involves getting a handle on the revenue and the costs — and doing the appropriate computations. In this chapter, you discover the basis of profit, revenue, cost, and breaking even — and how they all work together for you.

# Figuring Profit from Revenue and Cost

Profit is a good thing. It's what makes businesses run, become successful, and grow. In order to earn a profit, you have to have fewer expenses than you do income. Your cost of providing merchandise or a service has to be less than what you're paid to provide the goods or service.

Here's the general formula relating profit, revenue, and cost:

Profit = Revenue − Cost

## Understanding how volume affects profit, revenue, and cost

One of the simpler models or examples of profit, revenue, and cost is driven completely by the *volume* — the number of items produced and sold. In this model, the number of items produced and sold is the variable. The price being charged for an item and the cost to produce an item are constant numbers. This simple model doesn't really apply to big, complicated businesses, but it holds for some of the basic ventures.

For instance, if you can produce lemonade for $0.12 per cup — which includes the cost of the materials and the paper cup — and if you sell the lemonade for $0.50 per cup, this equation gives you your profit, $P$, when you sell $x$ cups of lemonade:

$$P(x) = 0.50x - 0.12x$$
$$= 0.38x$$

The notation $P(x)$ in the profit equation is called *function notation*. The notation $P(x)$ reads: *P of x*, where $x$ is the input amount used to evaluate the equation, $P$, to get an answer.

The variable $x$ represents the number of items, so in the revenue part of the equation it's the price multiplied by $x$. In the cost part of the equation, it's the cost per item multiplied by $x$. The difference is found with algebraic subtraction.

Continuing with the lemonade stand scenario, try out this example: What's the profit on 100 cups of lemonade if it costs you $0.12 for the materials per cup and you can sell the lemonade for $0.50 per cup? Also, how many cups of lemonade do you have to sell to make a profit of $50?

Using the formula $P(x) = 0.50x - 0.12x$, you get $P(x) = 0.38x$. If you let $x = 100$, you put the 100 in for the value of $x$, multiply 0.38 by 100 to get: $P(100) = 0.38(100) = 38$. So you make $38 for selling 100 cups of lemonade.

To determine how many cups of lemonade you need to sell to make a $50 profit, you replace $P(x)$ in the formula with 50, and then you solve for $x$, the number of cups. So your equation looks like this: $50 = 0.38x$.

Now you divide both sides of the equation by 0.38 to get about 131.6. You aren't going to sell anyone a part of a cup of lemonade, so round up to 132 cups. Using the formula, you get $P(132) = 0.38(132) = 50.16$. This amount exceeds $50, but if you use 131 in the formula instead, you get: $P(131) = 0.38(131) = 49.78$. Of course, this is under the target amount.

## Seeing how price is sensitive to demand

Many people are bargain hunters. They shop around until they find the best price for a particular item; some even wait until an item goes on sale before buying it. The merchant, on the other hand, wants to be competitive in pricing, but it also wants to make as much as the market will bear.

A common characteristic of price in relation to the number of items sold is this: the lower the price, the more the items tend to sell. It takes a fine balance, of course, between lowering or raising prices enough but not too much. The following sections show you how to find that balance.

### Charting the effects of lowering prices

A merchant can make use of an informal customer survey or perhaps the talents of a marketing firm to determine how to price an item. Also factored into setting the price is the cost to produce the item.

For example, say that a company usually sells 2,400 doodads a month when the price is $15 per doodad. The company determines that it can increase the number of doodads sold by 400 for each $0.50 drop in price. Table 18-1 illustrates the effect of this drop in price on the number of doodads sold.

| Table 18-1 | Determining Sales Due to Price Reductions | |
| --- | --- | --- |
| *Number of Reductions (n)* | *15 – 0.50n = Price* | *2,400 + 400n = Number Sold* |
| 0 | $15 – 0.50(0) = 15.00$ | $2{,}400 + 400(0) = 2{,}400$ |
| 1 | $15 – 0.50(1) = 14.50$ | $2{,}400 + 400(1) = 2{,}800$ |
| 2 | $15 – 0.50(2) = 14.00$ | $2{,}400 + 400(2) = 3{,}200$ |
| 3 | $15 – 0.50(3) = 13.50$ | $2{,}400 + 400(3) = 3{,}600$ |
| 4 | $15 – 0.50(4) = 13.00$ | $2{,}400 + 400(4) = 4{,}000$ |
| 5 | $15 – 0.50(5) = 12.50$ | $2{,}400 + 400(5) = 4{,}400$ |

The information in Table 18-1 is informative when it comes to determining the effects of lowering prices, but an equation is even more helpful. Why? Because it can be written into a spreadsheet, graphed, and worked into equations to analyze the situation. Read on to the next section to discover this helpful equation.

### Using an equation for demand

A *demand equation* is used to represent the effect of a drop in price. Demand equations are projections or predictions of what the sales volume will be based on the price or what the price should be based on estimated sales volume. These equations are produced by marketing analysts who study sales numbers and create models in the forms of formulas. A demand function is used instead of a set, unchanging price in the revenue part of a profit function. For example, instead of saying that the revenue is $R = 10x$, meaning that each item will sell for $10, you use the revenue function $R = px$, showing that the price, $p$, changes with the volume, $x$.

Consider the situation in the previous section where a company determines that it can increase the number of doodads sold by 400 for each drop of $0.50 in price. In that situation, the following demand equation can be used:

$$x = 2400 + 400\left(\frac{15 - p}{0.50}\right)$$

where $x$ is the number of items expected to be sold if the price is $p$. Because a profit equation uses the price being charged, $p$, and the number of items being sold, $x$, you solve the equation for $p$ like this:

$$x = 2400 + 400\left(\frac{15 - p}{0.50}\right)$$

$$x = 2400 + \overset{800}{\cancel{400}}\left(\frac{15 - p}{\underset{1}{\cancel{0.50}}}\right)$$

$$x = 2400 + 800(15 - p)$$

$$x = 2400 + 12000 - 800p$$

$$x = 14,400 - 800p$$

$$800p = 14,400 - x$$

$$p = \frac{14,400 - x}{800}$$

Using the equation to find the price, $p$, you see that when $x$ is the usual 2,400, the price is

$$p = \frac{14,400 - 2,400}{800} = \frac{12,000}{800} = 15$$

At a level of 2,800 sales, the price is

$$p = \frac{14,400 - 2,800}{800} = \frac{11,600}{800} = 14.50$$

These values are the same that you find in Table 18-1.

You can see from the demand equation that there's a limit on what $x$ can be. For instance, if you let $x$ be a number greater than 14,400, you get a negative price. You can specify that the equation applies only to volume within a certain range. You may say that $x$ must be less than 1 million or between 100 and 500, or some such constraints. Such guidelines are set to keep the price above a particular number so that the profit is a positive number.

The following example shows you how to use a demand equation to find a profit. Suppose that doodads sell for $15 regularly, but the company's marketing division has determined that doodads sell better if the price is reduced. The demand function that reflects an additional 400 sales for each $0.50 reduction is

$$p = \frac{14,400 - x}{800}$$

where $p$ is the price when $x$ doodads are sold. Constructing a doodad costs the company $8.50. What's the profit when the company sells 2,400 doodads? How about when it sells 3,600 doodads and 5,200 doodads?

To figure this problem, you simply use the profit equation (Profit = Revenue − Cost). You can fill in the revenue part of the formula by multiplying the price per doodad, $p$, by the number of items, $x$. Then you subtract the cost, 8.50$x$. Your equation looks like this: $P = px - 8.50x$. Now insert the demand equation for the $p$:

$$P = \left( \frac{14,400 - x}{800} \right) x - 8.50x$$

The profit equation can be simplified with algebra, but I'm just going to leave it for now to show the substitutions for $x$. I'll let $x = 2,400$, $3,600$, and $5,200$. Here are the resulting profits using each number:

$$P(2400) = \left( \frac{14,400 - 2400}{800} \right) 2400 - 8.50(2400) = 15,600$$

$$P(3600) = \left( \frac{14,400 - 3600}{800} \right) 3600 - 8.50(3600) = 18,000$$

$$P(5200) = \left( \frac{14,400 - 5200}{800} \right) 5200 - 8.50(5200) = 15,600$$

From these equations, you can see that the profit increases when the number of doodads sold goes from 2,400 to 3,600. Remember that the doodads cost $15 when 2,400 are sold. When 3,600 doodads are sold, you know that the price has been reduced to $13.50 each. But what happened with the sales level of 5,200? Shouldn't increased sales mean more profit? Nope! That's not the case when the price has been reduced too much. When 5,200 doodads are sold, the demand equation says that the price has been lowered to $11.50. The lower price has offset the increased sales.

## Sorting out variable and fixed costs

You can find two basic types of cost in production or services: fixed cost and variable cost. *Fixed cost* is the part of the total cost that's incurred whether you produce anything or not. Fixed cost may be for the use of a building or structure, payments to any salaried employees, insurance, and so on. In simpler models, the fixed cost is the *start-up cost.* This type of cost is incurred once no matter how many items are produced. For example, when you decide to sell imprinted T-shirts for a sports team, your start-up cost is for the screen-print template, which you pay for only once.

The *variable* cost is the part of the total cost that changes as the number of items being produced changes. When it comes to variable cost, the more items you produce, the more you have in cost.

With a lemonade stand, for example, the more cups of lemonade you put together to sell, the more it costs in sugar, lemons, and paper cups. With large companies, on the other hand, the cost may vary in another way: The amount charged per item may decrease (or the profit margin may increase). Why is this the case? Well, when supplies are bought in large quantities, discounts for volume can make the expenditure per item smaller.

### Figuring in fixed cost

The fixed cost of a small enterprise, such as the sale of fundraiser T-shirts or a stall at a flea market, is a set number that counts against the profit no matter how many items are sold. In a way, the fixed cost is the risk for being in a venture. You pay it whether you sell anything or not.

When including fixed cost in the profit formula, you add the fixed amount to the variable amount. Here are the formulas you may use:

Total Cost = Variable cost + Fixed cost

Total Cost = Price × number of items + Constant number

Try your hand at this problem: The booster club for the Bradley Braves decided to fashion a commemorative T-shirt to celebrate the team's success in the NCAA basketball tournament. The template for the screen printing costs $90. The T-shirts themselves cost $8 each, but it also costs $2 per shirt to get them printed. The booster club decides to sell the T-shirts for $12 each. What's the club's profit if it sells 300 T-shirts?

First you have to find the total cost for the T-shirts. This total includes both the variable and fixed costs. Start with the variable costs: $8 per shirt plus $2 per shirt for printing. So your total variable cost is $10 per shirt. In the cost formula, this variable part of the total cost is $10x$, where $x$ is the number of T-shirts. The fixed cost is $90. So the total cost is written like this: $C(x) = 10x + 90$.

The revenue part of the profit equation is written as $12x$, because the shirts will be sold for $12 each. By putting the revenue and cost portions into the profit equation and simplifying, you get

$$P(x) = 12x - (10x + 90)$$
$$= 12x - 10x - 90$$
$$= 2x - 90$$

Now replace the $x$ in the equation with 300, which is the amount the club anticipates to sell. This gives you

$$P(x) = 2(300) - 90$$
$$= 600 - 90$$
$$= 510$$

So the club will make $510 if it sells 300 T-shirts.

Notice that the club is in trouble if it doesn't sell at least 45 T-shirts. With fixed costs of $90, the club members need to sell at least $90 worth of T-shirts so they don't have a negative result in the profit formula. You solve for the number of items needed to break even by setting the profit function equal to 0 and solving for the variable. In this case, $P(x) = 0$ when $2x-90 = 0$. You get $2x = 90$, or $x = 45$.

Here's another example to get you up to speed: A plant food manufacturer has expenses involving mixing, packaging, and the monthly rent for the equipment. Rather than buy the machinery, the manufacturer pays $2,000 per month to rent the equipment needed to process the plant food. The manufacturer also pays $12,000 to produce 5,000 gallons of plant food. The plant food is mixed with water immediately before being packaged in 8-ounce bottles. During this packaging process, which costs $6,000, the manufacturer adds 4,000 gallons of water to the plant food. The 8-ounce bottles sell for $2.29 each. What profit function is used to determine the profit if the manufacturer sells 10,000 bottles of plant food in one month?

To find out, use the profit equation. The revenue part of this equation is easy: You simply multiply $2.29 by $x$, where $x$ is the number of bottles. The cost consists of the fixed part, $2,000 per month, and the variable part, which is the amount it costs to produce each bottle of food. To determine the cost per bottle, first figure out how many 8-ounce bottles are filled with the 5,000 gallons of plant food plus 4,000 gallons of water. *Tip:* One gallon = 4 quarts = 16 cups = 128 ounces; one cup = 8 ounces.

If you add the plant food and water, you get 9,000 gallons of fluid that fills the 8-ounce bottles. An 8-ounce bottle holds a cup of fluid, and 1 gallon is the same as 16 cups, so multiply the number of gallons by 16. You get $9,000 \times 16 = 144,000$ bottles. It costs $12,000 to produce the plant food and $6,000 to package the bottles, so a total of $18,000 is needed. To find the cost per bottle, divide $18,000 by 144,000 to get 0.125, or 12½ cents per bottle.

The profit function is written by subtracting the cost from the revenue. So for the plant food, your function looks like this:

$$P(x) = 2.29x - (0.125x + 2,000)$$
$$= 2.29x - 0.125x - 2000$$
$$= 2.165x - 2000$$

So the profit on 10,000 bottles in one month is

$$P(10,000) = 2.165(10,000) - 2000$$
$$= 21,650 - 2000 = 19,650$$

A $19,650 profit is a pretty decent one. Looks like a lot of money can be made in feeding plants.

### Examining how variable costs change with volume

The simplest method of figuring the cost of the materials and labor that are needed to produce an item is to take the total amount expended in materials and labor and divide it by the number of items. This computation gives you a fixed value that's then used in a profit function as the cost per item.

For example, if you determine that it costs $40 per item to produce a particular product, you use the number 40 as a multiplier of the variable that represents the number of items. In the profit equation, you can use something like $40x$ for the cost to show that each item of the $x$ number of items produced is multiplied by the individual cost.

Sometimes, though, the costs change with the number of items being produced. For instance, when a company buys supplies in bulk, it often gets a discount, which reduces the amount spent per item. This changing cost is represented by a function or relationship in the form of an equation. You start with the usual cost, such as $40 per item, and then subtract some amount as a reward for doing volume business.

For example, the $40 cost per item could be reduced by the fraction 0.0003 of a dollar, for every item you purchase. You could use this equation for the cost, $c$, per item: $c = 40 - 0.0003x$.

The greater the number of items, the less the cost. Why? Because something is always being subtracted from the base cost (in this case 40). With this particular example equation, not much is subtracted (unless $x$ is fairly large). But the numbers can eventually make a difference, and it may be necessary to put a cap or limit on the amount that can be subtracted. And remember, the reduced price applies to all the items purchased.

GO FIGURE

Ready for an example? Try this one: Suppose a ceramic business makes decorative jack-o'-lanterns every year and sells them for $55 each. The kiln-firing process during which all the jack-o'-lanterns are prepared costs $100 in electricity. Let's just say that the materials needed — clay and glazes — have a cost *per item* according to the equation $c = 40 - 0.0003x$, because the supplier rewards volume spending. What's the total cost for $x$ jack-o'-lanterns, and what's the profit if the business makes 100 of these decorations?

To solve this problem, remember that the total cost equals the variable cost plus the fixed cost. And the variable cost is the cost per jack-o'-lantern multiplied by the number of jack-o'-lanterns, $x$. So the variable cost is represented with $cx$. But you replace the $c$ with $40 - 0.0003x$ to get:

$$cx = (40 - 0.0003x)x$$
$$= 40x - 0.0003x^2$$

Now add the fixed cost of $100 to get the total cost for $x$ jack-o'-lanterns: $40x - 0.0003x^2 + 100$.

Great work! To find the profit that the business makes producing and selling 100 items, write the revenue as $55x$ and then subtract the cost. The profit is represented with this equation:

$$P(x) = 55x - (40x - 0.0003x^2 + 100)$$
$$= 55x - 40x + 0.0003x^2 - 100$$
$$= 15x - 0.0003x^2 - 100$$

To find the profit on 100 jack-o'-lanterns, replace the $x$ with 100, like so:

$$P(100) = 15(100) + 0.0003(100)^2 - 100$$
$$= 1{,}500 + 3 - 100 = 1{,}403$$

So, as you can see, the company makes over $1,400 on the jack-o'-lanterns.

REMEMBER

A cost-adjusting or changing formula like the one shown in this example usually has some range of values for which it can be used. If the number of items gets too large, the cost can become very small or even negative — which makes no sense. The supplier will have a chart or formula telling you the range of number of items that apply to the discount. For example, the supplier may say that the discount doesn't start until you've ordered at least 100 items and doesn't apply to orders exceeding 10,000 items. It's their call.

## Averaging the costs

One measure of how profits act is the *average cost* of the items produced. Here's the general formula to find the average cost:

Average cost = Total cost ÷ Number of items

So, for example, if the total cost of an item is $4,000, and you've produced 1,000 items, the average cost is $4 per item. Not too difficult, is it?

Good. Now, I want to give you an example of a legitimate cost function where the following apply:

The average cost per item when 50 items are produced is $24.

The average cost per item when 100 items are produced is $14.

The average cost per item when 200 items are produced is $9.

This is no magic trick. This example is an illustration of how the fixed cost factors into the total cost. It also shows that the greater the number of items produced, the more items there are to spread the fixed costs to.

Here's the math that corresponds to the numbers I gave: $C(x) = 4x + 1,000$. In this cost equation, the variable cost is represented with $4x$, where $x$ is the number of items. It costs $4 per item for materials and/or labor. The fixed cost is $1,000. When 50 items are produced, the $1,000 is divided evenly among the 50 items, each absorbing $20 of the fixed cost plus the $4. When 100 items are produced, each takes on $10 of the fixed cost plus $4. And when 200 items are produced, each has $5 of the fixed cost plus $4. As you see, the more items that are produced, the lower the average cost per item.

# What's It Gonna Be? Projecting Cost

When determining a company's profit, you use a formula that subtracts cost from revenue. This formula gives you the net value — which you hope is a positive number. Determining the revenue for this formula is relatively easy; the money comes in or promises are made, and this allows you to record the income for the products or service. Cost, on the other hand, is often a bit more elusive to predict than revenue. After all, you have to juggle many different types of costs, including

✔ Fixed and variable costs (see the previous section for details)

✔ Direct and indirect costs

✔ Differential and historical costs

The *projected cost* (what you estimate you'll have to provide to suppliers, labor force, and utility companies) is necessary when trying to determine a workable price to put in the revenue part of the profit equation. (See the earlier section, "Figuring Profit from Revenue and Cost," for more details on the profit equation.) Depending on your type of business, you generally break the various costs into smaller components or different categories. The many types of costs are often handled differently in the overall accounting processes. But the basic types, which I cover in the following sections, do have many similarities.

## Dividing up direct and indirect costs

*Direct cost* includes anything that can be specifically traceable or accountable to a particular product, service, or purpose. Here are some examples of direct costs:

- The materials needed to construct a house
- The labor used to install a fence
- The advertising bought to announce a bake sale

*Indirect cost* involves (surprise, surprise) costs that *aren't* directly traceable or accountable to a particular product or service. Instead, indirect cost may be associated with two or more different products or services of a company or business. The following are some examples of indirect cost:

- The cost of advertising for the different sale items in a store
- Salaries of managers and secretaries not directly involved in the production process
- Stationery and office supplies
- Insurance
- Heating and cooling

The list goes on and on. Just remember that the indirect costs get allocated to different items based on an overhead rate. An *overhead rate* is designed to assign a fair amount of the indirect costs to particular products based on some measures or amounts associated with the different products.

For example, an overhead rate may include the number of hours of labor for the different products, the number of hours of machine time, or even the level of direct costs. Each of these measures can then be used to determine how much of the indirect cost will be allocated to a particular product. The labor costs and amount of machine time are still assigned to the different products responsible as direct costs. The share of labor costs and machine time can be converted to percentages and used to assign or apportion the indirect costs to the different products in a fair and equitable manner.

If one product uses half of all the machine time, it could be assigned half of all the indirect costs. Formulas for assigning the indirect costs may vary, but they all revolve around divvying up percentages of the indirect cost so that the percentages all add up to 100%.

Try this example: Consider a company that makes four different products: A, B, C, and D. Here are the percentage breakdowns of each product:

**Product A** requires 10% of the labor and 20% of the machinery time, and it accounts for 20% of the total of all the direct costs.

**Product B** requires 30% of the labor and 20% machine time, and it accounts for 40% of the direct costs.

**Product C** uses 50% of the labor and 30% of the machine time, and it accounts for 20% of the direct costs.

**Product D** requires 10% of the labor and 30% of the machine time, and it accounts for 20% of the direct costs.

The total amount in indirect costs per year is $2,500,000. How much of the indirect cost should be assigned to each product? (*Tip:* When solving, assume that each category is equally weighted; in other words, machine time doesn't count more than labor.)

To find out how to assign the indirect costs, first add up all the percentages. Doing so will give you 300%. To get the fraction of the indirect costs that each product is responsible for, add up the percentages assigned to each product and divide by 300%. Refer to Table 18-2 for an organized approach to the math.

| Table 18-2 | | Assigning Percentages to Products | | | |
|---|---|---|---|---|---|
| Product | Labor Machinery | Time Direct | Costs | Total | Total ÷ 300% |
| A | 10% | 20% | 20% | 50% | ⅙ |
| B | 30% | 20% | 40% | 90% | ³⁄₁₀ |
| C | 50% | 30% | 20% | 100% | ⅓ |
| D | 10% | 30% | 20% | 60% | ⅕ |

Whoa! What in the world are those fractions doing there? They all have different denominators. How do you check to see if they add up correctly (to 1 or 100%)? It's easy! To add fractions together, simply find a common denominator for the fractions, change each fraction to an equivalent fraction with that denominator, and then add the numerators together and simplify. (Flip to Chapter 2 for more on fractions.) Here's what your work should look like:

$$\frac{1}{6} + \frac{3}{10} + \frac{1}{3} + \frac{1}{5}$$

$$= \frac{1}{6} \cdot \frac{5}{5} + \frac{3}{10} \cdot \frac{3}{3} + \frac{1}{3} \cdot \frac{10}{10} + \frac{1}{5} \cdot \frac{6}{6}$$

$$= \frac{5}{30} + \frac{9}{30} + \frac{10}{30} + \frac{6}{30} = \frac{30}{30} = 1$$

Now you have to find the amount that each product should be allocated in indirect costs. To do so, multiply the total indirect cost by the respective fraction for that product. Here's what I mean:

**Product A** $\quad \frac{1}{6}(2,500,000) = \frac{2,500,000}{6} \approx 416,666.67$

**Product B** $\quad \frac{3}{10}(2,500,000) = \frac{7,500,000}{10} = 750,000$

**Product C** $\quad \frac{1}{3}(2,500,000) = \frac{2,500,000}{3} \approx 833,333.33$

**Product D** $\quad \frac{1}{5}(2,500,000) = \frac{2,500,000}{5} = 500,000$

If you prefer not to work with fractions, you can do the computations using the percentages directly from the table. For each product, first you multiply $2,500,000 by its total percentage (without moving the decimal). Then you divide the result by 300. You'll be dealing with very large numbers, but your calculator should be able to take the heat for you. However, take heed: I really prefer not to work with such big numbers. It's way too easy to get lost in all those zeros.

# Weighing historical and differential cost

When determining how to price an item, you look at how much that item costs to produce. Well, you at least look at the item's production cost history — what it has been costing to produce it. Making plans for future production and sales, you look into the future for what it will cost you to continue the same process. *Historical cost* is what the cost has been in the past. *Differential cost* is what you predict the cost to be in the future.

Historical cost may be based on last month's or last year's figures. However, you may even go back several years to see if the cost of an item is holding steady. Usually, the cost of production increases over time. But on rare occasions, the cost may decrease because of some glut in the necessary materials or because of a change in technology. So when predicting profits and setting prices, you must use historical cost judiciously.

Differential cost is determined somewhat by changes in circumstances, and it's a prediction of what amount is necessary due to inflation, specification changes, availability of materials, or agreement with labor. Determining differential cost may seem somewhat like looking in a crystal ball. But don't worry; to make the process a bit more manageable and to provide a sense of security with the results, some tools are available to assist you in the process.

For instance, you can develop a formula or equation using the *high-low method* or the *method of least squares*. You can draw a line through a scatter plot and determine the equation of the line to use as the cost formula. Many calculators and spreadsheets give you pretty decent formulas, too, if you plug in the information you have at hand.

### Assuming a pattern in costs

One of the most prevalent assumptions in determining cost is that it will be *linear* and therefore follow a straight line. Except for a few little deviations, the increases (or decreases) in this cost assumption appear to follow a line that's upward or downward or fairly level — both in a picture or with a formula. A *linear function* is a frequently-used model in business and accounting.

Another assumption in cost behavior is that cost moves in steps. The *step function,* for instance, is horizontal or level until another expense, such as a new worker, is added. At that point, the cost jumps up and stays horizontal again until another addition comes along. The resulting graph looks like stair steps. Figure 18-1 shows you a linear model on the left and a step function on the right.

The last of the more popular assumptions is that cost follows a *curve.* However, the differences between a curve and a line are often so small that a line, which is easier to deal with, is usually satisfactory for approximations.

Linear cost assumptions top users' lists of function types because they're easier to write equations for, easier to deal with once the formula is written, and because, frequently, this is just how cost behaves. Complex cost formulas are created with technology — computers and calculators. You'll find that you can usually get by with the cost assumptions involving linearity, as shown in the following sections.

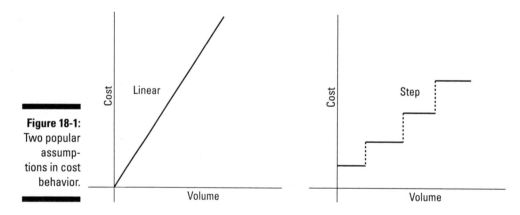

**Figure 18-1:**
Two popular assumptions in cost behavior.

### Developing a formula for a linear cost assumption

Using either historical or differential figures, the models or formulas for the cost of production based on the volume are found in several different ways. Your goal here is to write the equation of a linear function $C(x) = a + bx$, where $C$ is the total cost, $a$ is the fixed cost for some time period, and $bx$ is the variable cost (see the earlier section, "Sorting out variable and fixed costs," for more). The variable $b$ represents the cost per unit, and $x$ represents the number of units or volume of production.

The linear cost model is based on the slope-intercept form for the equation of a line. In algebra, you're presented with $y = mx + b$, where $m$ is the slope of the line (the way the line changes when $x$ changes) and $b$ is the $y$-intercept or initial value — it stays constant like the fixed cost.

To determine the values of $a$ and $b$ in the cost equation, $C(x) = a + bx$, your computation options include the following:

✔ Applying the high-low method to make a scatter plot of the actual costs

✔ Drawing a line through the points on a scatter plot

✔ Using the method of least squares

✔ Using just one year's or one month's worth of cost figures — one set of data — when that's all there is.

I explain how to use each of these options in the following sections.

Finding a decent cost equation is a matter of choice, judgment, and luck. You make the choice as to the method to use, you decide whether your resulting choice is reasonable, and, if you're lucky, the model will work for you when doing computations now and business in the future.

### Picking one representative value

If you've been in business for one year and anticipate that the cost of production isn't going to change, you'll probably use the fixed cost and cost per unit that you did last year to predict profit. The following example explains how.

Say, for instance, that a college club prepares goodie boxes for final exam time. The club sends notes to the parents of all students offering to prepare a box (on behalf of the paying parent) of healthy snacks, hot cocoa mix, coffee and tea, instant macaroni and cheese, crackers, and other goodies to help the student through final exams. Last year, the club was able to buy a box and fill it with goodies for $7.39 per box. The mailing and postage costs (stationery, printing, and stamps) came to a total of $473. Every parent got a letter, but not everyone ordered a goodie box. Using the figures from last year, how much will the group spend if it gets orders for 330 boxes this year?

To figure out the total expenses for this number of goodie boxes, you have to take into account the fixed and variable costs, which will be placed into the cost equation I introduce in the previous section: $C(x) = a + bx$. In this example, the fixed cost is $473, and the variable cost (the cost per box) is $7.39. So your cost equation is $C(x) = 473 + 7.39x$. The total cost for 330 boxes is found by replacing the $x$ in the equation with 330, like this:

$$C(330) = 473 + 7.39(330)$$
$$= 473 + 2{,}438.70$$
$$= 2{,}911.70$$

So it will cost the club almost $3,000 to do the mailing and fill 330 boxes. Using these figures, the club now has more information to help in determining the price it wants to charge for the boxes so it can make a decent profit.

### Using the high-low method

When a company has several years or several other periods of information on cost, it can decide that somewhere between the highest volume on record and the lowest volume on record works for an estimate of cost in the future. In this case, the company can choose to use the high-low method to determine a formula for the cost. The high-low method pretty much ignores all the data in between the highest and lowest volumes. The computation is much easier when you have just two points to consider. The missing numbers in the cost formula are computed algebraically and are assumed to represent what goes on with the cost of the product.

Table 18-3 shows the volume and total cost for a company over an eight-year period.

| Table 18-3 | Historical Costs in Actual Cost and Volume | |
| --- | --- | --- |
| Year | Actual Cost | Volume |
| 1999 | $1,200 | 1,000 |
| 2000 | $1,800 | 1,200 |
| 2001 | $1,500 | 900 |
| 2002 | $1,300 | 800 |
| 2003 | $1,500 | 1,100 |
| 2004 | $1,600 | 1,200 |
| 2005 | $1,700 | 1,400 |
| 2006 | $1,500 | 1,300 |

The high and low values (in volume) from the information given in Table 18-3 were in 2002 and 2005. The volume was 800 in 2002 and 1,400 in 2005.

### Resorting to the method of least squares

The *method of least squares* involves a formula used to find the best linear equation that's possible from the information you have at hand. Of course, if your information is no good, the resulting equation will be worthless. After all, garbage in, garbage out! But if you have accurate measures of volume and cost for a few years, you can use the least squares formula to find the best line for the information.

Here's what the least squares formula does: It gives you the equation of a line that's as close to each point in a scatter plot as possible. The line determined by the formula has the smallest differences in the distances from the points in the graph measured to a point on the line. Why is the formula named as it is? Because, built into the formula, the distances are squared so that points above the line (with a positive difference) and those below the line (with a negative difference) don't cancel each other out.

The best-fit linear equation, $y = a + bx$, for a set of points is found by solving the following system of equations for $a$ and $b$:

$$\begin{cases} na + b\left(\sum x\right) = \sum y \\ a\left(\sum x\right) + b\left(\sum x^2\right) = \sum xy \end{cases}$$

where $n$ is the number of points or data sets used, $\sum x$ is the sum of all the $x$ coordinates of the points, $\sum y$ is the sum of all the $y$ values of the points, $\sum x^2$ is the sum of all the squares of the $x$ values, and $\sum xy$ is the sum of the products of the $x$ and $y$ values.

The easiest way to deal with all the sums, products, and powers needed in the formula is to make a table. Table 18-4, which uses the information that's given in Table 18-3, shows an example of such a table.

| Table 18-4 | The Values to Use in the Least Squares Method | | |
|---|---|---|---|
| *Cost = y* | *Volume = x* | *$x^2$* | *xy* |
| 1,200 | 1,000 | 1,000,000 | 1,200,000 |
| 1,800 | 1,200 | 1,440,000 | 2,160,000 |
| 1,500 | 900 | 810,000 | 1,350,000 |
| 1,300 | 800 | 640,000 | 1,040,000 |
| 1,500 | 1,100 | 1,210,000 | 1,650,000 |
| 1,600 | 1,200 | 1,440,000 | 1,920,000 |
| 1,700 | 1,400 | 1,960,000 | 2,380,000 |
| 1,500 | 1,300 | 1,690,000 | 1,950,000 |
| **Sum** 12,100 | 8,900 | 10,190,000 | 13,650,000 |

Using the various sums in the least squares equations, you let $n = 8$, because you have eight data sets. You can see the sums of the various categories in Table 18-4. Fill in those various sums wherever they appear in the equations of the formula, like so:

$$8a + b(8,900) = 12,100$$

$$a(8,900) + b(10,190,000) = 13,650,000$$

To solve this system of equations for $a$ and $b$ (see Chapter 5 for more on solving systems of linear equations), you first simplify the second equation by dividing each term by 100 to get rid of some of those zeros. Then multiply the terms in the top equation by –89 and the terms in the bottom equation by 8. This multiplication gives you opposite numbers for the coefficients of $a$, so when you add the two equations together, you eliminate the $a$ terms and can solve for $b$. When you're finally ready to solve for $b$, you get $b \approx 0.654$. Now substitute that value of $b$ into the first equation and solve for $a$. Your math should look like this:

$$\begin{cases} 8a + b(8,900) = 12,100 \\ a(8,900) + b(10,190,000) = 13,650,000 \end{cases}$$

$$\begin{cases} 8a + 8,900b = 12,100 \\ 89a + 101,900b = 136,500 \end{cases}$$

$$\begin{cases} -712a - 792,100b = -1,076,900 \\ 712a + 815,200b = 1,092,000 \end{cases}$$

$$23,100b = 15,100$$

$$b = \frac{15,100}{23,100} \approx 0.654$$

$$8a + 8,900(0.654) = 12,100$$

$$8a = 12,100 - 5,820.6 = 6,279.4$$

$$a = \frac{6,279.4}{8} = 784.925$$

The best model for cost using the data in Table 18-4 is: $C(x) = 784.925 + 0.654x$. You get pretty much the same line using the different methods. Of course, the least squares line is the best measure, because it takes into account all the points. But this is where judgment comes in. Is all this computation worth the effort? Only you can decide that. When the scatter plot seems to have a distinctive flow in one direction, you probably won't get much difference in the lines when using the two different methods: high-low and least squares.

# Determining Break-Even Volume

Whether you like it or not, businesses are all created to make a profit. After all, it doesn't make sense to start up a business with the goal of losing money or going broke. Figuring out profit with a formula is simple enough. But coming up with the actual product and sales to make the profit happen is the biggest challenge.

The *break-even volume* (which is also sometimes called the *break-even point*) is the level of production at which the revenue and the cost are exactly the same. The profit at the break-even volume is zero. A profit of zero isn't exactly what you want, but at the break-even volume (or just beyond it) everything is all pluses: The difference between revenue and cost is a positive number.

Finding the break-even volume is relatively simple if you have a profit formula all worked out and ready to solve. You may have to come up with a cost formula to complete your profit equation (see the earlier section, "What's It Gonna Be? Projecting Cost," for details). After you have your profit equation, however, the break-even volume is relatively straightforward to solve for. Without a profit equation, you can resort to using a chart or a graph to estimate your break-even volume.

## Solving for break-even volume with a formula

You can solve for the break-even volume by setting the cost part of the profit equation equal to the revenue part of the profit equation and then solving for the volume, $x$. Or you can use the profit equation in its usual form, letting the profit equal 0; again, you solve for the volume, $x$.

Try solving for the break-even volume with this example: Say that a silk tie vendor purchases ties from his supplier through the mail. The vendor's cost for a shipment of ties includes the postage and handling charge of $47 plus the cost per tie, which is $13.80. He sells the ties for $29.95 each. What's his break-even volume?

The profit function for this vendor's venture is Profit = Revenue − Cost. After plugging in his specific numbers, you get this equation: Profit = $29.95x$ − $(47.00 + 13.80x)$, where $x$ is the number of ties. All you have to do now is let the profit equal 0 and then solve for the value of $x$, like this:

$$0 = 29.95x - (47.00 + 13.80x)$$
$$0 = 29.95x - 47.00 - 13.80x$$
$$0 = 16.15x - 47.00$$
$$47.00 = 16.15x$$
$$\frac{47.00}{16.15} = x$$
$$x \approx 2.910$$

The volume at break even, $x \approx 2.910$, means that selling three ties will do it — everything else is gravy. Just look at how the volume works in the profit function for two ties, three ties, and ten ties:

$$\text{Profit} = 29.95x - (47 + 13.80x)$$
$$= 16.15x - 47$$

$$\text{Profit (2 ties)} = 16.15(2) - 47$$
$$= 32.30 - 47$$
$$= -14.70$$

$$\text{Profit (3 ties)} = 16.15(3) - 47$$
$$= 48.45 - 47$$
$$= +1.45$$

$$\text{Profit (10 ties)} = 16.15(10) - 47$$
$$= 48.45 - 47$$
$$= +114.50$$

The profit with two ties is still negative, but three ties puts the vendor over the 2.910 break-even volume. And ten ties makes the venture look pretty good.

## Working out break-even volume from historical data

Even when an exact profit equation isn't available or hasn't yet been computed, you can make a decent estimate of the break-even volume from historical data. And even with two data sets you can find the point of intersection of revenue and cost.

Consider this example in which the AJ Company has information on its cost and revenue for the years 2004 and 2006. Using this information, it wants to estimate the volume at which it can expect to break even. In 2004, the company had a production and sales volume of 1,300 units for which the cost was $83,000 and the revenue was $104,000. In 2006, the volume was 2,500 units at a cost of $155,000 and revenue of $200,000. What can the AJ Company expect the break-even volume to be if this year's production is similar in the relationship between revenue and cost?

To solve this problem, you need to plot the two costs with their volumes and draw a line through them. On the same graph, plot the two revenues with their volumes and draw a line through them. Estimate where the lines cross, and you've found your break-even volume. From Figure 18-2, the break-even volume appears to be slightly more than 500 units.

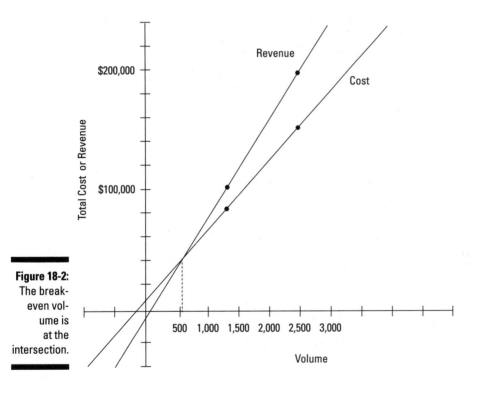

**Figure 18-2:**
The break-even volume is at the intersection.

# Chapter 19

# Accounting for Overhead and Depreciation

. . . . . . . . . . . . . . . . . . . . . . . . . . . . . . . . . . . . . . . . . . . . . . . . . . . . .

## In This Chapter

▶ Dealing with overhead by apportioning fairly

▶ Examining the many ways to calculate depreciation

. . . . . . . . . . . . . . . . . . . . . . . . . . . . . . . . . . . . . . . . . . . . . . . . . . . . .

*O*verhead, in business math, isn't the compartment above your airplane seat where you stow your carry-on luggage. Instead, *overhead* is the term used to describe a variety of different costs that are associated with a business but not necessarily with a particular product or product line in that business. Overhead may include any of the following:

- ✔ Order-getting costs, such as advertising, marketing research, and the salaries of salespersons

- ✔ Research and development of new products

- ✔ The salaries and supplies of administrative staff, such as managers and secretaries (these folks aren't necessarily assigned to one product, so the cost of employing them is considered overhead)

- ✔ Financial costs, such as interest on loans or dividends to owners

*Depreciation,* which is figured into the cost of doing business, is another type of nonmanufacturing cost. Computing the depreciation associated with an item allows you to determine the current value of that item. Depreciation is computed in many different ways; you choose the method that provides the best picture for you and gives you the greatest financial benefit.

This chapter provides several depreciation options and shows you the ins and outs of computing the depreciation and value of an asset. You'll also find apportioning scenarios, which show you how to divvy up overhead fairly.

# Keeping an Eye on Overhead Costs

Overhead costs often include marketing, advertising, postage, supplies, salaries, and other miscellaneous expenses that can't be delegated to a particular item or production line.

Overhead has to be paid and allocated fairly to the items or services being provided so that the profit from the different items can be determined and proper tax responsibilities executed.

Different overhead items are accounted for and apportioned in different ways. When many different costs are included in overhead, you may find that it's a challenge to determine a fair division. Even within an overhead area, different costs may need to be handled differently. I show you everything you need to know in the following sections.

## Working with order-getting costs

*Order-getting costs* are incurred in an effort to get orders for products or services. Order-getting costs may be associated with the revenue that comes from the particular orders, but not all activities and costs involved in order-getting necessarily result in sales. And often, multiple items are involved in a particular attempt at an order — whether they result in a sale or not.

Order-getting costs include the following:

- ✔ Advertising
- ✔ Commissions that are directly tied to particular items
- ✔ Marketing
- ✔ Office expenses
- ✔ Other sales promotions
- ✔ Salaries of salespersons who may sell one or several different products
- ✔ Salaries of secretaries
- ✔ Travel and entertainment expenses

Order-getting costs involve both the costs that are directly related to getting the orders and those that are intermingled with other departments or parts of the business. The cost of secretarial help, for instance, is apportioned fairly among the parts of the business that need the help of a secretary, and each part shares in the expense of having secretaries. The percentage share is determined for each shared cost category, and the total cost of all the categories is charged to order-getting.

To understand what I mean, consider some marketing and sales expenses incurred during the first two months of a year. Table 19-1 shows the expense and the amount per month. You'll notice that the salespersons' commissions for February are missing. The following example will show you how to calculate them.

| Table 19-1 | January and February Expenses | |
|---|---|---|
| **Area** | **January** | **February** |
| Sales supervisor | $4,320 | $4,560 |
| Salespersons' salaries | $9,000 | $9,000 |
| Salespersons' commissions | $6,000 | |
| Salespersons' travel and meals | $540 | $800 |
| Supplies – sales office | $230 | $275 |

If the sales supervisor gets a flat monthly salary plus 12% of the salespersons' commissions, what's that monthly salary, and what were the commissions in February? Also, are the travel and meal expenses proportionate with sales?

To answer all these questions, first determine the flat monthly salary of the sales supervisor. To do so, calculate what 12% of the commissions were in January by multiplying the commissions by 12% (in decimal form): $6,000 × 0.12 = $720. Now subtract the commission amount from the payment to the sales supervisor: $4,320 – $720 = $3,600.

If the sales supervisor gets $3,600 per month plus a percentage of the commissions, determine how much of the February payment to the supervisor is commission by subtracting the $3,600 from the amount paid in February: $4,560 – $3,600 = $960.

The $960 is 12% of the commissions, which means that you need to solve this equation: $12\%x = 960$, where the variable $x$ represents the commissions. If you divide both sides by 12% (or 0.12), you get that $x = 8,000$. So the salespersons' commissions in February were $8,000.

Finally, you want to determine whether the increased amount spent on travel and meals in February are consistent with the sales. In other words, did spending the additional money reap a proportionate amount of revenue? To find out, simply set up two fractions with the travel and meal expenses in the *numerator* (the top of the fraction) and the commissions in the *denominator* (the bottom).

You aren't shown the actual revenue figures, but if commissions are based on sales, you'll be comparing the same things. You just want to look at the percentages to see whether there's a big disparity between the expenses and results in these two months.

For January, you divide $540 by $6,000, and for February you divide $800 by $8,000. (You didn't have the commission amount until you did the figuring earlier.) Here's what the equations should look like:

$$\frac{540}{6000} = 0.09 = 9\% \text{ and } \frac{800}{8000} = 0.10 = 10\%$$

The percentages are different, but probably not significantly different. If you're in charge of revenue at this company, however, you may want to keep an eye on the numbers to see if a trend develops.

Give another example a try. Say that a company manufactures three products: A, B, and C. The company also has three supervisors — one for each product. One year, a $60,000 advertising expense was allocated to each product in proportion to the net sales of each. The next year, the advertising budget was increased to $90,000. Of this $90,000, a total of $50,000 was used for general advertising of the company to increase awareness of its commitment to the environment. The rest of the advertising budget was divvied up among the products as follows:

$16,000 for A

$18,000 for B

$6,000 for C

For cost purposes, the advertising expenses were allocated on the basis of sales. The sales and advertising expenses are shown in Table 19-2. Were the advertising expenses fairly divided?

| Table 19-2 | Sales and Advertising Expenses for A, B, C | | | |
|---|---|---|---|---|
| | **Total** | **A** | **B** | **C** |
| Net sales | $1,000,000 | $410,000 | $250,000 | $340,000 |
| Advertising cost | $90,000 | $37,000 | $25,000 | $28,000 |

You really have two things going on here: The amount allocated to each product to spend on advertising, and the amount being charged to each product for the cost of advertising. It could be that the best-selling product doesn't need as much advertising, but the amounts and percentages should still be presented and reviewed. Table 19-3 shows the three products, their percentage of total net sales, their percentage of budgeted money for advertising, and their percentage of the total advertising cost charged to them.

| Table 19-3 | | Proportions Allocated to A, B, C | |
| :--- | :--- | :--- | :--- |
| *Product* | *% Sales* | *% Advertising Budget* | *% Advertising Cost* |
| | Net sales<br>1,000,000 | Amount allocated<br>40,000 | Amount charged<br>90,000 |
| A | 41% | 40% | 41.1% |
| B | 25% | 45% | 27.8% |
| C | 34% | 15% | 31.1% |

Looking at Table 19-3, Product A seems to be in line. Its net sales, share of the advertising budget, and allocated advertising cost all are fairly close in number. Products B and C have some disparities, though. Product B, for instance, has 25% of the net sales and 45% of the advertising budget, and it's charged about 28% of the costs. If cost allocation is supposed to be based on sales, Product B is being charged more than it should be. On the other hand, Product B also is getting a huge amount of advertising money. Product C seems to be paying for more than its share of the advertising costs as well. The supervisors need to sit down and hash this out.

## *Deciding on allocation options*

Overhead comes in a variety of forms, and the costs associated with overhead are to be allocated to different products, product lines, or services in a way that fairly divides the costs. Different costs use different means of assigning the proportionate responsibility.

For instance, costs that are related to payroll or personnel can be divided based on hours of machine usage or total revenue of the product. Similarly, office supplies can be allocated based on hours or revenue. Financial costs, such as building loans, insurance, and utilities, can be divided based on space, revenue, or time.

When calculating overhead, a fair apportionment is determined, and then an overhead rate or percentage is assigned to the different products or services. The rate is usually revisited each year to determine whether any significant changes need to be made.

Consider, for example, a company that produces three different items: A, B, and C. The items are produced using two machines: Machine X and Machine Y. The labor force consists of 15 full-time employees. The total space for the production, storage, and office is 4,000 square feet. Table 19-4 shows the weekly allocation of machine time, labor, space, and yearly revenue for each product.

| Table 19-4 | Products A, B, and C | | |
|---|---|---|---|
| | **A** | **B** | **C** |
| Weekly hours on Machine X | 20 | 10 | 15 |
| Weekly hours on Machine Y | 10 | 15 | 15 |
| Weekly hours of labor | 240 | 200 | 160 |
| Storage space | 1,300 sq. ft. | 900 sq. ft. | 500 sq. ft. |
| Yearly revenue | $300,000 | $100,000 | $100,000 |

Using the information from Table 19-4, the overhead allocated to the different products can be computed. But when you calculate, do remember that different items produce different percentages and different shares of the categories. The following example problem will show you how to calculate the overhead allocations for this company.

The yearly cost of electricity for the business shown in Table 19-4 is $500,000. Using the total machinery hours, how is the cost divided among the three products? Is the allocation significantly different if you use hours of labor?

Product A uses a total of 30 hours on the two machines, while Products B and C use a total of 25 hours and 30 hours, respectively. If you divide the number of hours that each product uses the machines by 85 (the total hours on both machines), it turns out that Products A and C use 35.3% each; Product B uses 29.4%. If you multiply the $500,000 electric bill by the percentages, the allocation assigns $176,500 to Products A and C, and $147,000 to Product B.

If the number of hours of labor is used instead of machinery time, you have to divide the hours for each product by 600 (the total hours). In doing so, you find that Product A uses 40%, Product B uses 33.3%, and Product C uses 26.7%. These percentages change the amount of electricity charged to each product. For instance, now $200,000 is assigned to Product A, $166,500 is assigned to Product B, and $133,500 is assigned to Product C. As you can see, Product A now assumes much more of the cost of electricity.

Because the allotments come out rather differently, it's a judgment call as to which way to assign the overhead. Another option is to try to use several of the items with a weighted average as a multiplier. A *weighted average* is used to give some items in a list more value or worth than others. Weighted averages work something like diving or snowboard competitions; you get more points when the dive or maneuver is more difficult. To find a weighted average, multiply each item by the number assigned as its weight. Then divide the sum of all the multiplied numbers by the total of all the weights.

In this example, if you value Product A twice as much as you do Product B, and if you value Product C three times as much as you do Product B, you get this equation: 2A + 3C + B. The weights are 2, 3 and 1. The sum or total of all the weights is 6, so you simply divide the multiplications by 6: (2A+3C+ 1B) ÷ 6 gives you the weighted average.

Using all of the information in this example, you can now calculate the percentages for each category and product. Table 19-5 shows you the answers. Even though the storage space doesn't account for the total area available in the facility, the percentages reflect the part of the storage space that they occupy.

| Table 19-5 | Percentages of A, B, and C by Category | | |
|---|---|---|---|
| | A | B | C |
| Weekly hours on Machine X | 44.4% | 22.2% | 33.3% |
| Weekly hours on Machine Y | 25% | 37.5% | 37.5% |
| Weekly hours of labor | 40% | 33.3% | 26.7% |
| Storage space | 48.1% | 33.3% | 18.5% |
| Yearly revenue | 60% | 20% | 20% |

You may have noticed that the percentages for the weekly hours on Machine X and the storage space don't add up to exactly 100%. This discrepancy is due to rounding. If you're talking about huge amounts of money and want to avoid discrepancies, carry the decimals out several more places than just the one decimal place shown.

Now that you've been introduced to weighted averages, try out this example: Determine the percentage allocation for each product in Table 19-5 if you assume that

 ✔ The hours on the machines are weighted twice that of storage space.

 ✔ Labor is weighted three times as much as storage space.

 ✔ Revenue is weighted the same as storage space.

To compute each product's allocation, each percentage is multiplied by its respective weight factor: 2X, 2Y, 3Labor, 1Space, and 1Revenue. Then the total is divided by the total of the weightings, or multipliers (2 + 2 + 3 + 1 + 1 = 9). Here's what the calculations would look like for each product:

A: $2(44.4) + 2(25) + 3(40) + 1(48.1) + 1(60) = 366.9$

$$\frac{366.9}{9} = 40.8$$

B: $2(22.2) + 2(37.5) + 3(33.3) + 1(33.3) + 1(20) = 272.6$

$$\frac{272.6}{9} = 30.3$$

C: $2(33.3) + 2(37.5) + 3(26.7) + 1(18.5) + 1(20) = 260.2$

$$\frac{260.2}{9} = 28.9$$

So, using this weighted arrangement, Product A is assigned 40.8% of the overhead, Product B has 30.3%, and Product C is assigned 28.9%. Of course, you can play around with the numbers all you want to make the allocations come out the way you want them.

# Getting the Lowdown on Depreciation

Business assets, such as machinery, buildings, computers, and other equipment, are bound to one day grow old and become outdated. These physical assets lose value and their worth gradually gets used up. The procedure for determining the amount to be allocated to entries in valuation of capital equipment or property statements is referred to as *depreciation* (a reduction in value). A common example of this reduction in value involves cars. The value of a new car drops about 20% (or more!) as soon as you drive it off the dealer's lot. In five years, its value can drop around 35%.

A piece of machinery has an initial cost — what you pay at the time of purchase. After the *accumulated depreciation* (the sum of all the depreciation over time) is applied, the difference between the initial cost and the accumulated depreciation is the remaining value of the item. This value is often referred to as *salvage value* or *residual value*. Here's the formula:

Initial cost (purchase price) – Depreciation = Salvage value

When figuring depreciation, you have to determine the asset's estimated useful life (how long the asset can be used until it wears out or becomes outdated), its salvage value, and the accounting method that you're going to apply to compute depreciation over the lifetime of the asset. The methods most frequently applied to depreciation are

 ✔ The straight-line method of depreciation

 ✔ The sum-of-the-years'-digits method

 ✔ The declining balance method

I explain each of these methods in the following sections.

The sum-of-the-years'-digits and declining balance methods are considered to be *accelerated* methods, because they speed up or increase the amount of the depreciation early in the time period. Why would a company want to speed up the depreciation? It may need the additional tax break of being able to deduct more depreciation in the next few years rather than later.

## The straight-line method

The *straight-line method* of depreciation subtracts an equal amount from the value of the asset each year. When you look at a graph of the value of an asset for which the straight-line method is used, you see exactly what the method implies: a straight line. Figure 19-1 shows the value of a storage building that's determined with straight-line depreciation.

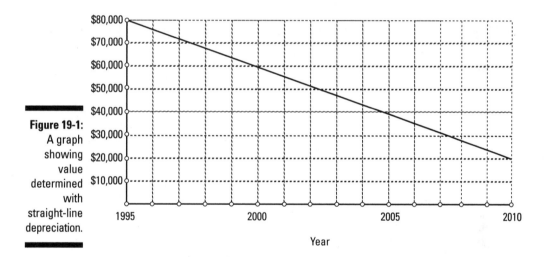

**Figure 19-1:** A graph showing value determined with straight-line depreciation.

Year

Figure 19-1 shows the initial value, useful life, and salvage value of the building. The initial value, in 1995, was $80,000. The salvage value in 2010 is $20,000. By subtracting the initial year from the last year shown on the chart, you can find the *useful life* of the building: 2010 − 1995 = 15 years.

Reading from the table, you see that the value in the year 2000 is $60,000, and the value in 2006 is about $37,000. Graphs are quick and easy to read, but a formula that gives you the exact value of the building during a particular year is better for the accountant (and the IRS).

The equation of a straight line when written in the *slope-intercept* form is

$$y = b + mx$$

where $b$ is the $y$-intercept (the point where the line crosses the vertical axis) and $m$ is the *slope* (the steepness or change). Applying the slope-intercept form to straight-line depreciation, the $y$-intercept, $b$, is the initial value of the asset; the slope, $m$, is the amount of yearly depreciation; and the input variable, $x$, is the number of years since purchase. The output variable, $y$, is the value of the asset. The following example will show you how to use the slope-intercept form to solve a straight-line depreciation problem.

Write an equation using straight-line depreciation to find the value of a building that's worth $80,000 in 1995 and $20,000 in 2010. Then use that equation to find the exact value of the building in the year 2006.

To start, use this format:

$$V(t) = V_0 - mt$$

where $V(t)$ is the value of the asset $t$ years since purchase, $V_0$ is the initial value, and $m$ is the yearly depreciation of the asset. The initial value is $80,000, so $V_0 = 80,000$. To find the amount per year of depreciation, divide the difference between the initial value and the salvage value by the number of years of useful life, like this:

$$m = \frac{\text{initial value} - \text{residual value}}{\text{number of years useful life}} = \frac{80,000 - 20,000}{15} = \frac{60,000}{15} = 4,000$$

Replacing the $m$ in the equation with 4,000, the formula to use for the value of the building in this situation is $V(t) = 80,000 - 4,000t$, where $t$ is the number of years since 1995. To find the value of the building in 2006, subtract to find the useful life: $2006 - 1995 = 11$ years. Replace the $t$ with 11 in the formula, and you get

$$V(11) = 80,000 - 4,000(11)$$
$$= 80,000 - 44,000$$
$$= 36,000$$

So you can see that the building is worth $36,000 in 2006.

You can make a chart or spreadsheet of the exact values of the building for the different years. Table 19-6 shows you the first five years for the value of a building whose initial value was $80,000 in 1995 and that depreciated by $4,000 per year.

| Table 19-6 | | The Value of the Building |
|---|---|---|
| *Year* | *t* | *80,000 – 4,000t* |
| 1995 | 0 | 80,000 – 0 = 80,000 |
| 1996 | 1 | 80,000 – 4,000 = 76,000 |
| 1997 | 2 | 80,000 – 8,000 = 72,000 |
| 1998 | 3 | 80,000 – 12,000 = 68,000 |
| 1999 | 4 | 80,000 – 16,000 = 64,000 |
| 2000 | 5 | 80,000 – 20,000 = 60,000 |

# The sum-of-the-years'-digits method

Some assets depreciate more rapidly in the first few years of use than they do later on. Vehicles, for instance, quickly lose a significant percentage of their value at the beginning of their useful life, and then their values stay steadier or decline slowly toward the end. One method used to accommodate this accelerated loss of value is the *sum-of-the-years'-digits method.*

A big difference between the sum-of-the-years'-digits method and other methods is that the computations with the sum-of-the-years'-digits method are made on the *depreciation base* (the initial value minus the salvage value), not on the initial cost of the item. Using the sum-of-the-years'-digits method, you create a series of fractions that add up to 1. The larger fractions come first in the list and then the fractions get progressively smaller from there. Each year's depreciation amount is determined by multiplying the depreciation base by the designated fraction.

The graph of the values of an item being depreciated by the sum-of-the-years'-digits method curves down sharply. Figure 19-2 shows an example graph of values that were computed using the sum-of-the-years'-digits method.

With the sum-of-the-years'-digits method, you take the number of years of estimated life of the asset and add that number to each positive whole number smaller than it. The sum of the numbers is the denominator (bottom) of the fractions to be used. The numerators (tops) of the fractions are the numbers that were added together. After subtracting the salvage value from the original value of the asset, you multiply the difference (the depreciation base) by the decreasing fractions to get the amount to be depreciated each year.

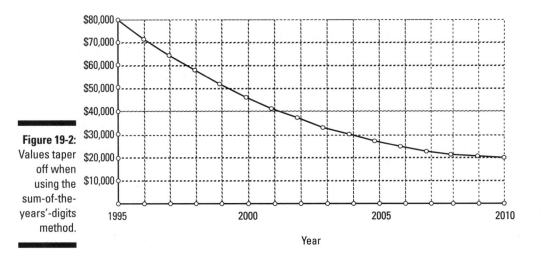

**Figure 19-2:**
Values taper
off when
using the
sum-of-the-
years'-digits
method.

So if the useful life of an asset is 5 years, you add 5 + 4 + 3 + 2 + 1 to get 15 for the denominators. Then you create the fractions that are used in figuring the yearly depreciation:

$$\frac{5}{15}, \frac{4}{15}, \frac{3}{15}, \frac{2}{15}, \frac{1}{15}$$

The sum of the first $n$ digits, $1 + 2 + 3 + \ldots + n$, is equal to

$$\frac{n(n+1)}{2}$$

So the sum of the first five digits, $1 + 2 + 3 + 4 + 5$, equals

$$\frac{5(5+1)}{2} = \frac{5(6)}{2} = \frac{30}{2} = 15$$

Ready for an example? Suppose a piece of equipment costs $90,000 and has a useful life of 10 years. The salvage value of the equipment is $13,000. Using the sum-of-the-years'-digits method, determine the amount of depreciation and the value of the equipment for each year of its useful life. (*Tip:* A spreadsheet is useful for these computations. Refer to Chapter 5 for more on using spreadsheets in your calculations.)

First you determine the fractions needed for the multiplying process. The sum of the digits 10 + 9 + 8 + . . . + 2 + 1 equals

$$\frac{10(10+1)}{2} = \frac{10(11)}{2} = \frac{110}{2} = 55$$

You create fractions with the number 55 in the denominators and the numbers 10, 9, 8, . . . 2, 1 in the numerators.

The depreciation base is $90,000 – $13,000 = $77,000. Make a table showing the amount being depreciated each year and the value of the asset in that same year. Table 19-7 is an example of such a table.

| Table 19-7 | | Using the Sum-of-the-Years'-Digits Method | |
|---|---|---|---|
| **Year** | **Multiplier** | **X $77,000** | **Value** |
| 0 | | | $90,000 |
| 1 | $\frac{10}{55}$ | $\frac{10}{55} \times 77,000 = 14,000$ | $90,000 – $14,000 = $76,000 |
| 2 | $\frac{9}{55}$ | $\frac{9}{55} \times 77,000 = 12,600$ | $76,000 – $12,600 = $63,400 |
| 3 | $\frac{8}{55}$ | $\frac{8}{55} \times 77,000 = 11,200$ | $63,400 – $11,200 = $52,200 |
| 4 | $\frac{7}{55}$ | $\frac{7}{55} \times 77,000 = 9,800$ | $52,200 – $9,800 = $42,400 |
| 5 | $\frac{6}{55}$ | $\frac{6}{55} \times 77,000 = 8,400$ | $42,400 – $8,400 = $34,000 |
| 6 | $\frac{5}{55}$ | $\frac{5}{55} \times 77,000 = 7,000$ | $34,000 – $7,000 = $27,000 |
| 7 | $\frac{4}{55}$ | $\frac{4}{55} \times 77,000 = 5,600$ | $27,000 – $5,600 = $21,400 |
| 8 | $\frac{3}{55}$ | $\frac{3}{55} \times 77,000 = 4,200$ | $21,400 – $4,200 = $17,200 |
| 9 | $\frac{2}{55}$ | $\frac{2}{55} \times 77,000 = 2,800$ | $17,200 – $2,800 = $14,400 |
| 10 | $\frac{1}{55}$ | $\frac{1}{55} \times 77,000 = 1,400$ | $14,400 – $1,400 = $13,000 |

Note that the final value comes out to be the salvage value. The sum-of-the-years'-digits methods subtracts parts of the depreciation base from the initial value until the final value is the salvage value.

# The declining balance method

Some assets lose a major portion of their value early in their estimated lives. Like the sum-of-the-years'-digits method (see the previous section), the *declining balance method* is another way to accelerate depreciation. However, the declining balance method is different from the sum-of-the-years'-digits method in two respects:

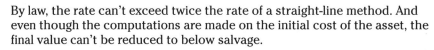

- ✔ The rate of depreciation is a uniform rate. (The rate doesn't change each year; it stays the same.)
- ✔ The computations are made on the *initial* value of the asset, not on the depreciation base.

By law, the rate can't exceed twice the rate of a straight-line method. And even though the computations are made on the initial cost of the asset, the final value can't be reduced to below salvage.

The declining balance method is accomplished using many possible multipliers. The annual depreciation can be multiplied by any number between 1 and 2, including the 2. You can multiply by 1.1, 1.2, 1.5, 1.75, 1.95, or anything, as long as it's not more than 2 (and less than 1). In the next sections, you see the different multipliers at work.

### Doubling your pleasure with the double declining balance method

In practice, the full extent of acceleration — twice — is most commonly used, because it creates the most rapid acceleration possible. When twice the straight-line method is employed, it's called the *double declining balance method.*

To compute depreciation using the double declining balance method, you follow these steps:

1. **Determine the annual percentage depreciation that you would use with the straight-line method.**

2. **Double the percentage from Step 1.**

3. **Multiply the asset's current value by the percentage to get the depreciation for that year.**

4. **To get the current value of the asset, subtract the depreciation from the current value.**

Try using the double declining balance method with this example: Imagine that a new truck costs $60,000 and has a useful life of 5 years. The salvage value of the truck at the end of 5 years is $8,000. Using the double declining balance method, determine the depreciation and value of the truck for its 5 years of useful life.

If you were to use the straight-line method over a 5-year period, you would depreciate ⅕ or 20% per year:

$$\frac{12,000}{60,000} = \frac{1}{5} = 0.20 = 20\%$$

So, doubling that 20% depreciation rate, you get 40% depreciation per year. The 40% first multiplies the initial value of the truck. Then, in future computations, the 40% multiplies the value of the truck from the previous year. The depreciation amount is subtracted from the previous year's value to get the new value. Table 19-8 shows you how the numbers work out in this method.

| Table 19-8 | Double Declining Balance Method | |
|---|---|---|
| *Year* | *Rate × Current Value = Depreciation* | *New Value* |
| 0 | | $60,000 |
| 1 | 40% × $60,000 = $24,000 | $60,000 − $24,000 = $36,000 |
| 2 | 40% × $36,000 = $14,400 | $36,000 − $14,400 = $21,600 |
| 3 | 40% × $21,600 = $8,640 | $21,600 − $8,640 = $12,960 |
| 4 | 40% × $12,960 = $5,184 | $12,960 − $5,184 = $7,776 |
| 5 | | 0 |

The value $7,776 (in the fourth year) is lower than the $8,000 salvage value. So only $4,960 is depreciated in the fourth year, and the value is recorded as $8,000 for the fourth year. You don't have any depreciation in the fifth year.

### Declining at a different rate

Multiplying a straight-line depreciation percentage or value by 2 is the maximum that's allowed, but what if you want to multiply by a number that accelerates the depreciation rate — but not quite as much as doubling it? For example, say that you want to use 1½ times or 1¾ times the rate. Well, guess what? You can do it! In fact, in Table 19-9, I show you what the depreciation schedule looks like if you use 1½ times the straight-line amount. The percentage amount you use for 1½ times is: 20% × 1½ = 30%.

So you'll be multiplying the value of the truck from the previous year by 30% to get the amount that's depreciated. Then you subtract the depreciation amount from the value of the truck that year to get the new value of the truck. In general, no matter what multiplier you use, the procedure to find the new, depreciated value is

% depreciation rate × Previous year's value = Amount of depreciation

Previous year's value − Amount of depreciation = New value

**Table 19-9    Accelerating the Depreciation Rate by 1½ Times**

| Year | Rate × Current Value = Depreciation | New Value |
|------|-------------------------------------|-----------|
| 0 | | $60,000 |
| 1 | 30% × $60,000 = $18,000 | $60,000 − $18,000 = $42,000 |
| 2 | 30% × $42,000 = $12,600 | $42,000 − $12,600 = $29,400 |
| 3 | 30% × $29,400 = $8,820 | $29,400 − $8,820 = $20,580 |
| 4 | 30% × $20,580 = $6,174 | $20,580 − $6,174 = $14,406 |
| 5 | 30% × $14,406 = $4,321.80 | $14,406 − $4,321.80 = $10,084.20 |

As you see in Table 19-9, the item isn't fully depreciated. The salvage value is $8,000, and this method doesn't reduce the value to salvage in the 5-year useful life span. The depreciation amount is accelerated at the beginning, but it tapers off toward the end. Can you accelerate the depreciation so that the value is fully depreciated at the end of its useful life? Yes you can! The following section tells you how.

### Determining a declining balance rate to fully depreciate

When using the declining balance method, you determine the remaining value of an asset for a particular year by subtracting the depreciation amount from the previous year's balance. So if the depreciation rate is 30%, you use this formula: $V - 30\%V = (1 - 30\%)V = (100\% - 30\%)V = 70\%V$. So you're left with $70\%V$. Every year, the value of the asset is 70% of what it was the previous year. The value changes each year, and the amount that's depreciated changes, but the percentages stay the same.

Here's the general formula for finding the salvage value for any percentage rate:

$$V_n = (1 - r)^n V_o$$

where $V_n$ is the salvage value after $n$ years of useful life, $r$ is the percentage rate, and $V_o$ is the original value of the asset. You want to figure out what rate allows the asset to be fully depreciated — and therefore have something to depreciate every year.

To find the rate you want, you have to divide the salvage value by the original value, take a root that's the same as the number of years of usable life, and then subtract the number you get from 1. You need a scientific calculator to do this computation.

To find a root on your calculator, change the value on the radical to its reciprocal and raise the <u>number</u> under the radical to that fractional power. For example, to find $\sqrt[5]{0.1333333}$, enter 0.1333333^(1/5) into your scientific calculator.

When solving the equation for the rate, $r$, your math looks like this:

$$V_n = (1 - r)^n V_o$$

$$\frac{V_n}{V_o} = (1 - r)^n$$

$$\sqrt[n]{\frac{V_n}{V_o}} = \sqrt[n]{(1 - r)^n}$$

$$\sqrt[n]{\frac{V_n}{V_o}} = 1 - r$$

$$r = 1 - \sqrt[n]{\frac{V_n}{V_o}}$$

Try your hand at an example: Use the formula to find a declining rate that fully depreciates a $60,000 truck that has a useful life of 5 years and a salvage value of $8,000.

Your math will look something like this:

$$r = 1 - \sqrt[n]{\frac{V_n}{V_o}}$$

$$r = 1 - \sqrt[5]{\frac{8,000}{60,000}} \approx 1 - \sqrt[5]{0.1333333} = 1 - 0.668325 \approx 0.332$$

The rate that works to fully depreciate the truck in five years is about 33.2%. That rate is slightly higher than the 30% used at 1½ times the straight-line rate; it's just enough more to do the full depreciation job.

Ready for another practice example? Use the formula for fully depreciating an asset that costs $80,000 and has a useful life of 20 years. The salvage value is $10,000.

Here's what your math should look like to solve this problem:

$$r = 1 - \sqrt[n]{\frac{V_n}{V_o}}$$

$$r = 1 - \sqrt[20]{\frac{10,000}{80,000}} \approx 1 - \sqrt[20]{0.125} \approx 0.098749537389$$

## Advertising on the Super Bowl

One important element in overhead expenses is advertising. While most companies try to spread their advertising and advertising expenses over the entire year, some put all their eggs in one basket: the Super Bowl basket. These folks hope that a flashy and memorable Super Bowl commercial will result in more customers and more revenue. With commercial spots commanding nearly 3 million dollars for 30 seconds (that's $100,000 per second), it's a real gamble that the Super Bowl commercial will be memorable enough and seen by enough people to have an impact. Advertisers are usually concerned that viewers who record programs are fast-forwarding through commercials during regular broadcasts, but this isn't the case with the Super Bowl. In this case, the commercials are sometimes more entertaining than the actual game. But that's still a lot of money to spend on advertising.

The rate is about 9.9%, which isn't quite double the straight-line rate. A 20-year useful life means that ¹⁄₂₀ corresponds to a 5% depreciation rate. The double-declining method would use twice 5%, which obviously would be 10%.

It's interesting how just a few decimal places can change the end value over a 20-year period. If the 10% rate is used, the value of the asset in the 20th year is $9,726 — lower than the salvage value. If you use 9.9%, the value in the 20th year comes out to be $9,945. It's still lower than the salvage value.

When including more decimal values, for instance 0.0987, or 9.87%, you get a value of $10,011, which is slightly higher than the salvage value. But if you include more decimal places and use 0.0987495 or 9.87495%, you get a value of $10,000 in the 20th year. (These numbers are easy to produce using a computer spreadsheet. See Chapter 5 for details.)

# Chapter 20

# Keeping Track of Inventory

● ● ● ● ● ● ● ● ● ● ● ● ● ● ● ● ● ● ● ● ● ● ● ● ● ● ● ● ● ● ● ● ● ● ● ● ● ● ● ● ● ● ● ●

● ● ● ● ● ● ● ● ● ● ● ● ● ● ● ● ● ● ● ● ● ● ● ● ● ● ● ● ● ● ● ● ● ● ● ● ● ● ● ● ● ● ● ●

A business has an *inventory* if it accumulates things such as raw materials used to produce products, supplies needed to run the business, partly finished goods, or finished goods. The object of the business is to sell the finished goods at a profit, so you'd think that the quicker the items move out the better. Sometimes, though, it pays to hold onto finished goods. Why? Well, it increases the size of the inventory and you can wait for the prices to increase. The hold versus sell proposition is a real balancing act for businesses, and sometimes it requires a lot of luck along with good business sense to make it all work.

Inventory can be accounted for using cost inventory or retail inventory. Both options are discussed and illustrated in this chapter so you can make a judgment as to which works best in your situation. The FIFO (first-in-first-out) and LIFO (last-in-first-out) methods, which are also discussed in this chapter, have implications involving when to take profits and how to reduce taxes.

## Controlling Inventory and Turnover

For small businesses, controlling inventory isn't terribly difficult — especially if the owner can simply walk into the storage area and count the number of boxes of supplies or the number of each item that has been produced. The larger the business, however, the bigger the challenge of keeping track of the items available in storage. Even more difficult is tracking what has already been loaded onto the trucks.

As a business grows, it becomes more and more important to keep track of inventory and how quickly the items being stored are sold. A measure of how quickly things move from inventory to sale and then are replenished is referred to as the *inventory turnover rate*.

In the following sections, I explain the differences between cost inventory and retail inventory, and I show you how to calculate each. I also introduce you to inventory turnover.

## Comparing cost inventory and retail inventory

Inventory is counted differently in different situations. The number of items is changed into monetary value by multiplying *how many* by *how much* in terms of money. Factor into that the choice between cost inventory and retail inventory and you find that you have options galore. As you can imagine, many decisions need to be made. In general, though, after you choose a particular inventory method, you stick with it and don't flip-flop around. You stick to one method so you don't run afoul of the tax people.

Here's the lowdown on the two primary inventory methods:

✔ **Cost inventory:** A *cost inventory* is used primarily by smaller retailers. The cost inventory gives an evaluation of the merchandise before any markup (in other words, before the goods are sold).

✔ **Retail inventory:** A *retail inventory* is used by larger stores to keep track of their merchandise. They find that sales, customer returns, discounts by employees, and shortages are easier to keep track of when the cost plus markup is assigned first. The cost inventory doesn't keep track of losses due to damage or mishandling as easily as retail inventory, so for large businesses, the retail inventory is generally preferred.

### Cost inventory: Keeping track of inventory with the cost of merchandise

The basis behind keeping track of a business's assets is to count up what's available at inventory. The company bookkeeper may keep several different types or categories of inventory, including raw products, partially finished goods, finished goods, office supplies, and so on. Whether the inventory is all lumped together or figured separately, the main process is the same: Count the beginning, factor in the additions and subtractions, and the net result is the end inventory.

Here's the formula for finding the inventory at the end of a period:

Inventory at period end = Beginning inventory + Purchases − Goods sold

Having a handle on inventory means keeping good track of purchases, sales, and adjustments due to loss or damage. The following example should help you understand how to do this by using the inventory formula.

Mother Nature's Bird Seed Shop had an inventory (at cost), on January 1, of $45,300. During the next three months, additional purchases of $13,000 and $675 were made. Also during those months, the cost of goods sold was $22,550. When the manager did a physical inventory at the end of the three months (actually counted every bag and box of every product), he determined that the inventory should be $36,825. Do the computed inventory and physical inventory agree?

To find out, first you need to determine the computed inventory by adding the purchases to the beginning inventory and then subtracting the sales: $45,300 + $13,000 + $675 − $22,550 = $36,425.

The physical inventory is $36,825, so you can see that the difference is $400. That is, the physical inventory is $400 *more* than what the computed inventory said it should be. One of many scenarios could be the culprit:

✔ The manager miscounted what's currently available.

✔ The inventory figures for the beginning of the year were wrong.

✔ Something happened during sales or delivery.

The manager, in this situation, will probably keep an eye on the figures in the future to see if the discrepancy resolves itself. Or he may just count the boxes again. How would you like to have that job?

### Retail inventory: Carrying inventory at retail

Large companies usually carry their inventories *at retail* rather than at cost (see the previous section to find out more about at cost inventory). When the companies need to report on their net worth, they do the necessary computations to reduce the inventories to cost using the percentage associated with each product's markup. With the inventory kept at retail, all purchases have the retail value calculated before inserting the numbers into the formula. Computing prices using markup percentages and determining cost from prices is covered in Chapter 17. The formulas you need are also given here.

To compute the retail price when given the cost of an item and the markup percentage, or to compute the original cost of an item when given its retail price and the markup, use these formulas:

$$\text{Retail price} = \frac{\text{Cost of item}}{100\% - \% \text{ markup}}$$

$$\text{Cost of item} = \text{Retail price} (100\% - \% \text{ markup})$$

Try out these formulas with the following example: The Man's Man Merchandise Mart had a suit inventory at retail of $19,500 at the beginning of the month. During the next two months, an additional $24,000 worth of suits was purchased at cost. Also during that two-month period, the store had sales of $37,900. If the standard markup on all items is 40%, what's the inventory at retail at the end of the two months?

To compute the value at retail of the newly purchased merchandise, use the formula to change $24,000 to its retail value with a 40% markup. Here's what your math should look like:

$$\text{Retail price} = \frac{\text{Cost of item}}{100\% - \%\text{ markup}} = \frac{24,000}{100\% - 40\%}$$

$$= \frac{24,000}{60\%} = \frac{24,000}{0.60} = \$40,000$$

Now use the $40,000 in purchases to compute the current inventory:

$$\text{Inventory} = \$19,500 + \$40,000 - \$37,900$$

$$= \$21,600$$

If the Man's Man Merchandise Mart needs to determine the value of the inventory at cost (so they can do a net worth report), you would work the formula like this:

$$\text{Cost of item} = 21,600\,(100\% - 40\%)$$

$$= 21,600(60\%)$$

$$= 21,600(0.60)$$

$$= 12,960$$

So, the merchandise at inventory has a cost of $12,960.

Keeping track of inventory at retail makes the computation easier when figuring in discounts, customer returns, and other variations. I show you how in the following example.

A local dairy store has a quick turnover due to the nature of the products it carries: Milk, eggs, and butter have a pretty short shelf life. The store's inventory at the beginning of the week (at retail) was $6,000. During the week, the net sales were $23,500. This total included $2,700 in sales to employees, who get a 10% discount. Purchases (at cost) were $18,000 during that week, and the markup was 20%. What was the ending inventory (at retail)?

The employees paid $2,700 for their merchandise, which was the price of the goods with a 10% discount; so employees pay 90% of the full cost. To get the price of the merchandise before the discount, divide $2,700 by 90% to get $3,000. The difference between the amount paid and the price of the merchandise is $300 — which must be subtracted from the inventory at retail.

To get the price of the goods purchased using the 20% markup, you subtract 100% – 20% to get the original value of 80% of what's being charged. Divide

$18,000 by 80% to get $22,500. So the inventory at the end of the week is figured with this formula:

> Inventory = Beginning amount – Net sales – Adjustment due to the employee discount + Purchased items (at retail)

Now you just have to plug in the correct numbers to get the answer:

> Inventory = $6,000 – $23,500 – $300 + $22,500
>
> = $4,700

## Estimating turnover with an inventory ratio

An *inventory turnover ratio* provides a business with an estimate of how quickly the current inventory is being sold and replaced by new inventory during a particular time period. In general, the higher the number obtained by the inventory turnover ratio, the more often the inventory is sold and replaced.

Inventory turnover is an important measure of management efficiency. For instance, if a business has too much inventory, it's probably tying up money that could be better used elsewhere. On the other hand, if the company doesn't have enough inventory, it may lose business if a customer wants immediate delivery when enough of the product isn't available to fill the order.

*Note:* An inventory that consists of fresh food will have a higher turnover than an inventory of equipment or goods that doesn't spoil. Also, a company may do a separate inventory turnover ratio for each different product — or it may decide to lump all the products together.

To calculate the inventory turnover ratio, use one of the following formulas:

$$\text{Inventory Turnover Ratio} = \frac{\text{Sales}}{\text{Inventory}}$$

$$\text{Inventory Turnover Ratio} = \frac{\text{Cost of Goods Sold}}{\text{Average Inventory}}$$

To get an average inventory, use this formula:

$$\text{Average Inventory} = \frac{\text{Beginning Inventory} + \text{Ending Inventory}}{2}$$

The examples in the following sections give you a chance to try your hand at these formulas.

### Working a few straightforward examples

The examples in this section are straightforward and easy. You just have to plug the numbers you have into the formulas. Grab some paper and try them out!

Ajax Company had a cost of goods sold of $3,387,250 during the past year. The average inventory for Ajax Company was $453,217. What was the inventory turnover ratio for last year?

Figuring out this answer is easy. Just plug the appropriate numbers into the inventory turnover ratio formula, like so:

$$\text{Inventory Turnover Ratio} = \frac{\text{Cost of Goods Sold}}{\text{Average Inventory}}$$
$$= \frac{3,387,250}{453,217} \approx 7.47$$

As you can see, the turnover ratio is approximately 7.47. This is a fine turnover ratio for many businesses. You'd expect a much higher turnover ratio in grocery stores or any business that deals with perishable or strictly seasonal products.

A small family business had a cost of sales of $75,000, a beginning inventory of $10,000, and an ending inventory of $8,000. What's the inventory turnover ratio?

First, find the average inventory:

$$\text{Average Inventory} = \frac{\text{Beginning Inventory} + \text{Ending Inventory}}{2}$$
$$= \frac{10,000 + 8,000}{2} = \frac{18,000}{2} = 9,000$$

Now, use the average inventory and the cost of sales to compute the inventory turnover ratio:

$$\text{Inventory Turnover Ratio} = \frac{\text{Cost of Goods Sold}}{\text{Average Inventory}}$$
$$= \frac{75,000}{9,000} \approx 8.33$$

So the turnover ratio is about 8.33. Whether this is a good turnover ratio depends on the business. What's going to be interesting to the small businessperson is how the turnover ratio changes and what seems to be influencing that change.

### Examples with a twist: Determining average inventory

One way to compute average inventory is to divide the beginning and ending amounts by 2. Another way to determine average inventory is to record the inventory periodically and use each of the figures in the average. Check out the following example to see what I mean.

A company that keeps its inventory at cost recorded the following inventory amounts:

January 1: $220,000

April 1: $240,000

July 1: $270,000

October 1: $330,000

December 31: $210,000

If this company had total retail sales of $2,100,000 and it uses a markup of 30%, what's this company's inventory turnover rate?

To find out, first calculate the average inventory by adding the five inventory amounts together and dividing by 5, like this:

$$\text{Average Inventory} = \frac{220,000 + 240,000 + 270,000 + 330,000 + 210,000}{5}$$

$$= \frac{1,270,000}{5} = 254,000$$

Now change the sales to their equivalent at cost using the number in the formula for the inventory turnover:

Cost = Retail price (100% − % markup)

= 2,100,000 (100% − 30%)

= 2,100,000 (70%)

= 2,100,000 (0.70)

= 1,470,000

Now, using the inventory turnover ratio, divide $1,470,000 by $254,000 to get the turnover rate:

$$\text{Inventory Turnover Ratio} = \frac{\text{Cost of Goods Sold}}{\text{Average Inventory}}$$

$$= \frac{1,470,000}{254,000} \approx 5.79$$

The inventory turnover is about 5.79. But what if the CFO decided that the inventory turnover should be computed at retail, rather than at cost? Would this make a difference in the reported inventory turnover rate? Look at the following example to find out.

# Inventory destroyed by the Great Chicago Fire

Montgomery Ward overcame many obstacles to make his mail-order business a success — not the least of which was having the first inventory of his fledgling business go up in flames in the Great Chicago Fire. Aaron Montgomery Ward started out as a traveling salesman, dealing in dry goods in the late 1800s. Ward was concerned about the plight of Americans living in the Midwest who were at the mercy of unscrupulous local retailers who took advantage of the remoteness of the areas and charged exorbitant prices for often poor-quality goods.

Montgomery determined that he would buy goods at low cost for cash and sell them to people through the mail. It took some time to convince backers to join his endeavor; he had no capital himself — just a vision. In 1872, Ward rented a small room on North Clark Street in Chicago and published the first mail-order catalog with more than 150 general merchandise items in it. Orders came by mail, and the purchases were delivered to the railroad station nearest the customer.

Business wasn't immediately successful, and Ward was not at all popular with the local merchants who had had a monopoly until that time. But the Montgomery Ward catalog came to be known as the "Wish Book." It was soon copied by Sears and others who saw the great promise of the mail-order trade.

A company keeps its inventory at cost and had an average inventory of $254,000 during the past year. This same company had net retail sales totaling $2,100,000. Determine its inventory turnover rate by changing the inventory to its equivalent retail amount (assuming a markup of 30%).

First, to change the inventory at cost to its equivalent at retail, use the following equation:

$$\text{retail} = \frac{\text{cost}}{100\% - \%\text{ markup}}$$

$$= \frac{254,000}{100\% - 30\%} = \frac{254,000}{70\%} = \frac{254,000}{0.70} \approx 362,857.14$$

Now, to determine the inventory turnover rate, use the inventory turnover ratio:

$$\text{Inventory Turnover Ratio} = \frac{\text{Cost of Goods Sold at Retail}}{\text{Average Inventory at Retail}}$$

$$= \frac{2,100,000}{362,857.14} \approx 5.79$$

As you can see, the rate comes out the same whether the figures are at cost or at retail. You just can't mix and match — you can't use at retail sales with at cost inventory.

# *FIFO and LIFO: Giving Order to Inventory*

Keeping track of items manufactured or purchased and then sold by a company can be a daunting task. Thank goodness for today's bar codes and hand-held devices that make counting and accounting easier and more efficient. And, of course, using technology means that you'll never have a mistake in your inventory. After all, technology is perfect. Yeah, right!

Okay, so keeping track of inventory still has its challenges in today's business environment. You may keep track of inventory with FIFO or maybe with LIFO. No, these aren't names for cute little puppy dogs. *FIFO* is the acronym for first-in-first-out, and *LIFO* stands for last-in-first-out. Most folks try to use FIFO when keeping track of products in their refrigerators and pantries. They use the older stuff first, before it turns green and nasty.

A company is using the FIFO system of inventory control when it sells to its customers the merchandise that it purchased first. The FIFO system works well for the retailer when costs and prices are rising because the computed profit is greater. The items purchased earlier cost the retailer less, and the price that customers are paying is higher.

A retailer uses the LIFO system when, perhaps for tax reasons, it doesn't want to show as great a profit. By using LIFO when prices are increasing, the difference between the cost and the price are closer, and the net profit is smaller.

Consider this scenario from a hardware store: If an order for leaf rakes comes in, the number of rakes and total cost for the rakes is removed from the inventory. Depending on whether you use FIFO or LIFO, the cost for the same number of rakes will differ — as will the profit. Table 20-1 shows purchases of leaf rakes by the hardware store over a six-month period.

| Table 20-1 | Purchases of Leaf Rakes | | |
|---|---|---|---|
| *Date* | *Number of Rakes* | *Price per Rake* | *Total Cost* |
| January 31 | 40 | $4.00 | $160.00 |
| February 28 | 24 | $4.25 | $102.00 |
| March 31 | 60 | $4.50 | $270.00 |
| April 30 | 72 | $4.40 | $316.80 |
| May 31 | 90 | $4.50 | $405.00 |
| June 30 | 48 | $4.60 | $220.80 |

In general, when figuring costs using either FIFO or LIFO, you remove as many items from the inventory listing as are needed to fill the order — taking all of the earliest (or latest, depending on the method) and moving through the inventory until the order is filled. You may have to take just part of one inventory line, if taking all of the items exceeds the order amount.

Using the information from Table 20-1, give the following example a shot. Suppose an order for 150 leaf rakes comes in, and the FIFO method is employed to determine the cost of those rakes. What is that cost?

Beginning with the first month, use the 40 rakes purchased in January, the 24 rakes purchased in February, the 60 rakes purchased in March, and 26 of the rakes purchased in April. (You only need 26 of the 72 rakes purchased in April to fill the order of 150 rakes.) Reading the totals for January, February, and March from the table, you add the total costs: $160 + $102 + $270 = $532. Now multiply the 26 rakes from April by $4.40 to get $114.40. Add this to the $532, and the total for all 150 rakes is: $532 + $114.40 = $646.40.

The business manager for the hardware store that received an order for 150 rakes wants to minimize the profit during this boom time, which happened to occur in the summer. She decides to figure the cost of the rakes using LIFO. What's the cost of the 150 rakes (using the figures in Table 20-1)?

Beginning with the last month, use the 48 rakes purchased in June, the 90 rakes purchased in May, and 12 of the rakes purchased in April to get to the 150 rakes needed to fill the order. From the table, the total purchases in June and May are: $220.80 + $405 = $625.80. Now multiply 12 by $4.40 for a cost of $52.80 in April. Add that to the earlier purchases, and the total is: $625.80 + $52.80 = $678.60. The cost is $32.20 more than that obtained using the FIFO method.

# Determining Economic Order Quantity

The *Economic Order Quantity,* or EOQ, is the optimum number of items to be manufactured or ordered each time a business produces products or places an order for products. The EOQ applies to retailers that need a steady supply of some item and that want to have enough of the item at hand but don't want to pay too much for shipping or storage. The EOQ also applies to manufacturers that don't want to pay too much for setup costs and storage of the items.

The balancing act here involves two opposing influences: the cost of ordering or manufacturing versus the cost of storage or carrying inventory. Both scenarios (a retailer purchasing and a manufacturer producing) assume some steady and predictable demand for the item or items being produced.

Figures 20-1 and 20-2 show two different options for manufacturing or ordering and the resulting inventory. Both graphs show a total of 2,000 items being manufactured or ordered. The graph in Figure 20-1 illustrates the number of items at inventory when there's only one manufacturing or order; the graph in Figure 20-2 shows the number of items at inventory when the manufacturing or ordering is done four separate times during the year.

GO FIGURE

A ceramic company makes decorative mugs with graphics depicting patriotic scenes. The company can count on orders for about 600 mugs each month throughout the year (for a total of about 7,200 each year). It costs $500 for the setup and labor to produce the mugs — no matter how many are produced at that time. The carrying cost (cost of storage, insurance, and interest) is $2 per mug per year. Which scenario is more economical: producing 600 mugs each month, producing 1,800 mugs every three months, or producing all 7,200 mugs at the beginning of the year?

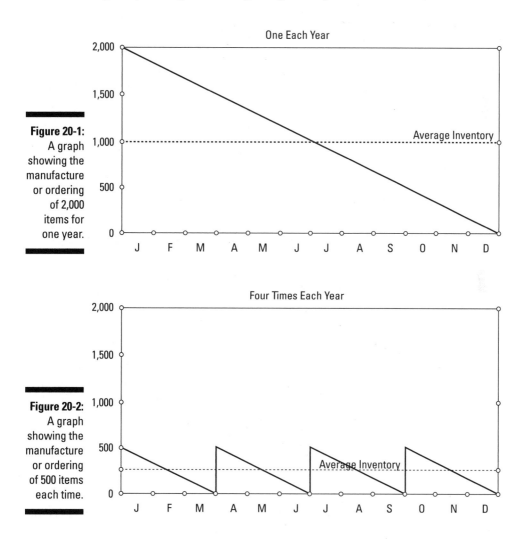

**Figure 20-1:** A graph showing the manufacture or ordering of 2,000 items for one year.

**Figure 20-2:** A graph showing the manufacture or ordering of 500 items each time.

Table 20-2 shows you the three different possibilities for the production schedule needed to produce 7,200 mugs per year. In each case, the carrying cost is equal to half the number produced during the time period. Why? Because half the production size is the average number of mugs in storage at any one time. And the cost of storage is for the entire year.

The average inventory is half the production size, because you figure that, right after production, you have all the mugs, and at the end of the time period, when you have to produce more mugs, you will be down to 0 mugs. So the average is the total amount plus 0 divided by 2 — the average. (Refer to Figure 20-2 for an illustration of the average inventory — how it's half the production level.)

| Table 20-2 | Determining the Cost of Production and Storage | | |
| --- | --- | --- | --- |
| Number of Mugs | Setup Cost | Carrying Cost Based on Average Inventory | Total Setup × Times per Year + Carrying Cost |
| 600 | $500 | 300 × $2 = $600 | $500 × 12 + $600 = $6,600 |
| 1,800 | $500 | 900 × $2 = $1,800 | $500 × 4 + $1800 = $3,800 |
| 7,200 | $500 | 3,600 × $2 = $7,200 | $500 × 1 + $7,200 = $7,700 |

Referring to Table 20-2, you see that the total cost first went down when the number of setups was reduced from 12 to 4, but then the total cost increased when the cost of storage became greater. The big balancing act of storing versus setups is boiled down to a formula that allows you to determine the best quantity to produce at one time — and the number of times per year to do the production.

The Economic Order Quantity, or EOQ, is estimated with the following formula:

$$Q = \sqrt{\frac{2SR}{CK}}$$

where $Q$ is the quantity you want (the number of items to produce or order at one time), $S$ is the setup cost or shipping cost per order, $R$ is the required number of items needed each year, $C$ is the cost per unit being manufactured or ordered, and $K$ is the inventory carrying charge given as a percentage of the cost or value of the item. The upcoming example shows you how to use this formula.

The ceramic company depicted in Table 20-2 can count on orders for about 600 mugs each month throughout the year. As noted in the earlier problem, it costs the company $500 for the setup and labor to produce mugs — no matter how many are produced at that time. The carrying cost is $2 per mug, which is determined by multiplying the cost per mug, $10, by the inventory carrying charge, 20%. Use the EOQ formula to determine the number of mugs that should be produced and how many times each year to make them.

Using the formula, you replace $S$ with $500, $R$ with 7,200, $C$ with $10, and $K$ with 0.20. You actually already have the value of $CK$ from the stated $2 carrying charge per mug. Now plug the numbers in like so:

$$Q = \sqrt{\frac{2(\$500)(7200)}{(\$10)(0.20)}} = \sqrt{\frac{7,200,000}{2}} = \sqrt{3,600,000} \approx 1,897.4$$

As you can see, that's almost 1,900 mugs per run. The company needs a total of 7,200 mugs for the year, and 1,900 doesn't divide 7,200 evenly, so consider two possibilities:

- ✔ **Option 1:** Produce 1,900 in each month of January, April, and July, and then 1,500 in October.

- ✔ **Option 2:** Produce graduated amounts of mugs with 1,920 in January, 1,880 in April, 1,840 in July, and the final 1,560 in October.

To compare the two options, refer to Table 20-3 for Option 1 and Table 20-4 for Option 2. The carrying charge changes with the number of mugs at inventory. In Option 1, the average of 1,900 mugs is being stored for three-fourths of the year; the average of 1,500 mugs is being stored for one-fourth of a year. In Option 2, the average inventory changes during each quarter year (beginning in January, April, July, and October), and that average number of mugs is stored for just its respective one-fourth of a year.

### Table 20-3: Option 1: Three Orders of 1,900 and One of 1,500

| Number of Mugs | Setup Cost | Carrying Cost Based on Average Inventory | Total Setup × Times per Year + Carrying Cost |
|---|---|---|---|
| 1,900 | $500 | 950 × 0.75 × $2 = $1,425 | $500 × 3 + $1,425 = $2,925 |
| 1,500 | $500 | 750 × 0.25 × $2 = $375 | $500 × 1 + $375 = $875 |
| | | | Total: $3,800 |

| Table 20-4 | | Option 2: Decreasing Values of Orders | |
|---|---|---|---|
| Number of Mugs | Setup Cost | Carrying Cost Based on Average Inventory | Total Setup × Times per Year + Carrying Cost |
| 1,920 | $500 | 960 × 0.25 × $2 = $480 | $500 + $480 = $980 |
| 1,880 | $500 | 940 × 0.25 × $2 = $470 | $500 + $470 = $970 |
| 1,840 | $500 | 920 × 0.25 × $2 = $460 | $500 + $460 = $960 |
| 1,560 | $500 | 780 × 0.25 × $2 = $390 | $500 + $390 = $890 |
| | | | **Total: $3,800** |

It appears that you can play around with the order sizes, but you don't really save anything by changing the order sizes. For simplicity, you may want to consider the simpler route and make the ordering process more consistent.

Try out one more example: Say that a specialty coffee shop orders its paper cups from a company in a neighboring state. Each order the shop places for cups has a flat service charge of $40; the actual shipping charges are included in the cost per cup. The cups cost $0.0325 each, and the coffee shop uses 900,000 cups annually, spread evenly over the entire year. If it costs the shop a 10% carrying charge (based on the cost of the cups) to store the cups, how many cups should the coffee shop order at a time? *Note:* Assume that the shop has storage room for up to 150,000 cups.

Using the EOQ formula, and letting $S$ be $40, $R$ be 900,000, $C$ be 0.0325, and $K$ be 10%, gives you:

$$Q = \sqrt{\frac{2(\$40)(900,000)}{(\$0.0325)(0.10)}} = \sqrt{\frac{72,000,000}{0.00325}} \approx \sqrt{2,215,384,615} \approx 148,842$$

That's almost 150,000 cups per order. Dividing 900,000 by 150,000, you get 6, so the coffee shop should place 6 orders of 150,000 cups each. If the shop tries to follow the value of the equation exactly, it would need to order 148,842 cups the first 5 times and then 155,790 on the 6th order. However, because the coffee shop doesn't have room for more than 150,000 cups, the 150,000 number is the best choice.

# Part VI

# Surviving the Math for Business Facilities and Operations

The 5th Wave          By Rich Tennant

"Can you explain your loan program again, this time without using the phrase 'yada, yada, yada?'"

# In this part . . .

Owning or managing a rental property has its challenges. For instance, you have to determine a fair rental amount, and then you actually have to collect that amount. And just when you think you're safe, loans, mortgages, utilities, and taxes get thrown into the mix when you determine revenue and cost. Last but not least, you have to calculate areas and perimeters in order to maintain your buildings. After all, you can't buy new carpet or have someone clean your building without knowing your square footage. As you can see, all the different facets of a rental business have their own special mathematical needs and processes. But don't worry! I have included everything you need to know in this part.

# Chapter 21

# Measuring Properties

. . . . . . . . . . . . . . . . . . . . . . . . . . . . . . . . . . . . . . . . . . . . . . . . .

## In This Chapter

▶ Calculating areas and perimeters

▶ Using square measurements

▶ Computing cost per square foot

▶ Determining areas of irregular shapes

▶ Taking advantage of surveyors' tools

. . . . . . . . . . . . . . . . . . . . . . . . . . . . . . . . . . . . . . . . . . . . . . . . .

Measurements tell you about the size of something. Say, a building lot, for example. And a Floridian's yardstick is the same as one belonging to someone from Montana. In fact, measurements are the same almost everywhere you go. So you can feel confident that your calculation of 6 square feet (sq. ft.) is the same as the next person's understanding of the same footage. Your challenge, instead, is to know how to do the computations of the area of a room or building so you can do comparisons with someone else.

Businesses both expand and contract. What I mean is that they sometimes need to add space, reconfigure existing space, rent out part of their space, or relocate to smaller quarters. And it's not just the inside areas that need to be measured. Many businesses have large parking lots that need to be resurfaced or cleared of snow; they also may have grounds to be mowed and landscaped. Many aspects of a business and the space it takes up involve knowing how to determine the amount of area needed.

Perfect squares are important when it comes to measurements, but so are irregular shapes. So this chapter shows you how to accommodate those irregular shapes in business — to think outside the proverbial square box. I show you how to compute square footage (even when the area being measured is not square) so you can use it to determine the cost of carpeting or marble flooring or how much heating capacity you need. Also, in this chapter, you discover how to measure like a surveyor: in metes and bounds. The information in this chapter allows you to measure and calculate on your own or verify what the Realtor, interior designer, or contractor tells you.

# Exploring Area and Perimeter Formulas for Different Figures

Several different area formulas are used throughout this chapter. In the later section, "Squaring Off with Square Measures," for instance, you see the formula for the area of a rectangle. In the section "Measuring Irregular Spaces," you become familiar with the formulas for the areas of triangles, trapezoids, and regular polygons (a *polygon* is a figure whose sides are segments).

This section helps you get started with basic area and perimeter tools. You find formulas for the commonly used figures found in buildings and plots of land. Even though every polygon can be broken into a number of triangles (see the later section "Trying out triangles") and the area can be computed using those triangles, this method should be saved for a last resort. Why? It's generally easier to use the specific formula designed for that specific type of figure.

In the following sections, I provide sketches of geometric figures, brief descriptions of those figures, and formulas for area and perimeter of the figures. The *perimeter* of a polygon or curved figure is the distance around the outside. The *area* of a figure is the number of square units you can fit inside it. You find some of the figures in this section later in the chapter, accompanying possible scenarios. Other figures and their formulas are offered here for your quick reference.

## Rectangles

A *rectangle* is a four-sided polygon with square corners (see Figure 21-1). The opposite sides, denoted by *l* for length and *w* for width, are parallel and equal in length. To find the perimeter of a rectangle, you can add up the four sides, or you can use this formula:

$$P = 2(l + w)$$

As you can see, this formula tells you to add the length to the width and then double the sum. To find the area of a rectangle, use this formula:

$$A = lw$$

Here you just have to multiply the length by the width. Pretty simple!

**Figure 21-1:**
A rectangle.

## Squares

A *square* is a four-sided polygon with square corners (see Figure 21-2). With this figure, all sides are the same measure, denoted *s*. The perimeter of a square is just 4 times the measure of a side:

$P = 4s$

The area of a square is the measure of a side squared:

$A = s^2$

**Figure 21-2:**
A square.

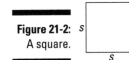 *s*

## Parallelograms

A *parallelogram* is a four-sided figure whose opposite sides, the length and width, are parallel and of the same measure. Think of it as being a squished rectangle — the angles are usually not 90°. You can see what a parallelogram looks like in Figure 21-3.

The perimeter of a parallelogram is found the same way as a rectangle's perimeter — just add up the sides. Or, if you'd like, you can use this formula:

$P = 2(l + w)$

To find the area, you multiply one of the sides by the distance between that side and the opposite side. You call the perpendicular distance between two sides of a parallelogram its height, *h*. So your formula looks like this:

$A = lh$

**Figure 21-3:**
A parallelo-
gram.

Here's an example to wrap your brain around: Say that a plot of land is shaped like a parallelogram. Walking around the perimeter of the plot, you measure 400 feet, 150 feet, 400 feet, and 150 feet. You've been told that the plot measures 40,000 sq. ft. What's the perimeter of the plot, and what's the height (the distance between the two longer sides)?

The perimeter is just the sum of the measures of the sides. You can either add the sides (400 + 150 + 400 + 150) or you can use the formula:

$$P = 2(400+150)$$
$$= 2(550)$$
$$= 1,100 \text{ feet}$$

If the area of the plot is 40,000 sq. ft., the distance between the two longer sides (the sides that measure 400 feet) must be 100 feet. The 100 feet is the *height*. Using the area formula, you see that 40,000 = 400 × 100.

## Triangles

A *triangle,* shown in Figure 21-4, is a three-sided polygon. There's no elegant way to find the perimeter — you just add up the lengths of the sides. You find the area of a triangle in one of two ways. You can use either of these formulas:

$$A = \frac{1}{2} bh$$
$$A = \sqrt{s(s-a)(s-b)(s-c)}$$

The first equation is the traditional formula where you multiply one side by the distance to the opposite corner and take half of that product (I refer to this as the *half-of formula*). The second formula is called *Heron's formula;* here you use the lengths of the sides and half the perimeter, *s*. I discuss both of these formulas in the section "Trying Out Triangles."

**Figure 21-4:**
A triangle.

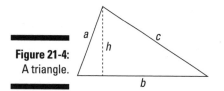

## Trapezoids

A *trapezoid* is a four-sided polygon with just two sides parallel (see Figure 21-5). The perimeter is found by adding up the lengths of the four sides:

$$P = a_1 + a_2 + b_1 + b_2$$

The area is found by multiplying the height, $h$, which is the distance between the two parallel sides, by the sum of the lengths of the two parallel sides and then taking half of the product. Here's the formula:

$$A = \frac{1}{2} h \left( b_1 + b_2 \right)$$

Refer to the section "Tracking trapezoids" for an example of finding the area of a trapezoid.

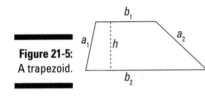

**Figure 21-5:**
A trapezoid.

## Regular polygons

A *regular polygon* is different from just any old polygon because all the sides are the same measure and all the angles are the same measure. Refer to Figure 21-6 to see a regular hexagon. The perimeter of a regular polygon is found by multiplying the measure of any side by the number of sides:

$$P = sn$$

The area of a regular polygon is found by multiplying the distance from a side to the center, denoted by $a$ (for apothem), by the length of one of the sides, $s$, by the number of sides, $n$, and then taking half of that product. In other words, the formula looks like this:

$$A = \frac{asn}{2}$$

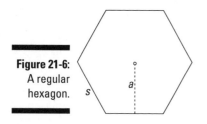

**Figure 21-6:**
A regular
hexagon.

# Circles

A *circle* is a curved shape where all the points on the figure are some set distance from the center (see Figure 21-7). The set distance is called the circle's *radius,* denoted by *r*. The distance all the way across a circle is called the *diameter.* The diameter is twice the radius. To find the perimeter of a circle (also called its *circumference*), you multiply 2 by 3.14159 (which is π) by the length of the radius. Here's the formula:

$$P = 2\pi r$$

To find the area of a circle, use

$$A = \pi r^2.$$

The value of π is approximately 3.14, 3.1416, 3.14159, or . . . . Actually, π (the Greek letter "pi") has no exact numerical value. Use as many or as few of the decimal places as you desire. The more digits you use, though, the more accurate your computation.

**Figure 21-7:**
A circle.

Here's an example: What are the perimeter and area of a circular garden if the distance all the way across the garden is 40 feet?

The diameter is twice the radius, so the radius of this circle is 20 feet. The perimeter of the garden is:

$$P = 2 \times 3.14159 \times 20$$

$$= 125.6636 \text{ feet}$$

The area is

$$A = 3.14159 \times 20^2$$

$$= 1256.\dot{6}36 \text{ sq. ft.}$$

You can now use these figures to determine how much border material to purchase and how much fertilizer is needed.

## *Ellipses*

An *ellipse* is an oval or egg-shaped figure (refer to Figure 21-8). Many door insets and mirrors are ellipses. The perimeter and area of an ellipse are found using two different measures, *a* and *b*. These two measures are: the longest distance inside the ellipse and the shortest distance across the inside. The two segments representing the distances are always perpendicular to one another. The perimeter, or distance around the outside of an ellipse, is found with this formula:

$$P = 2\pi \sqrt{\frac{a^2 + b^2}{8}}$$

The area is found with

$$A = \frac{\pi ab}{4}$$

**GO FIGURE**

Suppose an oval window in your front door is 3 feet long and 2 feet wide. How much trim must you buy to go around the outside of the window, and what's the area of the window?

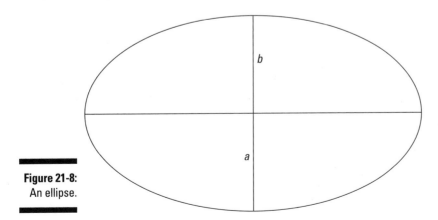

**Figure 21-8:**
An ellipse.

The trim around the outside is the perimeter. Here's how to figure the perimeter:

$$P = 2\,(3.14159)\sqrt{\frac{3^2 + 2^2}{8}} = 6.28318\sqrt{\frac{13}{8}} \approx 8.00951$$

So you need about 8 feet of trim. The area of the window is figured like this:

$$A = \frac{\pi\,(3)(2)}{4} \approx 4.71238$$

The area here is about 4.7 sq. ft.

# Squaring Off with Square Measurements

When you're told that an office or warehouse contains 2,400 square feet (sq. ft.) or that a parking area consists of 5 square miles (sq. mi.), you're being given a measurement of the building or area that's basically just a bunch of squares all stuck together. A room that has an area of 64 sq. ft. could be a square room, but it could also be rectangular or triangular. In fact, it may be a room that looks like an oddly shaped polygon. (A *polygon* is a geometric figure with line segments for its sides.)

Figure 21-9 shows you four different possibilities for a floor plan that measures 64 sq. ft. You could construct many more such rooms and offices in innumerable other shapes. Maybe you prefer hexagons or octagons or some other polygon with even more sides than eight. Notice that the triangular room has squares that are broken apart. Add the irregular pieces together, and you still get 64 sq. ft.

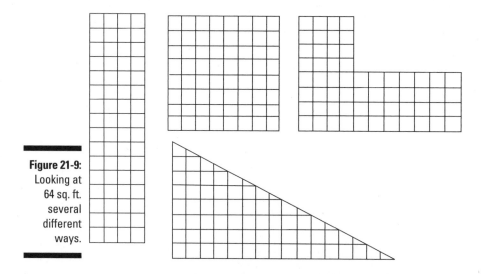

**Figure 21-9:**
Looking at
64 sq. ft.
several
different
ways.

## Figuring total area in various measurements

The total number of square feet in an office or warehouse is typically the sum of all the areas of useable space in that building. The useable area in a large office complex may include individual offices, secretarial work areas, mail-sorting rooms, lunch rooms, and storage spaces. A well-planned business also may have an outdoor walking track or picnic area.

Indoor rooms need flooring, paint, heating, and air conditioning. Outdoor areas need to be mowed, weeded, watered, and otherwise tended. All these activities revolve around determining the respective areas. In the following sections, I show you how to determine these areas, such as square footage, square yardage, and square acreage.

### Measuring in square feet and square yards

One of the more common measures of area is the square foot, which is often abbreviated as sq. ft. Floor tiles are often about 1 x 1 foot — making it easy to estimate the number of square feet in a particular room. Smaller tiles may be 1 x 1 inch. Carpeting is usually measured in square yards (sq. yd). Mixing and matching the measures makes it necessary to pay attention to details. An error in measurement can be costly in money, resources, or time. The upcoming example explains how to calculate total square footage and yardage.

Here's one tidbit that will help you with the example:

1 sq. yd. = 9 sq. ft. (3 ft. × 3 ft.)

Say that you've hired a new cleaning crew to take care of your office building. The total cost of cleaning is determined by how many square feet of carpeting and tile floor need to be cleaned. Your building has three levels, each with a total of 1,540 sq. ft. of useable space. The first level of your building contains four tiled bathrooms, each 3 sq. yd. in area, and a reception area that has a counter taking up 8 sq. ft. of the carpeted area. The second level has two tiled bathrooms, each 3 sq. yd. in area, and a tiled lunchroom with an area of 220 sq. ft. The third level has just two tiled bathrooms; one is 3 sq. yd., and the other (the executive bathroom) is 5 sq. yd. How many square feet of carpeting and how many square feet of tile are the cleaning crew responsible for?

To solve this problem, you need to first determine how many square feet of each level are carpeted by subtracting any tiled or covered areas. Add up the areas of all the carpeted and tiled rooms.

On the first level, you have four bathrooms, each 3 sq. yd. To get the square footage of one bathroom, multiply each side of this equation by 3: 1 sq. yd. = 9 sq. ft. You get: 3 sq. yd. = 27 sq. ft. Four bathrooms means that you multiply the 27 sq. ft. by 4, ending up with 108 sq. ft. of tiled floors in the bathrooms. The reception area takes up 8 sq. ft. of area, so the first level has 1,540 – 108 – 8 = 1,424 sq. ft. of carpeting.

The second level of the building has two bathrooms measuring 27 sq. ft. (the same computation as on the first level), which is a total of $27 \times 2 = 54$ sq. ft. The second level also has a tiled lunchroom with an area of 220 sq. ft. To get the amount of tiled floor, add 54 + 220 = 274 sq. ft. Now you're ready to find out how much carpeting the second level has: 1,540 – 274 = 1,266 sq. ft.

On the third level, the two bathrooms are tiled. One bathroom has the same 27 sq. ft. as the bathrooms on the first two levels. The second bathroom is 5 sq. yd., so multiply $9 \times 5 = 45$ sq. ft. To get the total amount of tiled floor on the third level, add 27 + 45 = 72. Now you can figure out the total amount of carpeting on that level: 1,540 – 72 = 1,468 sq. ft.

Add the areas of tiled floor for each level: 108 + 274 + 72 = 454 sq. ft. Then add the areas of carpeting for each level: 1,424 + 1,266 + 1,468 = 4,158 sq. ft. Great work! Now you're ready with the figures to negotiate with the cleaning crew.

### Determining acres

Large residential yards and parks are measured in acres rather than square feet, square yards, or square miles. Here are some famous examples of acreage:

- ✔ The White House lawn is about 18 acres.

- ✔ Hot Springs National Park in Arkansas is 5,760 acres.

- ✔ Yellowstone National Park — which is a part of three states — spans more than 2 million acres.

Here are a few things to keep in mind as you calculate acreage in the following example:

> 1 acre = 43,560 sq. ft.
>
> 1 sq. mi. = 640 acres = 27,878,400 sq. ft.

Also remember that the area of a rectangle is found by multiplying the length of the rectangle by the width of the rectangle.

Suppose you need to determine the size of a new parking lot. You start off by measuring with feet. Then you do some calculations to change to square feet, then to square miles, and finally to acres to see which measurement makes the most sense for your situation. Measuring the length of the rectangle along a road, you see that it measures 42,300 ft. And the width, measured along a road perpendicular to the first one, is 5,600 ft.

After you have your measurements in feet, you want to find the area of this rectangular plot of land in square feet. To do so, you multiply the length by the width: $42,300 \times 5,600 = 236,880,000$.

This total is more than 2 hundred million sq. ft. Do you see why measuring in acres or square miles makes more sense in this scenario?

To get your total into square miles, you have to divide the total number of square feet by 27,878,400 sq. ft.:

$$\frac{236,880,000}{27,878,400} \approx 8.4969$$

That's almost 8.5 sq. mi.

Now you're all set to change over to acres. To do so, multiply the square miles you just figured by 640 (the number of acres in a square mile): $8.4969 \times 640 \approx 5,438$ acres.

Another way to determine the acreage is to take the total number of square feet and divide by 43,560 (the number of square feet in an acre), like this:

$$\frac{236,880,000}{43,560} \approx 5,438$$

You get the same answer, of course. But you may find it easier to skip the calculation to change to square miles.

## Adjusting for increased area

Most businesses want to be efficient and save money, so they make the most of the rooms and area that they have in their buildings. But making employees happy to be at work counts for a lot, too. So, many business owners offer more pleasant work areas, break rooms, or other facilities to encourage more efficiency and loyalty.

The addition of a solarium, child-care area, or lounge enhances the quality of a workplace, and increases its value. But additions like these also introduce some other aspects that need to be dealt with: more area to pay taxes on, and more air volume to heat and cool.

## Computing percents of areas

Adding 300 sq. ft. onto an existing structure currently measuring 3,300 sq. ft. makes a much bigger impact than adding 300 sq. ft. onto a structure that measures 6,000 sq. ft. The percent increase gives a better measure and tells you more about the potential impact on utilities and taxes. The following example shows you exactly how to compute percents of area.

Say that a 12-x-25 foot solarium is added on the side of a building that originally had 3,300 square feet. What percentage of the total area will this solarium account for in the building? And by what percent was the area increased by adding the solarium?

A 12-x-25 foot room measures 300 sq. ft. (to get that answer, you simply multiply $12 \times 25$). The total area of the building plus the solarium is now: 3,300 + 300 = 3,600 sq. ft.

To find the percentage of the solarium, divide 300 by 3,600, write the answer as a decimal, and change the decimal to a percent by moving the decimal place. (Refer to Chapter 2 if you need more information on fractions, decimals, and percents.) Here's what your math should look like:

$$\frac{300}{3,600} = \frac{3}{36} = \frac{1}{12} \approx 0.0833$$

$$0.0833 = 8.33\% \approx 8\frac{1}{3}\%$$

As you can see, the solarium is a little more than 8% of the total area of the building. To determine the percent increase in the area of the building, divide the increase, 300 sq. ft., by the original square footage, 3,300 sq. ft, like this:

$$\frac{300}{3,300} = \frac{3}{33} = \frac{1}{11} \approx 0.0909$$

$$0.0909 = 9.09\% \approx 9\frac{1}{11}\%$$

The area of the building increased by a little more than 9%.

## Heating up and cooling off

The size of a company's furnace and air conditioner is determined not only by the size of the building but also by where in the world it's located. A company in the Upper Peninsula of Michigan, for instance, won't need air conditioning quite as much as a company in southern Florida. However, the company in Florida doesn't need to heat things up too much, either.

And what effect does an addition to a building have on the heating and air conditioning of the total building? If the heating and air conditioning needs increase at the same rate (percentage) that the size of the building increases, you can estimate the needs based on the new square footage. The following example shows you how.

Consider a small clothing store in Madison, Wisconsin, which measures 3,300 sq. ft. before the owners construct a 1,700 sq. ft. addition. If the original store needed 5.5 tons of cooling capacity and 145,200 Btus of heating capacity, what are the new requirements for the store plus the addition? Assume that the requirements for heating and cooling increase at the same rate as the floor space increases.

To determine the new requirements, create a proportion. (I show you how to handle ratios and proportions and simple algebraic equations in Chapter 4.) When solving the cooling problem, first put the old square footage over the new square footage in a fraction. The new square footage is 3,300 + 1,700 = 5,000 sq. ft. Now set that fraction equal to the old tonnage over the unknown tonnage, which you designate as $x$. Find the value of $x$ by cross-multiplying the numbers in the proportion and solving for $x$ in the equation. *Tip:* You can make life easier for yourself by first reducing the fraction on the left to make the numbers smaller. Your math should look like this:

$$\frac{3,300}{5,000} = \frac{5.5}{x}$$

$$\frac{33}{50} = \frac{5.5}{x}$$

$$33x = 5.5\,(50)$$

$$33x = 275$$

$$x = \frac{275}{33} \approx 8.33$$

It'll take about 8.3 tons of cooling to air-condition the whole store. What about the heating, you ask? Just set up another proportion using the original size of the store, the new size, and the original number of Btus; then let the new number of Btus be designated by $y$. Check out the answer:

$$\frac{3,300}{5,000} = \frac{145,200}{x}$$

$$\frac{33}{50} = \frac{145,200}{x}$$

$$33x = 145,200\,(50)$$

$$33x = 7,260,000$$

$$x = \frac{7,260,000}{33} = 220,000$$

The furnace or furnaces have to provide 220,000 Btus of heating capacity.

## *Taking frontage into consideration*

In Figure 21-9, you see four different ways to arrange 64 sq. ft. Those geometric figures could represent building lots, yards, patios, or ponds. One major consideration when computing the area of a building lot, park, or parking lot is the length of the *frontage* (the part or edge of a piece of land that lies next to public access) along a road, a lake, or another lot.

Having more lake frontage may be important to a restaurant that wants to take advantage of patio dining in nice weather and the gorgeous view during the spring and fall. Other considerations, of course, are that increased lake frontage means more upkeep. Even though Canada geese are lovely to look at, they leave behind rather messy and smelly souvenirs along the walkways. And don't forget that geese are prone to attacking the unwary stroller who may venture too close to goslings.

Suppose that a lakeside lot has lake frontage measuring 40 feet. If the rectangular lot has a total area of 8,400 sq. ft., what's the length of the lot?

The area of a rectangle is found by multiplying the length by the width. If you assume that the width is 40 feet, you can divide 8,400 by 40 to get 210 feet. Easy, wasn't it?

## *Calculating acreage*

Because large parks, grounds, and farms are so big, they're usually measured in acres rather than square feet or square yards. Here's some perspective to get you started: A football field is about 1⅓ acres.

The following equivalents can help you solve the following problem (and any real-life problem you're up against):

> 1 acre = 4,840 sq. yd.
>
> 1 acre = 43,560 sq. ft.
>
> 1 sq. mi. = 640 acres

If you're ready to work a math problem dealing with acres, check out the following example.

Say that a contractor wants to build a 35,000 sq. ft. office complex on 6 acres of land. The *footprint* (the amount of area taken up by the base of the structure) of that building is 22,000 sq. ft. The industrial park has a rule that a building's footprint can't exceed 10% of the total area of the lot. Is this plan going to work for the contractor?

---

# Oh, my aching acre

The word *acre* means, roughly, *open field*. An acre was supposed to be the amount of area or land that you could reasonably expect one man to plow in one day when his plow was hooked up to one ox.

An acre isn't really a square measure like a square foot or square yard. If you want a square piece of land that's exactly an acre, the square has to be about 208.7 feet on a side or about 69.6 yards on a side. Because of its origin, an acre isn't beholden to a square shape. After all, if an acre of land is long and narrow, it's easier to plow — because you don't have to turn your plow or tractor around as often.

---

You first need to convert your 6 acres to square feet using a proportion. Then you're ready to divide 22,000 by this total number of square feet. Change your decimal answer to a percent. (If you need to review fractions, decimals, and percents, flip to Chapter 2.) Here's what your math should look like:

$$\frac{1 \text{ acre}}{6 \text{ acres}} = \frac{43,560 \text{ sq. ft.}}{x \text{ sq. ft.}}$$

$$x = 6\,(43,560) = 261,360$$

$$\frac{22,000}{261,360} \approx 0.084175$$

This decimal is equivalent to about 8.4%. The contractor is fine.

# Determining Cost per Foot

When building a factory, store, or office, you want to start out with an approximate cost. This way you know what you're possibly getting yourself into before committing yourself to hiring the architect, buying the land, and so on. Of course, everything seems to cost more than you first figure, but a ballpark estimate at least helps you determine whether the project is even worth considering.

*Costs per square foot* are going to vary widely depending on what part of the country you live in, the quality of construction materials you intend to use, the cost of gasoline or diesel fuel, concrete availability, and sometimes other unpredictable factors. The examples in the following sections serve as general guidelines for estimating the costs and having intelligent conversations with Realtors and builders.

## Renting space

When renting an office or some type of retail or manufacturing space, the common denominator in determining the total cost of the area is the amount of floor space dedicated to particular purposes and the cost per unit of that area. Other factors include location, street accessibility, and parking. However, those items all help determine what the multiplier or cost per square unit becomes. The total price is still driven by area. The following examples give you a clue as to how area affects rent prices. I try to include different scenarios that may occur so that you can apply what you need from each to your own situation.

Imagine that an attorney wants to set up shop in a new city and is looking into renting an office in a particular downtown office building. If offices in that building go for $30 per square foot, and the lawyer has budgeted no more than $1,800 per month for rent, how large of an office can she afford?

To find out how much office the attorney can afford, you have to divide $1,800 by $30, which gives you 60 sq. ft. Guess her decision depends on how the office is divided and where the windows are.

Here's another example to consider: A manufacturing company is relocating to an area where the average cost of renting office space is $111 per sq. ft. The rental of industrial space is $140 per sq. ft. The company needs 5 offices that are at least 300 sq. ft. and 2 offices that are at least 500 sq. ft. It also needs 100,000 sq. ft. of industrial space. What can the company expect to be paying in monthly rent?

To solve, multiply each square footage requirement by the respective cost per square foot. Then add the products together. Here's how:

$$5 \text{ offices} \times 300 \text{ sq. ft.} \times \$111 = \$166,500$$
$$2 \text{ offices} \times 500 \text{ sq. ft.} \times \$111 = \$111,000$$
$$\underline{100,000 \text{ sq. ft.} \times \$140 = \$14,000,000}$$
$$\text{Total } \$14,277,500$$

That's more than $14 million just to rent the space. Might it be a better idea to build?

# Estimating building costs

A company can choose to rent office and manufacturing space, or it can decide to buy or build the structure it needs. If the company buys an existing structure, the cost is pretty clear-cut. Building a new structure, however, involves selecting a contractor and builder, and it requires some thought and preplanning. So you want to know the basics in terms of area and cost. This way you can select a contractor with confidence. After all, you don't want to come across any nasty surprises.

Estimating the cost of building a structure is tricky and can be fraught with pitfalls. So you have to keep a few things in mind. For example, when working with area, you need to consider the *finished square footage* (the number that includes all the livable area, from outside wall to outside wall). This number doesn't include garages or unfinished spaces, so the estimate is a bit misleading. You also need to consider the *constructed square footage,* which includes spaces that aren't lived in. The bottom line in cost is found by doing the actual figuring using material costs and labor. But a decent estimate is still helpful.

Estimating the cost correctly is important so that

 ✔ You don't overextend yourself.

 ✔ You can make budgeting plans for the future.

 ✔ You understand any tax implications.

Find out how to determine a building cost estimate with the following examples.

A two-story building for a dentist's office and laboratory is to be constructed on a slab with no garage and no basement. What's the estimated cost if the building is to measure 2,700 square feet and the estimated cost of construction is $123 per sq. ft.?

This problem is so easy! You simply multiply 2,700 by $123, which gives you $332,100 in building costs.

An office building with 2,700 sq. ft. of livable space is to have an unfinished basement measuring 1,200 sq. ft. The estimated total cost of the business and basement is $360,000. What's the cost per square foot?

Even though the building has 1,200 sq. ft. of basement, this isn't considered to be useable space, so you divide $360,000 by 2,700 to get $133.33 per sq. ft.

# Measuring Irregular Spaces

Wouldn't it be wonderful (or maybe extremely boring) if everything was rectangular in shape? A rectangle has a set width and length and no special character. No matter where you choose to walk from one end to the other of a rectangle, it'll be the same distance. Having a set measure is efficient, but it may not be appealing to the restaurant owner who wants nooks and crannies for intimate dining; to the church council that wants a large, open, octagonal area; or to the theatre manager who wants a trapezoidal auditorium that tapers in as it approaches the stage.

The advantage of having just rectangles — and those special rectangles, called *squares* — is that the computations for area and perimeter are relatively simple. To determine the area of a rectangle or square, you multiply the length by the width. To determine the perimeter, you add length and width together and then double the result. You can find many different area and perimeter formulas in the earlier section, "Exploring Area and Perimeter Formulas for Different Figures."

## Breaking spaces into rectangles

Many stores and office buildings have little alcoves or bends that form adjacent rooms. By creating interesting shapes for floor plans, you make determining area and perimeter (so that you can order carpeting or figure out the cost of cleaning materials) a bit of a challenge.

Most rooms in offices consist of areas with square corners or 90-degree angles. However, some creatively designed rooms may veer off in some other angles. A four-sided figure with all 90 degree angles for corners is a rectangle. And the rectangle is one of the easiest forms to use when figuring area and perimeter.

In Figure 21-10, I show you a large area and several options for breaking up the area into separate rectangles. When dividing an area into convenient rectangles, you can choose any number of combinations. You usually want to keep the number of rectangles as small as possible, though. After all, there's no need to make this more complicated than necessary.

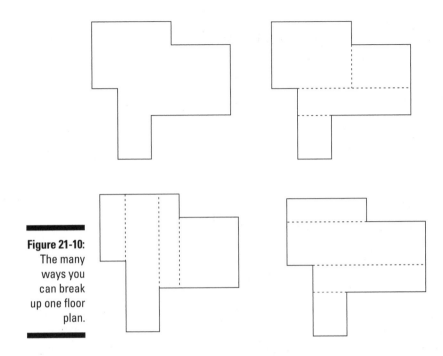

**Figure 21-10:**
The many
ways you
can break
up one floor
plan.

In Figure 21-11, you can see that I have chosen a particular division and have put in the dimensions of each individual rectangle. Did I just guess the numbers for the broken-up spaces? Of course not! These measurements need to be precise.

I'm assuming that the room has perfectly perpendicular corners and that the measures are correct. And you'll probably notice that the measures are all in whole feet. This probably isn't very realistic. Most rooms have measurements in feet *and* inches. But the whole feet will do to illustrate my point.

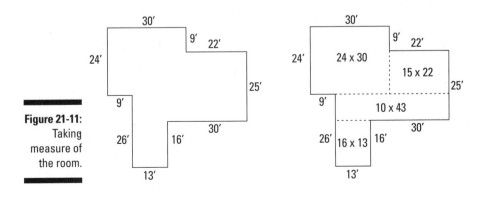

**Figure 21-11:**
Taking
measure of
the room.

As you can see in Figure 21-11, the upper left-hand rectangle has a measure of 24' x 30'. This division was the simplest, because the width and length are two entire walls. The upper right-hand rectangle has a measure of 15' x 22'. The 22-foot measure is pretty obvious. I get the 15-foot measure by subtracting the 9-foot measure that's along the right side of the first rectangle from 24 feet on the left side of that first rectangle. The middle rectangle that measures 10' x 43' takes a little more doing to figure out. I get the 10 foot measure by subtracting 16 feet from 26 feet. I get the 16 feet from the right side of the bottom rectangle and the 26 feet from the left side, opposite it. Whew! Only one more to explain: the bottom rectangle. This one is pretty easy. The 16 feet comes from the measure of the right side, and the 13 feet comes from the measure of the bottom.

Now you can compute the entire area. To do so, simply multiply each width and height and then add all the separate areas together, like this:

$$24 \times 30 = 720$$
$$15 \times 22 = 330$$
$$10 \times 43 = 430$$
$$\underline{16 \times 13 = 208}$$
$$1,688$$

So, the total area is 1,688 sq. ft. Great work!

## Trying out triangles

If your irregularly shaped room or parking area doesn't have all right angles for corners, you get to resort to more creative divisions of the entire area. Any *polygon* (a geometric figure with line segments for its sides) can be divided into a series of triangles.

For example, a four-sided figure divides into two triangles; a five-sided figure divides into three triangles; a six-sided figure divides into four triangles; and so on. The general rule is that the number of triangles you need is two less than the number of sides of the polygon.

Triangles work well when computing the area of an irregular polygon, because you can determine the area of a triangle using either the standard *half-of formula* or with *Heron's formula*.

Here's the half-of formula for finding the area of a triangle:

$$A = \frac{1}{2}bh$$

In this formula, *b* is the measure of the *base* of the triangle (one of the sides) and *h* is the measure of the distance drawn from the *vertex* (corner) that's opposite the base down to the base. The *h* (height) is always perpendicular to the base. (If you want see a labeled triangle, check out Figure 21-4 earlier in this chapter.)

This is Heron's formula for finding the area of a triangle:

$$A = \sqrt{s(s-a)(s-b)(s-c)}$$

In this formula, *a*, *b*, and *c* are the measures of the three sides of the triangle and *s* is the *semi-perimeter* (half the perimeter).

In Figure 21-12, I show you one way to divide up an irregular area using triangles.

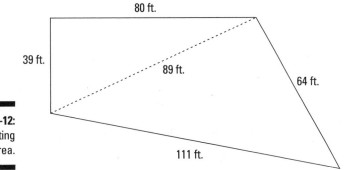

80 ft.

39 ft.

89 ft.

64 ft.

111 ft.

**Figure 21-12:** Triangulating the area.

Now I want to show you how to find the area of this polygon using the two formulas I just introduced. The area in Figure 21-12 is a *quadrilateral* (a four-sided polygon). Only one corner is square (90°), so I divide the area into two triangles. The top triangle is called a *right* triangle, because it contains a right, or 90°, angle.

To find the area of the top triangle (the right triangle) in Figure 21-12, use the half-of formula that I introduce earlier in this section. The base, *b*, can be the side measuring 39 feet, and the height, *h*, is the side measuring 80 feet. Here's what the formula looks like with all the numbers plugged in:

$$A = \frac{1}{2}(39)(80) = 1,560 \text{ sq. ft.}$$

**TIP** The half-of formula for the area of a triangle is easier and quicker to use, so it's my choice when possible. I could have used Heron's formula on this triangle, too, but I don't use it unless necessary. The bottom triangle in Figure 21-12 isn't a right triangle. So, to find the area, use Heron's formula. You first find the perimeter, *P*, which is the sum of the measures of the sides of the triangle:

$$P = 89 + 64 + 111$$
$$= 264$$

The *semi-perimeter, s,* is half that amount, so $s = 132$. Now use the formula to solve for the area:

$$A = \sqrt{132(132 - 89)(132 - 64)(132 - 111)}$$
$$= \sqrt{132(43)(68)(21)}$$
$$= \sqrt{8,105,328} \approx 2,846.986 \text{ sq. ft.}$$

Add the two areas of the two triangles together to get the total area of the polygon: $1,560 + 2,846.986 = 4,406.986$ sq. ft.

## Tracking trapezoids

Some areas in buildings, parking lots, or other such structures are made up of rectangles with triangles seemingly attached to either end. This shape resembles a *trapezoid.* To see what a trapezoidal area looks like, check out Figure 21-13.

To determine the total area of such a structure, you can use the technique involving dividing the area into triangles, as I show you in the earlier section "Trying out triangles," or you can take advantage of the formula for the area of a trapezoid to save yourself some time.

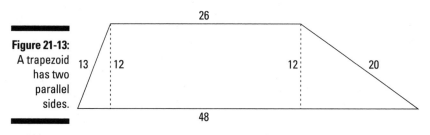

**Figure 21-13:**
A trapezoid has two parallel sides.

The area of a trapezoid that's *h* units between the two parallel bases is found with the following formula:

$$A = \frac{1}{2}h(b_1 + b_2)$$

The two $b$'s with the subscripts represent the lengths of the two parallel bases. Using the previous formula, you can find the area of the trapezoid in Figure 21-13. Here's what the math looks like:

$$A = \frac{1}{2}(12)(26 + 48) = 6(74) = 444$$

So, as you can see, the area is 444 square units.

# Describing Property with Metes and Bounds

*Metes and bounds* (not to be confused with *leaps and bounds*) is a surveyor's description of a piece of property. Using metes and bounds when measuring property dates back to colonial times. The United States inherited the term from England, which has used metes and bounds for hundreds of years.

Metes and bounds is a way of describing someone's land or property by way of a landmark and compass directions. For example, a listing in metes and bounds may start with something like this: "Beginning at the corner of Main and University, running N30°E 300 feet to a point, then S63°E 150 feet to a point . . . ." The surveyor's description in metes and bounds begins at some specific point, proceeds around the boundaries of a property with references to direction and some linear measurement, and always ends up back where he or she started.

## Surveyors' directions

Before you can deal with both the directions and distances in metes and bounds, you need to acquaint yourself with surveyors' measures. It's important for you to note that the directions are given in terms of being east or west of north or south. For example, a direction that starts out with N for north has you pointing upward on a piece of paper. Then you lean either left or right by a particular number of degrees, depending on whether the second direction is west or east. So the angle N30°E means 30 degrees east of due north. This would be pointing at about 1:00 on a clock. The angle S60°W means 60 degrees away from due south or about 8:00 on a clock. Figure 21-14 shows you some compass degree headings to help you with the directions.

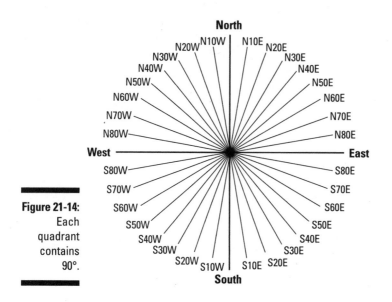

**Figure 21-14:**
Each
quadrant
contains
90°.

## Measuring a boundary

Consider the following description, which uses metes and bounds to describe property. Then refer to Figure 21-15 for a diagram of what the description represents.

Start at the corner of Sugar Plum Avenue and Maple Drive and proceed N66°E for 678 feet, where you should see a large boulder. From there, go S54°E for 850 feet where there's a large tree stump. Now go S44°W for 743 feet to an abandoned well. From there go N25°W for 560 feet to a wild raspberry thicket. (Leave the raspberries alone!) Return to the starting point by going N60°W for 572 feet.

After describing this plot in metes and bounds, you then have a diagram, which helps you calculate certain measurements. For example, you can figure the perimeter of this plot by adding up the measures between the points. You can also figure the area by dividing the area up into three triangles, finding the area of each triangle, and adding the areas together.

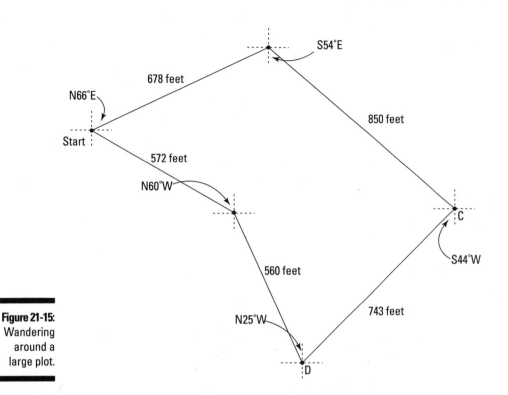

**Figure 21-15:**
Wandering
around a
large plot.

# Understanding the Rectangular Survey System

The *rectangular survey system,* also called the *government survey system,* has been used in North America since the 1700s. The rectangular survey system establishes units of land that are approximately square — 24 miles on a side. The measure is considered approximate because the system is based on meridians that run north and south, converging at the poles. The meridians are closer together as you get closer to the poles. In a square plot that's 24 miles on a side, you aren't going to see all that much convergence or narrowing. However, there is some, and the error has to be accounted for. Any discrepancies in measurement, due to the convergence of the meridians and any other built-in errors, are usually absorbed into sections along the north and west boundary of the square. The 24-mile square is divided into 16 more squares, each measuring 6 miles on a side. The 16 squares are then subdivided into 1-mile square sections.

## Basing measures on meridians

The rectangular survey system is implemented for a particular area by establishing some starting point determined by astronomical measures of latitude and longitude. For instance, using the starting point, a *principal meridian* is determined (running north and south), and then a *base line* (running east and west) is set. Running parallel to the principal meridian are *guide meridians,* each of which are 24 miles east and west of the principal meridian or another guide meridian. And then there are *standard parallels* that run parallel to the base line. These are also 24 miles apart.

Figure 21-16 shows you a principal meridian, base line, and some guide meridians and standard parallels. The darkened section is then shown in more detail in Figure 21-17.

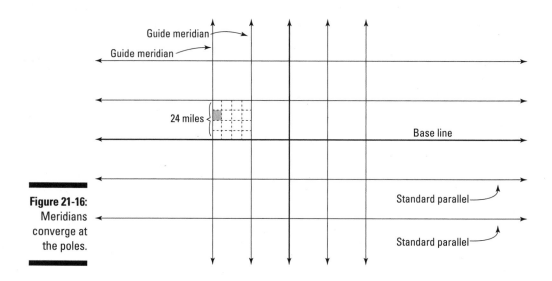

**Figure 21-16:**
Meridians
converge at
the poles.

## Subdividing the 24-mile square

Each 24-mile square is subdivided. Then it's identified by *ranges,* which are 6-mile-wide strips running vertically (north and south), and by *townships,* which are 6-mile-wide strips running horizontally (east and west).

The squares that are 24 miles on a side are divided into 16 squares that are 6 miles on a side. Each of these squares is subdivided into 36 parts, each of which is a 1-mile square called a *section*. Each section contains 640 acres. Figure 21-17 shows you how a section may be broken up into various plots containing different numbers of acres.

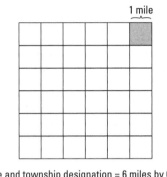

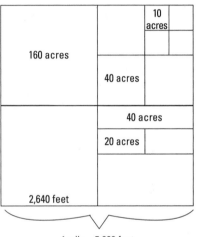

**Figure 21-17:** Acres making up pieces of square miles.

Range and township designation = 6 miles by 6 miles

1 mile = 5,280 feet

Using the rectangular survey system, a particular plot of land can be described by its principal meridian and guide meridian, then by ranges and townships, then by sections, and finally by the portion of the section in which it lies.

# Chapter 22

# Taking Out Mortgages and Property-Related Loans

. . . . . . . . . . . . . . . . . . . . . . . . . . . . . . . . . . . . . . . . . . . . .

## In This Chapter

▶ Taking care of closing costs

▶ Using different methods to amortize loans

▶ Considering alternate payment schedules for mortgages

▶ Determining borrowing power

▶ Surveying alternatives to amortized loans and mortgages

. . . . . . . . . . . . . . . . . . . . . . . . . . . . . . . . . . . . . . . . . . . . .

*A* *mortgage* on a building or property is a promise to pay for that building or property over a period of time. In other words, it's a way of making a purchase without having to pay for the whole thing immediately. What makes a mortgage different from other types of loans or debts is that the building or property acts as the security; an actual physical entity is involved. The mortgage is both the process used to produce the loan and the loan itself — the meaning is dual.

Many different variables figure into the amount and term of a mortgage, and the amount of it is usually some percentage of the value of the property. A down payment on the property reduces the amount of money that has to be loaned. The time agreed on for repayment is usually between 20 or 30 years — but even that can change, depending on the circumstances and the financial situation of the person borrowing the money.

Some mortgage-holders like to pay ahead — to finish up early and save money on interest. Others need a contract for deed. The implications of paying ahead or changing the method of payment aren't always clear. Don't worry, though. This chapter shows you how to better understand the mathematics of the lending process related to mortgages and other property loans.

For instance, in this chapter I solve the mysteries involved in closing on a mortgage and computing mortgage payments and total interest. You see how to accurately make use of appropriate formulas or tables (you get to choose your favorite method). I also discuss other types of property loans.

# Closing In on Closing Costs

When finally securing a mortgage on a property, the prospective buyer comes to a settlement meeting or *closing.* At this settlement meeting, the borrower has to deal with any *closing costs* or other terms.

A closing in money-speak is when all the papers are finally signed, the promises are officially and formally made, and the buyers get to move their stuff into their building or build on their newly acquired property.

Closing costs vary from mortgage to mortgage, from lender to lender, and from situation to situation. Even more confusing is the fact that different closing costs are called different things by different institutions. Some lenders even resort to padding their closing fees because they've had to reduce their rates to be competitive. Top it all off with the fact that errors in computations can occur, and whew! With all these possibilities coming at you, you need to be prepared to sort through all the information and computations to be sure that what you see is what you want.

The following are some common closing costs:

- ✔ Points (percent of loan amount)
- ✔ Appraisal fee
- ✔ Title search
- ✔ Survey fee (dependent on property size)
- ✔ Recording and transfer fees
- ✔ Stamp tax (based on amount borrowed)
- ✔ Attorney fees
- ✔ Prorated interest
- ✔ Prepaid property taxes

I use this example throughout the section to explain some of the most common closing costs and how to compute them: Suppose you're purchasing a $300,000 office building. You made a contract deposit of $10,000 at the time of the offer. (By the way, a *contract deposit* is an amount of money that holds or saves the property for you. It's your guarantee that you'll buy the place, and the seller's promise not to sell the property to someone else.) The total down payment is 15% of the sale price. Assume that you're planning to take out a 30-year, 5.5% fixed rate loan.

## Dealing with down payments

A *down payment* is a certain percentage of the amount being paid for a building. The greater the down payment, the less to be paid back later — which in turn reduces the amount that needs to be paid in interest over the years.

In the example involving the $300,000 building (check out the beginning of this section for the details), the down payment is $300,000 × 0.15 = $45,000. Because you made a contract deposit of $10,000 so that the seller would commit to selling to you, you have to subtract $45,000 – 10,000 = $35,000, which tells you the amount of the additional down payment that's due at the time of the closing. The down payment includes your initial deposit plus whatever is needed to complete the down payment. This way, you don't have to come up with the entire down payment until the time of the closing. The loan amount is then the difference between the sale price and the entire down payment: $300,000 – 45,000 = $255,000.

## Paying down using points

In the mortgage world, *points* represent additional money that's paid to the lender at the time of the closing. The more points you pay, the more you can lower the interest rate. The amounts of the points are percentages of the amount of the loan. For example, 2.5 points translates into 2.5% of the loan amount. And each point you pay on a 30-year loan reduces your interest rate by about one-eighth of a percentage point.

Consider the example I introduce at the beginning of the section. The $300,000 building with a down payment of $45,000 has a loan amount of $255,000. You're paying 2.5 in points at closing. So to compute the additional amount that you'll be paying, multiply the loan amount by the decimal form of the points percentage: $255,000 × 0.025 = $6,375.

You planned to pay, before points, 5.5% interest. However, each point you pay reduces the interest rate by about one-eighth of a percentage point. One-eighth point is 0.125, or 0.125%. Multiplying that by 2.5, you see that the reduction in the interest rate is 0.3125%:

$$\frac{1}{8} = .125$$
$$2.5\,(.125) = .3125$$

This lowers the interest rate on the loan to 5.5 – 0.3125 = 5.1875%. This may not seem like much, but on a 30-year loan, that amounts to significant savings. (Refer to the later section, "Showing the effect of lowering the interest rate," for more information.)

## Considering appraisal fees

An *independent appraiser* (someone not connected to the lender or the borrower) determines the market value of a building or piece of property. That appraiser charges an *appraisal fee* for her work. The fee usually depends on the price of the building. For instance, assume that the appraiser in the example (at the beginning of the section) charges a flat fee of $200 plus $25 for each $10,000 the home costs in excess of $200,000. So for the building that costs $300,000, the $25 extra is charged on $100,000 — which is 10 multiples of $10,000. So the charge is $200 + 10($25) = $200 + 250 = $450.

## Prorating property tax

Property taxes are generally collected by the local governing body once or twice each year. The seller of property owes taxes for the number of days or months he owned the property during a particular tax period, and the buyer owes taxes on the remainder of the tax period. To split up the tax liability fairly, you have to deal with some prorating. For instance, if a property tax payment is due on September 30 and the closing is earlier in the year on April 10, the buyer is responsible for the property taxes from April 10 through September 30, and the seller is responsible for the previous months.

Consider again the $300,000 office building. The tax rate on this building is 6.75% of the assessed value. And the assessed value is ⅓ of the value of the building. To determine the property tax on the building, multiply $300,000 × ⅓ × 0.0675 = $6,750. If the closing is on April 10, and the property taxes are due on September 30, how much is the buyer responsible for paying?

Five and two-thirds months separate April 10 and September 30. So first you need to figure out what fraction of a year 5⅔ months is. To do so, divide by 12, and then multiply that fraction by the annual property taxes, like so:

$$\frac{5\frac{2}{3}}{12} = \frac{\frac{17}{3}}{12} = \frac{17}{3} \cdot \frac{1}{12} = \frac{17}{36}$$

$$\frac{17}{36}(6750) = \frac{17}{\underset{2}{\cancel{36}}} \cdot \frac{\overset{375}{\cancel{6750}}}{1} = \frac{6375}{2} = 3187.50$$

As you can see, the buyer owes $3,187.50 of the taxes and the seller owes the difference, $6750 − $3187.50 = $3,562.50.

Often, the closing occurs before the tax bills are available. A common practice to handle this situation is to estimate the total amount of the tax by figuring 110% of the previous year's tax bill. Check out the following example to see how to compute taxes this way.

**GO FIGURE**

What's the estimate of the coming year's property tax if last year's was $6,750 and you're using 110% as a best guess?

To find out, you multiply $6,750 by 1.10 to get $7,425.

# Amortizing Loans with Three Different Methods

An *amortized loan* is one in which regular, set payments are made and the interest on the loan is figured on what's left to be repaid. Even though the *principal* (the amount still owed) is decreasing, the payment stays the same, because that payment is more of an averaging-out of all the payments — from when the principal is at the highest until the loan is almost completely repaid. You can determine the monthly payment of an amortized loan using an online calculator, a table of values, or a good, old-fashioned formula. I explain each of these methods in the following sections.

By the way, the term *mortgage* is only applied to loans for buildings or property. So, while a mortgage is a loan, a loan isn't necessarily a mortgage. However, payments and interest on both are figured the same way.

## Taking advantage of online calculators

An online calculator is a handy tool for figuring mortgage payments. The beauty of online calculators is that they take all the mystery and scariness out of computing mortgage payments, mortgage interest, points, and so on.

Most financial institutions with Web sites offer you the convenience of doing computations using their particular calculator. You enter the amount of the loan and any other pertinent information, and the calculator will almost immediately produce the payment amount and interest rates available. You can play around with the numbers — see what happens if you want to borrow more or make a bigger down payment. You get to do your homework before actually going to the institution to arrange for a loan.

Be aware, though, that you often have to *register* or provide some sort of personal information when you use these Web sites. You may get e-mails or phone calls from the institution, offering to do business with you.

## Consulting mortgage payment tables

Visiting an online mortgage calculator (see the previous section) is probably the quickest method of figuring monthly mortgage payments, but using a table or chart is quick and efficient as well. It would take a book the size of a dictionary to list all the possible loan amounts and corresponding interest rates. But you don't want to lug around a book. You want something you can carry in your pocket or wallet. That's why folks started constructing tables showing certain interest rates, the number of years of borrowing, and the amount of the payment per thousand dollars of the loan.

When using a payment table to determine your monthly mortgage payment amount, you first find the rate down the side of the table and the number of years across the top. Find the entry corresponding to the rate and years you want and then multiply by the number of thousands of dollars that you want to borrow.

In Table 22-1, I show you the monthly payment it would take to amortize a $1,000 loan. I include interest rates from 4% to 9% in increments of a quarter of a percent, and loans for 10, 15, 20, 25, and 30 years.

| Table 22-1 | Monthly Payment to Amortize a Loan of $1,000 | | | | |
|---|---|---|---|---|---|
| Interest Rate | 10 Years | 15 Years | 20 Years | 25 Years | 30 Years |
| 4 | 10.124514 | 7.396879 | 6.059803 | 5.278368 | 4.774153 |
| 4¼ | 10.243753 | 7.522784 | 6.192345 | 5.417381 | 4.919399 |
| 4½ | 10.363841 | 7.649933 | 6.326494 | 5.558325 | 5.066853 |
| 4¾ | 10.484774 | 7.778319 | 6.462236 | 5.701174 | 5.216473 |
| 5 | 10.606552 | 7.907936 | 6.599557 | 5.845900 | 5.368216 |
| 5¼ | 10.729170 | 8.038777 | 6.738442 | 5.992477 | 5.522037 |
| 5½ | 10.852628 | 8.170835 | 6.878873 | 6.140875 | 5.677890 |
| 5¾ | 10.976922 | 8.304101 | 7.020835 | 6.291064 | 5.835729 |
| 6 | 11.102050 | 8.438568 | 7.164311 | 6.443014 | 5.995505 |
| 6¼ | 11.228010 | 8.574229 | 7.309282 | 6.596694 | 6.157172 |
| 6½ | 11.354798 | 8.711074 | 7.455731 | 6.752072 | 6.320680 |
| 6¾ | 11.482411 | 8.849095 | 7.603640 | 6.909115 | 6.485981 |

| Interest Rate | 10 Years | 15 Years | 20 Years | 25 Years | 30 Years |
|---|---|---|---|---|---|
| 7 | 11.610848 | 8.988283 | 7.752989 | 7.067792 | 6.653025 |
| 7¼ | 11.740104 | 9.128629 | 7.903760 | 7.228069 | 6.821763 |
| 7½ | 11.870177 | 9.270124 | 8.055932 | 7.389912 | 6.992145 |
| 7¾ | 12.001063 | 9.412758 | 8.209486 | 7.553288 | 7.164122 |
| 8 | 12.132759 | 9.556521 | 8.364401 | 7.718162 | 7.337646 |
| 8¼ | 12.265263 | 9.701404 | 8.520657 | 7.884501 | 7.512666 |
| 8½ | 12.398569 | 9.847396 | 8.678232 | 8.052271 | 7.689135 |
| 8¾ | 12.532675 | 9.994487 | 8.837107 | 8.221436 | 7.867004 |
| 9 | 12.667577 | 10.142666 | 8.997260 | 8.391964 | 8.046226 |

Use Table 22-1 to determine the monthly payment of a $200,000 loan at 6¾% interest for 25 years.

To do so, find the row for 6¾% interest. Then scan across until you hit the column for 25 years. At that intersecting cell of the table, you should find the number 6.909115. That's the amount of the payment — about $6.91 — that you'd have to pay every month for 25 years, to repay a loan of $1,000.

Now, to determine the monthly payment for a $200,000 loan, multiply 6.909115 by 200 to get $1,381.823, or about $1,382 each month. (You multiply by 200, because the loan is for 200 *thousand*.) Pretty snazzy table, isn't it?

### Interpolating for interest

Say that you want to figure the payment on a $150,000 loan at 5⅛% interest for 20 years. What do you do in this situation where the interest rate that you want is some amount between two of the rates given on the table? You interpolate! *Interpolating,* as used in mathematics, simply means finding some intermediate or middle value between two consecutive values that appear in a table. It's essentially finding a number between two other numbers. Interpolation isn't as good as using a formula or a given table value, but it provides a decent estimate of the number you want.

Use Table 22-1 to determine the monthly payment on a $150,000 loan at 5⅛% for 20 years.

To do this, you first determine that 5⅛% is halfway between 5% and 5¼%. Now you know that you want the payment value that's halfway between the two given values on the table. To determine that midpoint, find the difference between the two payment values, take half of that difference, and then add it to the 5% amount. Here's what your math should look like:

$$5\frac{1}{4}\% \text{ for } 20 \text{ years} = 6.738442$$
$$5\% \text{ for } 20 \text{ years} = \underline{6.599557}$$
$$\text{Difference} = 0.138885$$
$$\text{Half the difference} = \frac{0.138885}{2} = 0.0694425$$
$$\text{Add half difference to 5\%: } 6.599557$$
$$+ 0.0694425$$
$$6.6689995$$

The payment per $1,000 is 6.6689995, so the payment on a loan of $150,000 is $150 \times 6.6689995 = \$1,000.349925$, or about $1,000.35 per month. Later in this chapter, I use a formula to determine the payment in this example. This way you can compare the amounts to see how well interpolation works. Don't forget that I said it was a decent estimate. So take it for what it's worth.

In the previous problem, the percentage needed was halfway between the values on the table. What if the percentage you want isn't quite so cooperative? For example, say you want to use Table 22-1 to figure the monthly payment on a loan of $100,000 at 7.3125% for 30 years.

The first challenge is to determine where 7.3125 belongs in a table that has fractional percentage rates. The number 7¼ = 7.25 and 7½ = 7.5. (Refer to Chapter 2 for a complete explanation of changing fractions to decimals.) So where does 7.3125 belong between these two numbers? Is it halfway? One-third of the way? To determine how far 7.3125 is from 7.25 and 7.5 in the table, find the percent difference of 7.3125 between the two table entries. (You can find more information on percent increases and decreases in Chapter 3.) Your work should look like this:

$$\frac{\text{difference between 7.3125 and 7.25}}{\text{difference between 7.5 and 7.25}} = \frac{7.3125 - 7.25}{7.5 - 7.25}$$
$$= \frac{0.0625}{0.25}$$
$$= .25 \text{ or } 25\%$$

The rate 7.3125% is one-quarter of the way between 7¼% and 7½%. So, to find the monthly payment from the table, you find ¼, or 25%, of the payment amount between 7¼% and 7½%. Then you add that number to the given payment amount at 7¼%. Here's what your work should look like:

$$7\tfrac{1}{2}\% \text{ for } 30 \text{ years} = 6.992145$$

$$7\tfrac{1}{4}\% \text{ for } 30 \text{ years} = \underline{6.827163}$$

$$\text{Difference} = 0.170382$$

$$\tfrac{1}{4} \text{ the difference} = \frac{0.170382}{4} = 0.0425955$$

Add difference to $7\tfrac{1}{4}\%$: 6.821763

$$+\ \underline{0.0425955}$$
$$6.8643585$$

Finally, to determine the payment on a loan of $100,000, you multiply 6.8643585 by 100, which gives you a payment of about $686.44 per month. Great work!

### *Interpolating for time*

The entries in Table 22-1 show increments of 5 years going from 10 years to 30 years. It would be nice if all loan lengths fell into neat little categories like this, but they don't. Loans are made in other, not so round, yearly amounts. But don't worry. You can compute estimates for years other than multiples of five in much the same way as you compute estimates for interest rates between those on the table. Check out the following example for some practice.

Suppose a dentist wants to take out a $60,000 loan at 8% interest and have it paid off the year that she retires, which is 12 years from now. What will the monthly payments be?

A loan term of 12 years is obviously between 10 years and 15 years. But you need to know exactly how far between the two entries 12 years is. To find out, you take the difference between 12 years and 10 years, 12 − 10 = 2, and then divide that by the difference between 15 years and 10 years:

$$\frac{12-10}{15-10} = \frac{2}{5} = .40 \text{ or } 40\%$$

As you can see, 12 years is 40% of the way between the two column entries. Now you have to find 40% of the difference between a loan for 10 years and a loan for 15 years. The payments decrease as the number of years increases, so the 40% difference is subtracted from the payment at the 10-year level:

$$8\% \text{ for } 10 \text{ years} = 12.132759$$

$$8\% \text{ for } 15 \text{ years} = \underline{9.556521}$$

$$\text{Difference} = 2.576238$$

40% of the difference = $0.40\,(2.576238) = 1.0304952$

Subtract difference from 10 years: 12.132759

$$-\ \underline{1.0304952}$$
$$11.1022638$$

The monthly payment is now found by multiplying 11.1022638 by 60 (for $60,000): 11.1022638 × 60 = 666.13584. So the dentist's payments are about $666.14 per month.

## Working your brain with an old-fashioned formula

You can use an online calculator or a set of tables to find a particular mortgage payment (see both methods in action earlier in the chapter). But sometimes you may want to do your own calculating with a simple, scientific calculator and the appropriate formula. I show you how in this section.

The formula for determining the monthly payment, $R$, on an amortized loan of $P$ dollars for $t$ years at an interest rate $r$ is

$$R = \frac{P\left(\frac{r}{12}\right)}{1 - \left(1 + \frac{r}{12}\right)^{-12t}}$$

If the payments are in an increment other than monthly, replace the 12's in the formula with the number of times each year the payments are made.

In the earlier section "Interpolating for interest," I show you how to use Table 22-1 to estimate the monthly payment for a loan of $150,000 at 5⅛% for 20 years. I use the formula here to do the same computation.

The loan amount, $P$, is 150,000; the interest rate, $r$, is 5⅛ %, or 0.05125; and the number of years, $t$, is 20. So now you just plug all of the numbers into the formula, like so

$$R = \frac{150,000\left(\frac{0.05125}{12}\right)}{1 - \left(1 + \frac{0.05125}{12}\right)^{-12(20)}}$$

Whoa! You have to put all that in your little scientific calculator? Well, not all at once. I take you through the steps that work best. (In Chapter 5, I give explanations on how to solve formulas, so refer to that chapter if you need more information than I give here.)

The first step in using the formula is to find the values in the parentheses. You can then substitute the computed numbers into the parentheses as part of simplifying the whole thing. This step looks like this:

$$\left(\frac{0.05125}{12} \approx 0.0042708\right) \text{ and } \left(1 + \frac{0.05125}{12} \approx 1.0042708\right)$$

So the formula now reads

$$R = \frac{150,000\,(0.0042708)}{1 - (1.0042708)^{-12\,(20)}}$$

Next, you deal with the pesky exponent in the denominator (the bottom part of the fraction). Multiply –12 by 20 to get –240. Then raise 1.4270833 to the –240th power. Subtract that answer from 1. Your work looks like this:

$$(1.0042708)^{-240} \approx 0.3595833$$

$$R = \frac{150,000\,(0.0042708)}{1 - 0.3595833} = \frac{150,000\,(0.0042708)}{0.6404167}$$

**WARNING!**

When raising to a negative power on your calculator, be sure to put that negative number in parentheses. Your entry will look like this: ^(–240).

Now multiply the numbers in the numerator (the top part of the fraction) and divide by the number in the denominator to get the monthly payment:

$$R = \frac{150,000\,(0.0042708)}{0.6404167} = \frac{640.62}{0.6404167} \approx 1,000.317450$$

Your payment comes out to about $1,000.32 per month. This solution is just three cents different from the one where I use interpolation in a table. Not too bad.

# Going Off Schedule with Amortized Loans and Mortgages

Amortized loans are the standard for building mortgages and other property-related loans. You determine the monthly or yearly (or some other periodic) payment using either tables or a formula — or you can hit the Internet for an online calculator. (I show you how to use each of these methods in the earlier section, "Amortizing Loans with Three Different Methods.")

But what if you want to go *off schedule?* What if you want to pay ahead on your mortgage or make some adjustment to the regular payments? What's the end result? Do you save money? And, if you save money by paying ahead or changing the pattern, is the savings enough to even bother with?

In this section, I show you how to determine what amount of a property mortgage payment is actually interest, and I explain what effects changing your payment can make on the end result. Finally, I help you explore the effect of fiddling with the interest rate. For instance, did you know that by paying a larger down payment or paying points to the lender, you can realize a significant savings? That's right! Read on for details. (For more information on loans that aren't used for property-related expenses, flip to Chapter 12.)

# Determining how much of your payment is interest

Lending institutions are required to send out reports to loan-holders telling them how much interest they paid during the previous year. This practice helps taxpayers determine how much they can deduct for expenses when figuring their income tax. The amount paid in interest changes every year with an amortized loan. Why? Because the principal (the amount of the loan) keeps decreasing as payments are made. So more interest is paid at the beginning of a loan payback than at the end. (Check out Chapter 9 for more information on simple and compound interest.)

### Figuring total interest paid

By increasing the number of years needed to pay back a property mortgage, you decrease the monthly payments. This perk is great for the borrower when money is tight and he doesn't want to put as much money toward the expenses. But how does this affect the total amount of money paid back by the borrower? Will the amount of interest being charged increase significantly? Is the amount of the increase great enough to encourage you to look at other options?

Say that a business owner wants to move into a bigger building, so he borrows $130,000 at 4¼% for 30 years. In this case, his monthly payment is $639.52. What's the total amount that the business owner is paying back over the entire 30-year period, and how much of that is interest?

A monthly payment of $639.52 for 30 years amounts to $639.52 × 30 × 12 = $230,227.20 as a total payment. If the amount borrowed was $130,000, then $100,227.20 was paid in interest. (You find the interest by subtracting the amount borrowed from the total payback: $230,227.20 − $130,000 = $100,227.20.) That's a lot of money!

Compare that to the same business owner borrowing the same $130,000 at 4¼% for 25 years. The monthly payment goes up to $704.26. The total amount paid over the 25-year period is $704.26 × 25 × 12 = $211,278. The interest paid is $211,278 − 130,000 = $81,278. This is still a lot of money, but it's almost $20,000 less than the amount that's paid over the 30 years.

### Computing unpaid balance

After a certain number of mortgage payments, a borrower wants to know the amount of the *unpaid balance* — how much is left of the original loan and how much of the interest is being computed on. Knowing this information is important for tax purposes and for planning on what can be deducted as overhead. It also helps the borrower determine whether she needs to take out an additional loan based on the equity in the property so she can build an addition and expand her business.

The amount of the unpaid balance of a loan that has a monthly payment of $R$ dollars at a rate of $r$ percent for $t$ years if $x$ payments have already been made is found with the following formula:

$$A = R \left[ \frac{1 - \left(1 + \dfrac{r}{12}\right)^{-(12t - x)}}{\dfrac{r}{12}} \right]$$

This formula looks hairy, but don't worry. The following example will help you work through the steps.

A small business owner has made 24 monthly payments on an $80,000 mortgage at 5% interest over 25 years. Her monthly payments are $467.67. How much does she have left to pay on her loan? With this figure, she can make plans for an expansion.

To find out, plug your numbers into the formula, like this:

$$A = 467.67 \left[ \frac{1 - \left(1 + \dfrac{0.05}{12}\right)^{-(12(25) - 24)}}{\dfrac{0.05}{12}} \right]$$

And, showing you the computation in steps:

You first divide the 0.05 by 12 for both the numerator and denominator; in the exponent, you multiply the 12 by the 25.

Next, you do the subtraction in the exponent to get –276.

Now you add the 1 to the decimal in the parentheses and raise the result to the –276 power.

Subtract the result from the parentheses from the 1.

Now simplify the fraction in the bracket by dividing, and finally, multiply the result of the division by the payment amount. Whew!

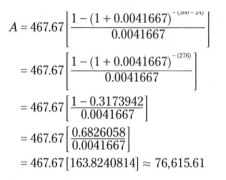

$$A = 467.67 \left[ \frac{1 - (1 + 0.0041667)^{-(300 - 24)}}{0.0041667} \right]$$

$$= 467.67 \left[ \frac{1 - (1 + 0.0041667)^{-(276)}}{0.0041667} \right]$$

$$= 467.67 \left[ \frac{1 - 0.3173942}{0.0041667} \right]$$

$$= 467.67 \left[ \frac{0.6826058}{0.0041667} \right]$$

$$= 467.67 [163.8240814] \approx 76,615.61$$

The balance is about $76,615.61 on the original $80,000 loan. Even though she has made payments totaling $467.67 × 24 = $11,224.08, her loan balance has been reduced by a little more than $3,000. This example illustrates how most of the first payments on an amortized loan are interest.

### Finding out how much of a particular payment is interest

Each monthly payment of an amortized loan has different amounts devoted to reducing the principal and to paying interest. The percentage of interest in the first few payments is very high, and the percentage decreases with each payment. The high interest payments have a tax advantage to a new business. How? Well, it allows it to have more to deduct at tax time. To determine just how much of a particular payment is interest and how much is being applied to the principal, you need to know what the balance is at a particular point in time. (In the section "Computing unpaid balance," you see how to come up with this number.)

After you know your balance, you can use the following formula to figure the amount of interest in a monthly payment:

$$I = Pr \left( \frac{1}{12} \right)$$

This formula is just the basic interest formula, $I = Prt$, where $P$ is the principal, $r$ is the rate of interest, and $t$ is the amount of time (in this case for one month, so it's divided by 12).

Suppose Jack, a budding entrepreneur, wants to purchase a building for his pool-cleaning business. He takes out a loan for $180,000 over 30 years at 7% interest. His mortgage terms say that he will pay monthly payments. How much of the payment is interest after one year, and how much of the payment is interest after 20 years?

First determine the amount of the monthly payment. Using Table 22-1, you do this by multiplying 6.653025 × 180 = $1,197.54. (If you need help using a table to determine the monthly payment, spend some time with the earlier section, "Consulting mortgage payment tables.")

Next, compute the remaining balance after 1 year (12 payments) and after 20 years (240 payments). Here's what your math should look like:

$$1\text{ year:} A = 1197.54 \left[ \frac{1 - \left(1 + \frac{0.07}{12}\right)^{-(12(30)) - 12}}{\frac{0.07}{12}} \right] = 178{,}170.87$$

$$20\text{ years:} A = 1197.54 \left[ \frac{1 - \left(1 + \frac{0.07}{12}\right)^{-(12(30)) - 240}}{\frac{0.07}{12}} \right] = 103{,}139.75$$

After 1 year, the principal is reduced by only about $2,000, and after 20 years, the principal is down to about $100,000.

Now you just have to compute the amount of interest being paid at each level. You do this using the formula I introduced earlier in this section:

$$1\text{ year: } I = 178{,}170.87\,(0.07)\left(\frac{1}{12}\right) = 1{,}039.33$$

$$20\text{ years: } I = 103{,}139.75\,(0.07)\left(\frac{1}{12}\right) = 601.65$$

So, after 1 year of paying almost $1,200 per month, the amount of the payment going to interest is about $1,039 of that $1,200. After 20 years, the amount of interest goes down to about $600 — or half the payment.

## Altering the payments

After a loan repayment schedule has been established, the borrower knows what payment is required each month in order to pay off the loan on time. Savvy borrowers realize that an awful lot of money goes into paying interest, so they'll consider some alternatives — some ways to reduce the total amount paid back.

Of course, it's often to the borrower's advantage to have a lot of interest payment to deduct. Why? Well, the interest on a loan is a deduction that offsets the amount of income tax to be paid. The whole process is a bit of a balancing act between taking advantage of the tax laws and not going overboard with the amount of money being paid back. (You can read more in Chapter 12.)

For those of you who want to reduce the total amount of money paid back on an amortized loan, you have several options. All the options, which I explain in the following sections, revolve around changing the amount of payment, changing the number of times payments are made, or changing the interest rate.

A borrower can change the amount of each of his or her payments by either increasing the amount by a certain number of dollars or by paying half the amount twice each month. Both of these methods affect the total amount paid — one more significantly than the other.

### Increasing the dollar amount

A person who borrows $90,000 at 6¼% over a period of 25 years has monthly payments of $593.70. (The earlier section, "Amortizing Loans with Three Different Methods," shows you how to figure monthly payments.) Increasing the amount of the monthly payment will decrease the number of years of repayment and the total payout. How? Any additional money made in a payment is applied directly to reducing the principal. If the principal is lower, then the amount of interest on that lower principal is also smaller. This has a snowball effect — reducing the principal and the interest month after month. (In Chapter 12, you find an example of the effect that making larger loan payments can have.)

How long will it take a person with a $90,000 mortgage at 6¼% for 25 years to repay her loan if she makes $600 payments (instead of the scheduled $593.70)? And how much will she save in interest? What about payments of $625? How will that increase in payment affect the length of time and the amount of interest?

To be honest, a nice spreadsheet does this figuring pretty well (see Chapter 5 for details), but I have a dandy formula for you to use so you can compute the amounts with a scientific calculator.

To determine the number of months needed, $n$, to pay off an amortized loan whose principal is $P$, when the rate of interest is $r$ percent, and you're making payments of $R$ dollars, use the following formula:

$$n = \frac{\ln R - \ln\left(R - P\left(\frac{r}{12}\right)\right)}{\ln\left(1 + \frac{r}{12}\right)}$$

The notation *ln* in the formula represents the *natural logarithm.* It's a function that allows you to solve for a variable that's in the exponent of a number. Your scientific calculator has a *log* button and an *ln* button. The log button is for base 10 numbers, and the ln button is for base $e$ numbers. The number $e$ has a value of approximately 2.71818. You "hit" the *ln* button, then the number that you want evaluated, and then *enter.*

So, to find the number of months it will take to repay the original 25-year mortgage if the payments are increased to either $600 or $625, you just plug the numbers into the formula and solve:

$$\text{Payments of \$600: } n = \frac{\ln 600 - \ln\left(600 - 90{,}000\left(\frac{0.0625}{12}\right)\right)}{\ln\left(1 + \frac{0.0625}{12}\right)}$$

$$\approx \frac{6.3969297 - 4.8771039}{0.0051948} \approx 292.567$$

$$\text{Payments of \$625: } n = \frac{\ln 625 - \ln\left(625 - 90{,}000\left(\frac{0.0625}{12}\right)\right)}{\ln\left(1 + \frac{0.0625}{12}\right)}$$

$$\approx \frac{6.4377516 - 5.0514573}{0.0051948} \approx 266.862$$

You work from the inside of the parentheses outward. First divide the 0.0625 by 12 in both the numerator and denominator. In the numerator, take that result, multiply by 90,000, and then subtract the product from 600. Find the *ln* of the difference. In the denominator, you add 1 to the division result and find the *ln* of that sum. The computations are all based on the *order of operations* in formulas which I discuss in great detail in Chapter 5.

Increasing the payments to $600 makes the number of months equal to 292.567, which, when divided by 12 months, comes out to about 24.38 years. That's shortening the loan-repayment by about 7½ months. Increasing the payments to $625 reduces the loan-repayment period to 22.239 years, or about 22 years, 3 months. The time isn't decreased much with the additional $7, but it's cut by almost three years with payments of $625.

Now, to compare the amount of interest paid, look at the total amount paid back and the amount of interest paid for each of the three loan payment amounts:

$$\$593.70 \times 25 \text{ years} \times 12 = \$178{,}110$$
$$\$178{,}110 - 90{,}000 = \$88{,}110 \text{ interest}$$
$$\$600.00 \times 24.38 \text{ years} \times 12 \text{ years} = \$175{,}536$$
$$\$175{,}536 - 90{,}000 = \$85{,}536 \text{ interest}$$
$$\$625.00 \times 22.24 \text{ years} \times 12 \text{ years} = \$166{,}800$$
$$\$166{,}800 - 90{,}000 = \$76{,}800 \text{ interest}$$

For each case, find the total payback by multiplying the amount of the payment times the number of years times 12 months. Then subtract the amount of the loan from that total payback.

Even a small addition of $7 per month results in a couple thousand dollars saved in interest payments. Increasing the payment by $32 per month saves over $11,000 in interest.

### Paying more often, but paying half as much

You wouldn't think that paying twice as often — and half as much — would make much difference in the total amount repaid on a loan. But it does. If you don't believe me, check out the following example.

Consider the borrower in the previous section who borrowed $90,000 at 6¼% interest over 25 years. The monthly payment comes out to be $593.70, and half of that is $296.85. Paying $296.85 twice a month results in $n$ payments:

$$n = \frac{\ln 296.85 - \ln\left(296.85 - 90{,}000\right)\left(\frac{0.0625}{24}\right)}{\ln\left(1 + \frac{0.0625}{24}\right)}$$

$$\approx \frac{5.6932270 - 4.1347665}{0.0026008} \approx 599.224$$

Remember that the 6.25% rate is divided by 24 instead of 12, because the borrower pays two payments per month. Now take the number of payments, 599.224 and divide by 24 to get 24.968 years. That's almost the same 25 years. So how does the total amount repaid change?

To find out, you multiply the payment amount times the number of years times 24 (the number of payments in 12 months). Here's what the math looks like:

$$\$296.85 \times 24.968 \text{ years} \times 24 = \$177{,}882$$

$$\$177{,}882 - 90{,}000 = \$87{,}882 \text{ interest}$$

Even this method saves a couple hundred dollars in interest.

### Changing the number of payments

A popular method used to repay mortgages on a regular schedule and cut down the interest paid is to make more payments. If you make the scheduled payment every four weeks instead of every month, by the end of the year you will have made one extra payment. For businesses that have unusually good months or other sources of income, this method makes a lot of sense. See how the math works out in the following example.

Suppose an advertising firm bought a business retreat in Colorado for holding client conferences and took out a mortgage of $420,000 at 4½% interest for 20 years. The scheduled monthly payments are $2,657.13. The company decides to make payments every four weeks. How will this change the length of the repayment and the amount of interest paid?

To find out how many payments need to be made, you use the formula for determining that number. Your math will look like this:

$$n = \frac{\ln 2{,}657.13 - \ln\left(2{,}657.13 - 420{,}000\left(\frac{0.045}{13}\right)\right)}{\ln\left(1 + \frac{0.045}{13}\right)}$$

$$\approx \frac{7.885002 - 7.0928096}{0.0034556} \approx 229.249$$

At 13 payments per year, 229.249 divided by 13 is about 17.635 years rather than the 20-year term of the mortgage. And look at the savings when it comes to interest:

$$\$2{,}657.13 \times 20 \text{ years} \times 12 = \$637{,}711.20$$
$$\$637.711.20 - 420{,}000 = \$217{,}711.20 \text{ interest}$$
$$\$2{,}657.13 \times 17.635 \text{ years} \times 13 = \$609{,}160.34$$
$$\$609{,}160.34 - 420{,}000 = \$189{,}160.34 \text{ interest}$$

You find the amount of interest paid by determining the total amount of the payback and subtracting the amount borrowed. The two computations show you the total paybacks for 20 years and 17.635 years and the amount of interest in each case.

If you subtract the two resulting amounts ($217,711.20 – $189,160.34), you can see that the borrower will save more than $28,000 in interest.

### Showing the effect of lowering the interest rate

In the earlier section, "Paying down using points," I indicate that by paying more points at closing you save money in the long run. The scenario in that section is that a loan of $255,000 at 5.5% interest was to be paid back in 30 years. Paying $6,375 in points reduced the interest rate to 5.1875%. Do the borrowers really make up the extra $6,375 paid up front if they lower the interest rate? To compare, you have to find the total payback over 30 years.

What's the total payback of a $255,000 mortgage at 5.5% compared to the total for the same amount at 5.1875% when both are for 30 years? Is the difference greater than $6,375?

Using Table 22-1, the monthly payments for the loan at 5.5% are $1,447.86. Next, using the formula for monthly payments for the amortized loan at 5.1875%, you get $1,398.26. I got to use the easier method (the table) with the first computation because the interest rate is a nice, round number. For the second computation, I had to dig out my calculator. (To get more information on the methods to figure monthly payments, check out the earlier section, "Amortizing Loans with Three Different Methods.")

The difference between the two payments is $49.60. Multiply that difference by 360 (12 payments per year for 30 years) and the total difference is $17,856. This total difference is almost three times the extra amount paid in points. Not too shabby!

# Talking about Borrowing Power

Most small businesses have to borrow money at some time or another. Borrowing money isn't a bad thing, but it has to be done carefully and with great forethought. Businesses can borrow money using any of the following methods:

- ✔ **A line of credit:** Allows the business to borrow funds as needed up to a prescribed limit.
- ✔ **A term loan:** This loan can be short term or long term and is usually amortized.
- ✔ **A credit card:** Lets the business purchase merchandise and temporarily postpone payment.

When a business borrows money, it has to establish that the current and expected income provides enough cash flow to make the loan payments. The business should make a yearly model or projection, because a company may have peaks and valleys in its income during the course of 12 months.

Suppose, for example, a business wants to borrow money to purchase a new assembling machine. The machine will increase the total income of the company and allow for future growth into other markets. After doing a study on the effect of adding the new machine and the projected increases in income, the business owner determines that the company can allocate $12,500 per month to repaying a loan for the purchase of the equipment. If the business can afford to make monthly payments of $12,500 to pay back the loan, how much money can it borrow if the interest rate is 8¾%? Compute this for both 10 years and 15 years.

Using Table 22-1, the payment per $1,000 borrowed at 8¾% for 10 years is 12.532675. If you figure a monthly payment by multiplying the number of thousands of dollars of the loan by the value in the table, then you need to divide the payment by the number in the table to get the number of thousands of dollars of the loan.

For instance, if you divide $12,500 by 12.532675, you get 997.392815, or approximately $997,000 that can be borrowed. Using the same process for 15 years, divide $12,500 by 9.994487 to get 1250.689505, which is $1.25 million. If the machinery has a long life expectancy, the loan seems like a good idea.

# Investigating Alternative Loans

Two types of loans that require special handling are *contracts for deed* and *construction loans*. Not all states encourage or even make it easy to purchase property with a contract for deed. In fact, they often make it very difficult. A contract for deed puts the responsibility for the amount of the loan on an individual, the current owner of the property. Lending institutions are usually more prepared to determine a person's ability to repay a loan than an individual who may have a personal connection (a close friend or relative). A contract for deed may be the only way to make a sale possible, though.

A construction loan is pretty standard fare for both businesses and developers. A construction loan has a predetermined limit, but the total amount isn't borrowed all at the same time. Over a period of some months or a year, the construction costs accumulate until a traditional loan comes into effect.

## Taking out a contract for deed

A *contract for deed* is a contract between the owner of some property and the person who wants to buy that property. This type of loan is different from a mortgage with a bank or lending institution because the seller is taking on the position of lender. The seller and buyer agree to a price and a payment schedule, and the buyer makes the payments directly to the seller.

A contract for deed is often used if a particular buyer either doesn't qualify for a conventional loan or if there just isn't time for all the paperwork and processes necessary with a lending institution. Beware, though. There may be a reason that a business doesn't qualify for a conventional loan.

Some creative payment plans are often instituted with a contract for deed. Such plans include interest-only plans and short balloon plans. I explain each in the following sections.

### Interest-only contracts

With an *interest-only contract,* the buyer is paying the interest on the amount and not making any dent in the principal. The interest-only part usually takes place for a specified amount of time. This type of contract may be great tax-wise, but it leaves the buyer with the full amount to pay back eventually. However, interest-only contracts also allow the buyer to "own" the property for a while.

A small business owner is anxious to sell his shop so he can move to another state. He sells his business on a contract for deed for $85,000. He's currently making mortgage payments on a conventional $60,000 loan at 8¾% for 25 years. The contract for deed that he's agreed to is an interest-only contract for

2 years at 6¼%; at the end of the 2 years, the buyer will pay the full amount — probably by just getting a conventional loan. How much will the monthly interest payments be, and will the buyer's payments cover the current owner's monthly mortgage payments?

The interest-only payments on $85,000 are computed by multiplying the $85,000 by the interest rate of 6¼% by ¹⁄₁₂. Each month, the interest is figured on the same amount, because the principal isn't being reduced. Here's what the math looks like:

$$\$85,000 \times 0.0625 \times \frac{1}{12} = \$442.71$$

Next figure out the payments that the current owner is making. You can do this by using Table 22-1. You multiply 8.221436 (the monthly payment) by 60 (the table gives the amount per thousand, and the loan is for 60 thousand) to get $493.29. (To get more information on using tables to find monthly payments, check out the earlier section, "Consulting mortgage payment tables.")

The owner will be paying out more on his loan than he'll be getting from the buyer. Sometimes, however, this is acceptable, especially if the owner is in a hurry.

### Balloon payment contracts

A *balloon payment contract* allows the borrower to pay back a loan, usually using an amortization schedule for a 25- or 30-year loan, for a certain number of years — say 5 or 10 years. At the end of the specified time, the entire balance becomes payable. The buyer either has to take out another loan to repay the balance — or come up with the entire amount owed.

Balloon contracts are great for fledgling businesses — making the payments smaller at the beginning. But this is really a gamble — gambling that the situation will be better in the five or ten years and that the borrower will be in a position to pay the greater amount of money.

Suppose a couple purchases a fast-food franchise and makes arrangements to borrow $450,000 at 4½% for 6 months. At the end of six months they'll make a balloon payment — pay off the balance of the loan (they won the lottery and will have the check by then). The payments during the 6 months are based on a 30-year amortized loan. What's the total amount that the couple will pay before making the final, balloon payment, and how much will the balloon payment be?

Using Table 22-1 and 30 years, you get monthly payments of $2,280.08. Multiply the monthly payment by six, and their total output comes to $13,680.50. (For more information on using tables to find monthly payments, flip to the earlier section, "Consulting mortgage payment tables.")

The unpaid balance is found using the formula I introduce in the earlier section, "Computing unpaid balance." The formula is

$$A = R \left[ \frac{1 - \left(1 + \dfrac{r}{12}\right)^{-(12t - x)}}{\dfrac{r}{12}} \right]$$

where monthly payments are $R$ dollars at a rate of $r$ percent for $t$ years and $x$ payments have already been made.

After you plug in your numbers, your math should look like this:

$$A = R \left[ \frac{\left(1 + \dfrac{r}{12}\right)^{12t} - 1}{\dfrac{r}{12}} \right]$$

So, as you can see, the couple will have to pay $446,410.23; their 6 months of payments decreased the principal by about $3,600.

## Building a construction loan

REMEMBER

A *construction loan* usually has two parts: an interest-only loan until the construction project is finished and a conventional mortgage loan as soon as the business owner takes occupancy. The lender needs to know the plans for construction — the amount of money needed and the timeline. The details of the loan are worked out based on the construction plans. This type of loan makes a lot of sense to the builder, because only the amount of money advanced to a particular point is creating interest. The end total is agreed upon, and only the amount needed at a particular time is released.

Say, for example, that a $300,000 construction loan is agreed on in which the construction company pays interest only at 6% for the first 6 months and then 5¾% when the building is occupied and the full amount comes due. Assuming that the average balance of the borrowed amount is $150,000 during the 6-month construction period, estimate the amount of interest paid during that first 6 months.

Even though the money needed for construction is probably doled out in uneven increments during the 6-month period, an estimate is made using the average amount of $150,000. Just use the simple interest formula, $I = prt$, to get

$I = 150,000(0.06)(½)$

$= \$4,500$

You can read more about simple and compound interest in Chapter 9.

# Part VII
# The Part of Tens

The 5th Wave          By Rich Tennant

"I don't like the numbers on this report. Send it down to Marketing and see what they can do with it."

# In this part . . .

You can't beat the traditional *For Dummies* Part of Tens! This part includes two quick and easy chapters that outline top ten lists on some fun but important business math stuff. For example, I include a chapter on ten neat tricks for dealing with numbers. Use them to speed up your computations or to amaze your cohorts.

You'll also find a chapter on what to look for in financial reports, which are those lengthy and complicated reports that include details on the different facets and phases of a company's situation. In this chapter, I provide you with ten tips that help you zoom in on what's needed when confronted with a financial report.

# Chapter 23

# Ten Tips for Leasing and Managing Rental Property

*T*he biggest question that pops up when you're faced with the prospect of needing space is whether to rent or to buy. The plusses and minuses of both options are special to your particular circumstances. And, if you're in the business of renting property to others, you need to know how to best arrange for a fair, profitable, and manageable situation using a lease.

Leases come in all shapes and sizes. Okay, they're all printed on paper, but the terms and specifics of a lease vary as much as the types of properties they represent. Because of all these variations, it's always a good idea to get legal counsel and tax advice before entering into a lease agreement. However, you can do all the legwork and be more assured of having the arrangement that you want if you take time to acquaint yourself with the options.

In this chapter, you find all sorts of information about leases. I describe the most common types of leases: gross leases, net leases, percentage leases, and step leases. I also explain some of the ways that you can protect your lease, including clauses, allowances, and security deposits. The different clauses that can be added to a lease spell out some special terms and protect both the renter and the property owner. You don't want any nasty surprises, so it pays to understand what's involved with a particular lease and its clauses.

Each type of lease has its own particular advantages and disadvantages. Each type of lease can be modified to fit the situation. However, do remember that when bound by any of these arrangements, both the landlord and the tenant need to be aware of their responsibilities under the terms of the arrangement — and they must make good on them. Complete and accurate records need be kept and regular accounting and audits need be scheduled.

# Getting the Full Treatment with a Gross Lease

With the terms of a *gross lease,* the landlord receives the gross amount of money in rent and pays all expenses associated with the property. If you're entering into a gross lease as a landlord, you agree to be responsible for all the expenses that are normally associated with ownership: taxes, maintenance, utilities, insurance, and repairs.

On the other hand, if you're the renter in a gross lease, you have the advantage of a set monthly payment for all the expenses of housing your business. In other words, you don't have to worry about fluctuating utility charges, changes in your share of the common areas, or increases in taxes.

As you probably know, property owners or landlords are interested in having their investments in a property grow. They also want to receive a steady income from that property. Setting a rental amount that provides the income and is at a level that attracts good tenants is essential to financial success.

When determining a fair rental amount, the landlord takes into account all of the usual expenses and determines an amount that provides a profit. When more than one tenant is involved, the landlord must determine each tenant's share of any common areas. This share is referred to as the *common area maintenance,* or CAM. Other provisions provided in a gross lease include clauses involving escalation of rent and expense stop provisions. (You can find more on these provisions later in this chapter. See the sections, "Stepping It Up with a Step Lease" and "Inserting Expense Provisions into Your Lease."

A gross lease is fairly rare. One reason that a gross lease is rarely used is because the different expenses associated with the rental of part of a building can vary greatly with different tenants' usage. For instance, if each office has a thermostat and heating and cooling isn't limited or controlled, a tenant may set the temperature to excessively high temperatures in the winter or excessively cool temperatures in the summer. One tenant may use more in heat, lights, and water than another, but the kicker is that they all pay the same amount. This is why an *expense stop provision* is often used with gross leases.

Similarly, with a gross lease, the expenses are based on projected costs for garbage and taxes. Parking lot maintenance may be contracted out with a set amount for a year, but you can't control what governmental bodies do with the garbage pickup fees and real estate tax rates. Increases in the real estate tax don't usually take effect immediately, so there's time to adjust accordingly. However, the garbage doesn't wait. An *operating expense escalation clause* may be added to a gross lease to take care of such contingencies.

# *Using a Single Net Lease to Get Your Tenant to Share the Expenses*

If you thought that a net lease dealt with fishing or butterfly collecting, you're going to be slightly disappointed. Instead, with a *net lease,* the landlord receives a net sum and the tenant pays for some of the expenses. These expenses, which are paid by the tenant, can include utilities, repairs, insurance, or taxes. Different types of net leases are defined by the expenses that are or are not included.

A *single net lease,* for example, is a net lease where the tenant pays a monthly amount in rent and the property taxes as well. With this type of lease, the property owner/landlord is still responsible for all the other expenses, such as utilities, maintenance, and so on. For instance, the landlord can include a charge of $200 per month toward the taxes, $600 per quarter, or some such arrangement. If the landlord deposits this tax payment into a separate account, the interest can be earned on the payments — quite a bonus!

Even though the tenant is paying the taxes, it's best to have the tax payment go through the owner or landlord. That way, the property owner is sure that the taxes get paid on time. Cities and counties get a bit testy about late or unpaid property taxes. If your tenant neglects to pay the taxes on time, you'll come across penalties and other inconveniences (such as having the tax bill sold at auction).

# *Signing a Double or Triple Net Lease*

A *double net lease* is a lease where the tenant pays the monthly set amount along with the property taxes and property insurance. With a *triple net lease,* the tenant pays the property taxes, property insurance, the maintenance, and the monthly rent. In both situations, the landlord is still responsible for all the other expenses.

The double and triple net leases are conventional arrangements. But there's no reason that you can't get creative and set up a *quadruple net lease,* a *quintuple net lease,* or *sextuple net lease.* (Okay, I went a bit overboard here; this is beginning to sound like a discussion of multiple births.) In any case, a properly executed lease can be set up to meet the needs and desires of the landlord and the tenant. Spelling out all the details in a lease helps avoid confusion as to who's responsible for what.

# Trying Out a Percentage Lease

A *percentage lease* is arranged to benefit both the landlord and the renter. The landlord benefits by sharing in the business tenant's good profits, and the renter benefits when business is in a slump or down-period. Percentage leases make sense for businesses that fluctuate with the seasons, but they also make financial planning difficult. Another downside is this: Even though a landlord receives more money with a percentage lease when the tenant's business is good, the additional revenue may be offset by greater expenses, such as utilities, if a net lease is in place.

The terms of a percentage lease may require the renter to pay a certain flat rate per month plus 2 percent of the gross sales for the month. This arrangement works if the added amount is sufficient to cover the added expense of the increased sales. And a possibility is a rate schedule that includes increases or decreases determined by the particular level of volume.

# Stepping It Up with a Step Lease

A *step lease* (or *escalation lease*) makes provisions for the amount of the lease to increase over a period of time — usually each year, but other arrangements, such as quarterly or monthly, can also be used.

Over the course of the lease term, the rent can increase at predetermined amounts. These amounts are negotiated at the beginning of the lease term. With a step lease, the landlord accepts smaller amounts of money at the beginning of the term, expecting to benefit from the success of the enterprise after it takes hold. At the same time, the building is appreciating in value, and the landlord has a tenant locked in. With a step lease, each step may be the same over the term of the lease, or it may have unequal steps that reach the same level in the end.

Real estate values increase over time, and inflation continues to affect the cost of doing business, so landlords expect to get their fair share. The main reason a building owner may use a step lease is to allow the renter and owner to enter into a multiyear contract while, at the same time, making accommodations for yearly increases.

Another reason an owner may choose a step lease is to benefit the tenant. When a business is first starting up, the business owner prefers to reinvest money into the business to stimulate its growth (rather than give profits to the owner or landlord). A landlord can go along with this and protect himself with a larger security deposit. The step lease, in this situation, allows the landlord to reap the benefits as the new business takes hold and becomes more profitable.

# Inserting Expense Provisions into Your Lease

When you're entering into a lease as a landlord, you have to be sure to protect yourself. You can do so by adding specific clauses and provisions. For example, an *expense stop provision* protects the landlord from excessive expenses. And an *allowance per foot clause* can secure the promise of renovations and cooperation while the renovations or changes are being made.

Expense stop provisions are negotiated when the terms of the lease are being set. A landlord and renter will negotiate how the issue is to affect the terms of the lease using historical information regarding the expected expenses for a particular item.

With the allowance per foot clause, either the improvements are added to the rental amount because of the increase in utility of the property, or the amount is decreased due to expected inconveniences. Sometimes, the landlord pays for the improvement while the tenant decides how the improvement is made — within the allowance.

A gross lease requires a landlord to pay all the expenses — such as taxes, maintenance, and utilities (see the earlier section "Getting the Full Treatment with a Gross Lease"). However, a landlord can protect himself from an overly high utility expense (whether it's due to abuse by a tenant, unseasonably cold or warm weather, or an unexpected utility rate hike) by inserting an expense stop provision in the lease. The provision should protect the owner, but should also be fair to the tenant. Utility companies usually offer level payment plans to even out spikes in the cost.

# Including an Allowance for Improvements in Your Lease

When negotiating for commercial space, you sometimes want to include improvements to the property as part of the contract. In some cases, those improvements are the landlord's responsibility, and sometimes they're the tenant's responsibility. When the tenant takes on the improvements, he or she may be able to negotiate a reduction in the rental amount. Of course, the landlord wants to have some control over the quality of the work being done and expects all permanent fixtures to be left in place when the tenant leaves.

It's often more advantageous for the landlord to take on the improvements, because the landlord has more of a vested interest in the improvement on his property. One example of such a lease is when a tenant agrees to a 5-year step lease with a particular cost per square foot of space. If the tenant requests some improvements to the property, her proposal may include an extension in the length of the lease. In this case, the owner of the property needs to weigh the advantage of having the renter for a longer period of time — at her terms — versus the cost of the improvement.

# Protecting Your Lease with a Security Deposit

A *security deposit* basically acts to cover the landlord's expenses should a tenant violate the lease agreement by breaking the lease (not renting for the full term that was agreed upon) or by not keeping the property in good condition — if the wear and tear is greater than what's normally expected during the time period of the lease.

When a business has little or no credit history, the landlord may require a larger security deposit because of the increased risk of default or violation. On the other hand, if the credit history of a business is strong, the landlord may waive the security deposit requirement altogether or require only a nominal amount. After a tenant establishes himself as being reliable, he can sometimes negotiate an agreement that allows him to reclaim part or all of the security deposit.

Assuming all the conditions are met, the challenge here is determining how much of the security deposit should be returned. Does the added time cover the cost of the improvements? Is it worth the security of having the tenant for more time? Other agreements may include having the landlord share the interest that's being earned on the security deposit; this would take some negotiating.

# Adding a Sublease Clause to Your Commercial Lease

Because commercial leases are typically much longer than residential leases, it's usually a good idea to have a *sublease clause* in your rental contract. A sublease clause defines whether, as a renter, you're allowed to find another renter to take your place — with the same terms that you've agreed on with the landlord. After all, if you need to move to other quarters, you'd probably prefer to sublease the current area instead of breaking the contract and paying the penalties. Most leases contain a clause stating that the landlord must approve any subtenants; this type of clause also covers all the legal ramifications in terms of your responsibility and the owner's responsibility.

# Deciding Whether You Should Renew, Renegotiate, or Break a Lease

Some tenants want to renew their leases at the end of the current lease term. Others need to break their lease midway through and not complete the full term. And other times, a tenant may want to sublease a property as a way of fulfilling the lease obligation. Sometimes a landlord may even be the one to break a lease.

All these various contingencies, such as breaking the lease, subleasing, or even renewing, should be spelled out in the original lease. That way if one of these occurrences needs to be dealt with, the terms of the lease give direction on how to handle them.

Remember that the length of a commercial lease is usually much longer than a residential lease. That means it's common to see commercial leases that are up to five or ten years in length as a normal course of business; this length can add stability and security. But, depending on the type of business you run and the projected growth, locking yourself into a long lease can cause a conundrum for your planning. So if you think that your business has the potential to outgrow the current space, you need to plan ahead for expansion.

One way to provide for the changes in your situation is to renegotiate the terms of your lease, taking into account your new needs. This is often the more acceptable situation — the least disruptive to business. If good terms can't be set, or if the current facilities can't be altered to meet your needs, you may have to go through the expense of breaking your lease — and thereby paying a hefty fine to get out of that contract.

For example, consider a company that's currently halfway through the third year of a 5-year lease. If that company decides that it needs to expand, it has to determine whether it's better to increase the space at a possible premium rate (if the landlord wants to take advantage of the situation) or to break the lease, pay the penalty, and move elsewhere. Even though there are multiple expenses involved in relocating, it may still be the preferred option.

# Chapter 24

# Ten Things to Watch Out for When Reading Financial Reports

Companies — especially those that are publicly traded — produce many different financial statements each year so that their managers, owners, stockholders, and the IRS have access to the information necessary to make decisions. Each of the different financial reports has its own specifications and particular format, and even those formats change with the type of company and the situation.

This chapter doesn't educate you on how to read the different financial forms. Instead, what you find here are some of the things to watch out for to determine whether more study is necessary if a report isn't as clear and forthright as you think it should be. If you're interested in understanding how to read financial reports, you can get an explanation with lots of detail in *Reading Financial Reports For Dummies* (Wiley).

## Apologies for Shortfalls

A corporation's annual report offers the board of directors the opportunity to strut their stuff and summarize the past year. They also get to use the report to show their strategies for the coming year. So if you're reading the report and come across an apology (veiled or outright) in the chairman's letter that a particular lofty goal wasn't quite met because of unforeseen events, you want to investigate further to see just how far the goal was missed by and how unlikely the event in question may have been. If missing the goal has significant impact on the future of the company, then you need to keep an eye on future developments.

# Blurring in the Overall Picture

A company is most happy to report record revenues and successfully-met sales goals, but is the annual report clear on where that income is being generated? How and where is the revenue coming from during the year? Are the different divisions that make up the company clearly defined? It's fine that revenue is coming in, but you want to know where the revenue is coming from and if the result can be replicated.

# Wishy-Washy Descriptions of Trends

The management of a company uses various records to determine the financial situation both past and present. So you should be able to find documentation concerning the financial trends of the company — whether the financials are good news or bad news. And that documentation should be written in a clear and honest discourse. If the report is wishy-washy and uninformative, it's your job to find out more. You don't want vague generalities about the finances; you want cold, hard facts. If you're being given noninformation, you need to ask questions. Are the details available, or are you gently being told that you just don't need to know?

# An Overabundance of Footnotes

Footnotes are designed to clarify information in a financial report. Beware of an overabundance of footnotes or of vague references in footnotes. Why? Because a report is meant to be informative and instructive. Footnotes are extra bits of information or clarification and shouldn't contain the meat of the report. And remember that any time information is pulled away from the body of a report, you have an interruption of the flow of material. This interruption isn't good because it impedes your ability to understand the report and see the whole picture.

# Listings of the Directors

You'll find a listing of the board of directors in a company's annual report. Determine how many directors are on the board. Do you really think that the number of directors currently on the board is enough? Or do you think that there are too many? You also may want to read through the directors' brief biographies so you can get an idea of who's sitting on the board and what their agendas may be. By looking at their histories, do you think that the directors represent a wide range of experience and corporate responsibility, or are they all apparent buddies of the president? A wide range of experience provides a more global perspective — not everyone is looking at the company with the same preparation and background. Buddies of the president may be more forgiving of errors and missteps; you want directors who are more objective.

A family-owned business may have an overabundance of people with the same last name serving on the board. Keeping it in the family may be fine, but you need to delve further if you think that undue influence could be hurting the company.

# Uncollected Revenues

In a company's balance statement, you'll find entries for the *receivables,* which is the income not yet collected for goods purchased or services provided. Does the amount in accounts receivable seem a bit high compared to previous periods of time or comparable companies? Does the company seem to have an awful lot of uncollected revenue? If so, the company may be exposing itself to potentially bad debts at a higher-than-needed scale. Ask questions of the company representatives to see if they have plausible reasons for the situation.

# A High Working Capital

When reading a company's cash flow statement, be wary of especially large amounts of working capital (money just sitting there — items produced and warehoused but not sold). A big number in the working capital entry may mean that inventory isn't being moved out as quickly as it should be. And if inventory isn't moving out at an appropriate rate, then sales are probably stagnating, and that will affect revenue in the future.

# Long-Term Debt

Many companies use loans to help their businesses grow. And why not? Borrowing makes good business sense as long as the amount is reasonable and the projected payback is doable. Having said that, beware of a company with a large amount of long-term debt. Having to pay interest on long-term debt tends to cut into a company's cash flow.

# Low Profit Margins

A *profit margin* is the difference between the revenue generated and the cost required to generate that revenue. In general, the larger the profit margin, the safer the investment. On the other hand, a low margin can be risky. For example, a company operating at a 4% profit margin is quickly in trouble if the sales decline by 4% or more. You should expect the profit margin to increase over time, which is accomplished by increasing sales at a faster rate than the expenses incurred while generating those sales.

# Warning Words

Financial reports are full of numbers, and unfortunately, numbers can lie. In Chapter 8, for example, I show you how to lie with statistics. (Okay, I don't show you *how* to actually lie, but I do show you how it can be done.) Numbers can be slippery fellows, but so can words. So watch for the hidden meanings when you see words and phrases like *challenging, undertaking corrective measures, adjustments,* and *significant difficulties.* When you see *challenging,* does it mean something that the company is looking forward to facing or something that has dragged the financial situation downward? *Corrective measures* may be slight adjustments to take care of changes in the financial climate, or they may be huge layoffs or firing of employees who have done some harm. *Adjustments* can mean just about anything — look for what the adjustments refer to in the financial picture. And *significant difficulties* isn't included in a financial statement unless something significant really has happened. Delve further — something is afoot.

You want the reports to be honest and to include the bad news along with the good, but be wary of the company trying to gloss over certain pertinent information.

# Index

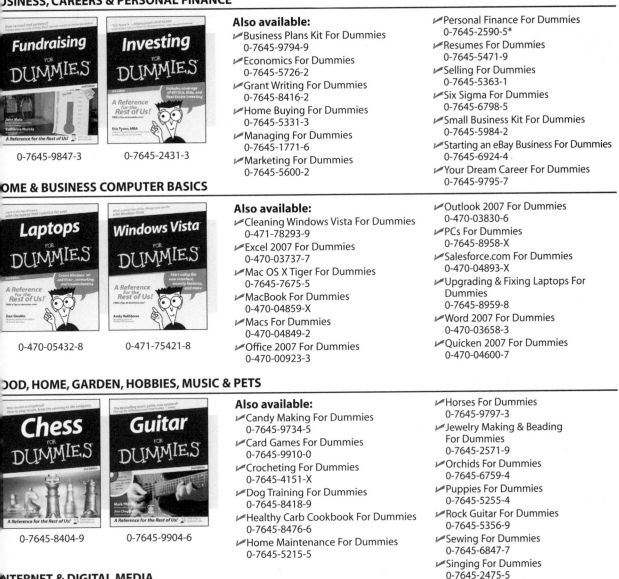

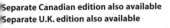

## SPORTS, FITNESS, PARENTING, RELIGION & SPIRITUALITY

0-471-76871-5

0-7645-7841-3

**Also available:**
- ✓ Catholicism For Dummies
  0-7645-5391-7
- ✓ Exercise Balls For Dummies
  0-7645-5623-1
- ✓ Fitness For Dummies
  0-7645-7851-0
- ✓ Football For Dummies
  0-7645-3936-1
- ✓ Judaism For Dummies
  0-7645-5299-6
- ✓ Potty Training For Dummies
  0-7645-5417-4
- ✓ Buddhism For Dummies
  0-7645-5359-3

- ✓ Pregnancy For Dummies
  0-7645-4483-7 †
- ✓ Ten Minute Tone-Ups For Dummies
  0-7645-7207-5
- ✓ NASCAR For Dummies
  0-7645-7681-X
- ✓ Religion For Dummies
  0-7645-5264-3
- ✓ Soccer For Dummies
  0-7645-5229-5
- ✓ Women in the Bible For Dummies
  0-7645-8475-8

## TRAVEL

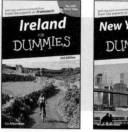

0-7645-7749-2

0-7645-6945-7

**Also available:**
- ✓ Alaska For Dummies
  0-7645-7746-8
- ✓ Cruise Vacations For Dummies
  0-7645-6941-4
- ✓ England For Dummies
  0-7645-4276-1
- ✓ Europe For Dummies
  0-7645-7529-5
- ✓ Germany For Dummies
  0-7645-7823-5
- ✓ Hawaii For Dummies
  0-7645-7402-7

- ✓ Italy For Dummies
  0-7645-7386-1
- ✓ Las Vegas For Dummies
  0-7645-7382-9
- ✓ London For Dummies
  0-7645-4277-X
- ✓ Paris For Dummies
  0-7645-7630-5
- ✓ RV Vacations For Dummies
  0-7645-4442-X
- ✓ Walt Disney World & Orlando
  For Dummies
  0-7645-9660-8

## GRAPHICS, DESIGN & WEB DEVELOPMENT

0-7645-8815-X

0-7645-9571-7

**Also available:**
- ✓ 3D Game Animation For Dummies
  0-7645-8789-7
- ✓ AutoCAD 2006 For Dummies
  0-7645-8925-3
- ✓ Building a Web Site For Dummies
  0-7645-7144-3
- ✓ Creating Web Pages For Dummies
  0-470-08030-2
- ✓ Creating Web Pages All-in-One Desk
  Reference For Dummies
  0-7645-4345-8
- ✓ Dreamweaver 8 For Dummies
  0-7645-9649-7

- ✓ InDesign CS2 For Dummies
  0-7645-9572-5
- ✓ Macromedia Flash 8 For Dummies
  0-7645-9691-8
- ✓ Photoshop CS2 and Digital
  Photography For Dummies
  0-7645-9580-6
- ✓ Photoshop Elements 4 For Dummies
  0-471-77483-9
- ✓ Syndicating Web Sites with RSS Feeds
  For Dummies
  0-7645-8848-6
- ✓ Yahoo! SiteBuilder For Dummies
  0-7645-9800-7

## NETWORKING, SECURITY, PROGRAMMING & DATABASES

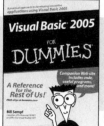

0-7645-7728-X

0-471-74940-0

**Also available:**
- ✓ Access 2007 For Dummies
  0-470-04612-0
- ✓ ASP.NET 2 For Dummies
  0-7645-7907-X
- ✓ C# 2005 For Dummies
  0-7645-9704-3
- ✓ Hacking For Dummies
  0-470-05235-X
- ✓ Hacking Wireless Networks
  For Dummies
  0-7645-9730-2
- ✓ Java For Dummies
  0-470-08716-1

- ✓ Microsoft SQL Server 2005 For Dummies
  0-7645-7755-7
- ✓ Networking All-in-One Desk Reference
  For Dummies
  0-7645-9939-9
- ✓ Preventing Identity Theft For Dummies
  0-7645-7336-5
- ✓ Telecom For Dummies
  0-471-77085-X
- ✓ Visual Studio 2005 All-in-One Desk
  Reference For Dummies
  0-7645-9775-2
- ✓ XML For Dummies
  0-7645-8845-1

## HEALTH & SELF-HELP

0-7645-8450-2

0-7645-4149-8

**Also available:**

- Bipolar Disorder For Dummies
  0-7645-8451-0
- Chemotherapy and Radiation
  For Dummies
  0-7645-7832-4
- Controlling Cholesterol For Dummies
  0-7645-5440-9
- Diabetes For Dummies
  0-7645-6820-5* †
- Divorce For Dummies
  0-7645-8417-0 †

- Fibromyalgia For Dummies
  0-7645-5441-7
- Low-Calorie Dieting For Dummies
  0-7645-9905-4
- Meditation For Dummies
  0-471-77774-9
- Osteoporosis For Dummies
  0-7645-7621-6
- Overcoming Anxiety For Dummies
  0-7645-5447-6
- Reiki For Dummies
  0-7645-9907-0
- Stress Management For Dummies
  0-7645-5144-2

## EDUCATION, HISTORY, REFERENCE & TEST PREPARATION

0-7645-8381-6

0-7645-9554-7

**Also available:**

- The ACT For Dummies
  0-7645-9652-7
- Algebra For Dummies
  0-7645-5325-9
- Algebra Workbook For Dummies
  0-7645-8467-7
- Astronomy For Dummies
  0-7645-8465-0
- Calculus For Dummies
  0-7645-2498-4
- Chemistry For Dummies
  0-7645-5430-1
- Forensics For Dummies
  0-7645-5580-4

- Freemasons For Dummies
  0-7645-9796-5
- French For Dummies
  0-7645-5193-0
- Geometry For Dummies
  0-7645-5324-0
- Organic Chemistry I For Dummies
  0-7645-6902-3
- The SAT I For Dummies
  0-7645-7193-1
- Spanish For Dummies
  0-7645-5194-9
- Statistics For Dummies
  0-7645-5423-9

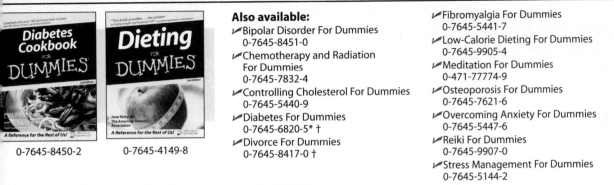

* Separate Canadian edition also available
† Separate U.K. edition also available

Available wherever books are sold. For more information or to order direct: U.S. customers visit www.dummies.com or call 1-877-762-2974.
U.K. customers visit www.wileyeurope.com or call 0800 243407. Canadian customers visit www.wiley.ca or call 1-800-567-4797.